GSFLOW—Coupled Ground-Water and Surface-Water Flow Model Based on the Integration of the Precipitation-Runoff Modeling System (PRMS) and the Modular Ground-Water Flow Model (MODFLOW-2005)

By Steven L. Markstrom, Richard G. Niswonger, R. Steven Regan, David E. Prudic, and Paul M. Barlow

Chapter 1 of
Section D, Ground-Water/Surface-Water
Book 6, Modeling Techniques

Techniques and Methods 6–D1

U.S. Department of the Interior
U.S. Geological Survey

U.S. Department of the Interior
DIRK KEMPTHORNE, Secretary

U.S. Geological Survey
Mark D. Myers, Director

U.S. Geological Survey, Reston, Virginia: 2008

For product and ordering information:
World Wide Web: http://www.usgs.gov/pubprod
Telephone: 1-888-ASK-USGS

For more information on the USGS--the Federal source for science about the Earth, its natural and living resources, natural hazards, and the environment:
World Wide Web: http://www.usgs.gov
Telephone: 1-888-ASK-USGS

Suggested citation:
Markstrom, S.L., Niswonger, R.G., Regan, R.S., Prudic, D.E., and Barlow, P.M., 2008, GSFLOW—Coupled ground-water and surface-water flow model based on the integration of the Precipitation-Runoff Modeling System (PRMS) and the Modular Ground-Water Flow Model (MODFLOW-2005): U.S. Geological Survey Techniques and Methods 6-D1, 240 p.

Contents

Contents—Continued

Figures

Figures—Continued

Figures—Continued

Tables

Conversion Factors and Datums

Conversion Factors

Multiply	By	To obtain
cubic meter (m³)	35.31	cubic foot (ft³)
cubic meter (m³)	264.2	gallon (gal)
cubic meter per day (m³/d)	35.31	cubic foot per day (ft³/d)
cubic meter per day (m³/d)	264.2	gallon per day (gal/d)
cubic meter per second (m³/s)	70.07	acre-foot per day (acre-ft/d)
cubic meter per second (m³/s)	35.31	cubic foot per second (ft³/s)
gram (g)	0.03527	ounce, avoirdupois (oz)
gram per cubic centimeter (g/cm³)	62.4220	pound per cubic foot (lb/ft³)
kilogram (kg)	2.205	pound, avoirdupois (lb)
kilometer (km)	0.6214	mile (mi)
kilopascal (kPa)	0.2961	inch of mercury at 60°F (in Hg)
kilopascal (kPa)	0.1450	pound per square inch (lb/ft²)
liter (L)	0.2642	gallon (gal)
meter (m)	3.281	foot (ft)
meter per day (m/d)	3.281	foot per day (ft/d)
meter per kilometer (m/km)	5.27983	foot per mile (ft/mi)
meter per second (m/s)	3.281	foot per second (ft/s)
meter squared per day (m²/d)	10.76	foot squared per day (ft²/d)
millimeter (mm)	0.03937	inch (in.)
square meter (m²)	10.76	square foot (ft²)
square kilometer (km²)	247.1	acre
square kilometer (km²)	0.3861	square mile (mi²)

Temperature in degrees Celsius (°C) may be converted to degrees Fahrenheit (°F) as follows:

°F= (1.8×°C) +32

Temperature in degrees Fahrenheit (°F) may be converted to degrees Celsius (°C) as follows:

°C= (°F-32)/1.8

Datums

Vertical coordinate information is referenced to the North American Vertical Datum of 1988 (NAVD 88)

Horizontal coordinate information is referenced to North American Datum of 1983 (NAD 83)

Altitude, as used in this report, refers to distance above the vertical datum.

Preface

This report describes the U.S. Geological Survey coupled Ground-water and Surface-water Flow model (GSFLOW). The performance of the program has been tested in a variety of applications. Future applications, however, might reveal errors that were not detected in the test simulations. Users are requested to send notification of any errors found in this report or the model program to:

Office of Ground Water
U.S. Geological Survey
411 National Center
Reston, VA 20192
(703) 648-5001

The latest version of the model program and this report can be obtained using the Internet at address: http://water.usgs.gov/software/.

GSFLOW—Coupled Ground-Water and Surface-Water Flow Model Based on the Integration of the Precipitation-Runoff Modeling System (PRMS) and the Modular Ground-Water Flow Model (MODFLOW-2005)

By Steven L. Markstrom, Richard G. Niswonger, R. Steven Regan, David E. Prudic, and Paul M. Barlow

Abstract

The need to assess the effects of variability in climate, biota, geology, and human activities on water availability and flow requires the development of models that couple two or more components of the hydrologic cycle. An integrated hydrologic model called GSFLOW (**G**round-water and **S**urface-water **FLOW**) was developed to simulate coupled ground-water and surface-water resources. The new model is based on the integration of the U.S. Geological Survey Precipitation-Runoff Modeling System (PRMS) and the U.S. Geological Survey Modular Ground-Water Flow Model (MODFLOW). Additional model components were developed, and existing components were modified, to facilitate integration of the models. Methods were developed to route flow among the PRMS Hydrologic Response Units (HRUs) and between the HRUs and the MODFLOW finite-difference cells. This report describes the organization, concepts, design, and mathematical formulation of all GSFLOW model components. An important aspect of the integrated model design is its ability to conserve water mass and to provide comprehensive water budgets for a location of interest. This report includes descriptions of how water budgets are calculated for the integrated model and for individual model components. GSFLOW provides a robust modeling system for simulating flow through the hydrologic cycle, while allowing for future enhancements to incorporate other simulation techniques.

Introduction

Hydrologic models of ground-water and surface-water systems traditionally have been developed with a focus on either the ground-water or surface-water resource. For example, models used to evaluate the effects of pumping and surface-water interactions on ground-water resources typically estimate recharge using long-term average precipitation and evaporation, without considering flow through the soil and unsaturated zones (Cosner and Harsh, 1978; Morgan, 1988; Danskin, 1998; Lindgren and Landon, 1999). Similarly, models used to evaluate the effects of climate variability on surface-water resources typically represent ground-water flow to streams using empirical and(or) simplified equations that do not consider the effects of local geology and surface-water interactions (Beven and others, 1984; Jeton, 1999; Zarriello and Ries, 2000; Ely, 2006).

Integrated Hydrologic Models

Integrated hydrologic models are useful for analyzing complex water-resources problems faced by society because they can consider feedback processes that affect the timing and rates of evapotranspiration, surface runoff, soil-zone flow, and ground-water interactions (fig. 1). Two general approaches have been used to design integrated models. One approach is referred to as "fully integrated" whereby equations governing surface and subsurface flows are solved simultaneously (VanderKwaak, 1999; Panday and Huyakorn, 2004). The second approach partitions the surface and subsurface systems into separate regions and the governing equations that describe flow in each region are integrated (coupled) using iterative solution methods (Smith and Hebbert, 1983; Refsgaard and Storm, 1995; Sophocleous and Perkins, 2000). This second approach is called the "coupled regions" approach.

The fully integrated approach uses the three-dimensional form of Richards' equation to simulate unsaturated and saturated flow (Freeze, 1971; VanderKwaak, 1999; Panday and Huyakorn, 2004; Thoms and others, 2006). Models that use the three-dimensional form of Richards' equation require much finer spatial grids and smaller time steps than typically are used to simulate saturated flow, which limits their applicability for simulating flow through regional hydrologic systems that encompass hundreds to thousands of square kilometers and simulation time periods of months to decades.

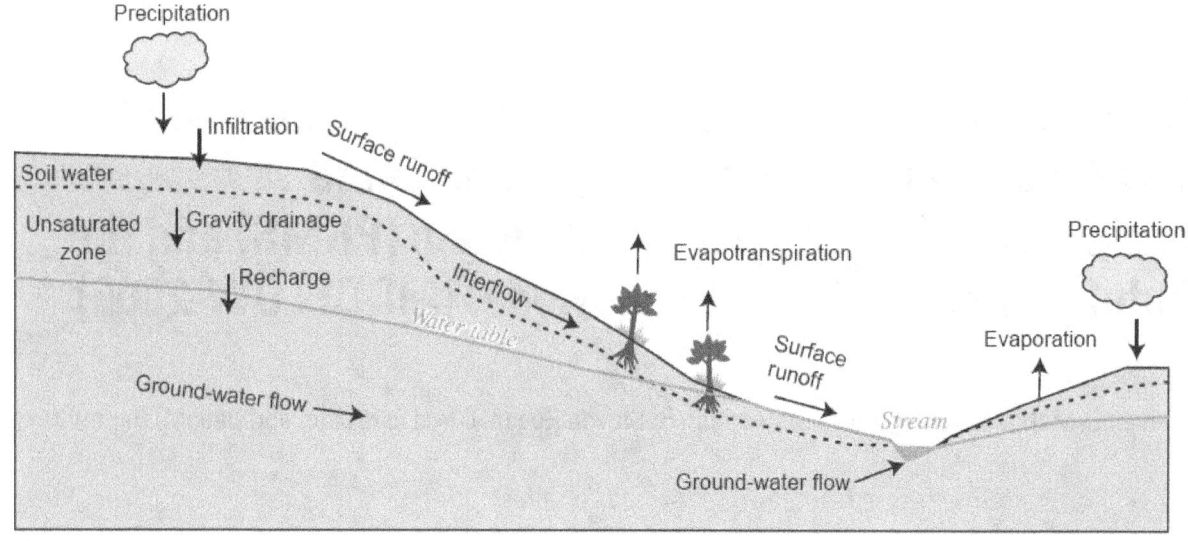

EXPLANATION

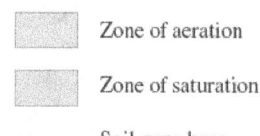   Zone of aeration

Zone of saturation

· · · · ·   Soil-zone base

Figure 1.   Distribution, flow, and interaction of water on the land and in the subsurface.

Thus, coupled models that do not use the three-dimensional form of Richards' equation may be better suited for simulating flow through regional hydrologic systems.

An efficient approach for simulating unsaturated flow in integrated models takes advantage of the fact that the dominant direction of flow within the unsaturated zone usually is vertical when averaged over large areas (Mantoglou, 1992; Chen and others, 1994; Harter and Hopmans, 2004). Accordingly, equations can be used to simulate flow and storage in the soil, unsaturated, and saturated zones separately, such that model efficiency is improved with some sacrifice of model accuracy. For this approach, separate equations are used to simulate horizontal and vertical flow through the soil zone, gravity-driven vertical flow through the unsaturated zone, and three-dimensional ground-water flow through the saturated zone. This coupled-regions approach was the approach used in the development of the GSFLOW (Ground-water/Surface-water **FLOW**) model described in this report and outlined by Fenske and Prudic (1998).

GSFLOW was developed to simulate coupled ground-water/surface-water flow in one or more watersheds by simultaneously simulating flow across the land surface and within subsurface saturated and unsaturated materials.

GSFLOW can be used to evaluate the effects of such factors as land-use change, climate variability, and ground-water withdrawals on surface and subsurface flow. The model was designed to simulate the most important processes affecting surface-water and ground-water flow using a numerically efficient algorithm. The model incorporates well documented methods for simulating runoff and infiltration from precipitation, as well as the interaction of surface water with ground water in watersheds that range from a few square kilometers to several thousand square kilometers, and for time periods that range from months to several decades.

GSFLOW version 1.0 (referred to as GSFLOW in the remainder of the report) is an integration of the U.S. Geological Survey (USGS) Precipitation-Runoff Modeling System (PRMS; Leavesley and others, 1983; Leavesley and others, 2005) with the 2005 version of the USGS Modular Ground-Water Flow Model (MODFLOW-2005; Harbaugh, 2005). PRMS and MODFLOW have similar modular programming methods, which allow for their integration while retaining independence that permits substitution of additional PRMS modules and MODFLOW packages. Both models have a long history of support and development.

PRMS was selected because it: (1) can simulate land-surface hydrologic processes of evapotranspiration, runoff, infiltration, and interflow by balancing energy and mass budgets of the plant canopy, snowpack, and soil zone on the basis of distributed climate information (temperature, precipitation, and solar radiation); (2) can be used to analyze the effects of urbanization on the spatial distribution of ground-water recharge (Vaccaro, 1992; Steuer and Hunt, 2001); (3) can be used with other models for water-resources management and forecasting (Fulp and others, 1995; Wilby and others, 1999; Berris and others, 2001; Hay and others, 2002; Mastin and Vaccaro, 2002; Hay and Clark, 2003; Clark and Hay, 2004); and (4) has a modular design that allows for selection of alternative hydrologic process algorithms among existing or easily added modules. Some examples of the use of PRMS in hydrologic studies are the simulation of sediment production for semi-arid watersheds (Rankl, 1987); heat and water transfer for seasonally frozen soils (Emerson, 1991); use of radar data to specify rainfall input (Yates and others, 2000), streamflow and wetland storage (Vining, 2002); and flow-frequency characteristics (Olson, 2002).

MODFLOW was selected because it is one of the most widely used and tested ground-water flow codes, and is capable of simulating: (1) three-dimensional saturated ground-water flow and storage; (2) one-dimensional unsaturated flow and ground-water discharge to the land surface (Niswonger and others, 2006a); and (3) ground-water interactions with streams (Prudic, 1989; Swain and Wexler, 1996; Jobson and Harbaugh, 1999; Prudic and others, 2004; Niswonger and Prudic, 2005) and lakes (Cheng and Anderson, 1993; Council, 1998; Merritt and Konikow, 2000). Moreover, MODFLOW has been coupled with other precipitation-runoff models (Ross and others, 1997; Hunt and Steuer, 2000; Sophocleous and Perkins, 2000; Nishikawa and others, 2005; Said and others, 2005) and with solute-transport models (Konikow and others, 1996; Guo and Langevin, 2002; Langevin and others, 2003), and has sensitivity and parameter estimation capabilities (Hill and others, 2000).

The initial version of GSFLOW does not include all capabilities of the PRMS and MODFLOW models. Additional capabilities likely will be added to GSFLOW over time. Moreover, future versions of GSFLOW can include alternative models to PRMS and MODFLOW for simulating surface- and(or) ground-water flow. Additional models can be integrated to simulate other environmental and anthropogenic processes, such as water quality, ecology, geochemistry, and management strategies, and pre- and post-processors will likely be developed for GSFLOW to facilitate the use of GSFLOW by practicing hydrologists.

Purpose and Scope

This report describes GSFLOW, version 1.0. The report first describes the design of GSFLOW and includes descriptions of PRMS and MODFLOW-2005 and how they were integrated for GSFLOW. The equations and order of calculations used in GSFLOW are presented in section "Computations of Flow." Inflows, outflows, and storages are discussed in section "Water Budgets." Important assumptions used in the development of GSFLOW and how these assumptions limit the applicability of GSFLOW are described in section "Assumptions and Limitations." The section "Input and Output Files" describes data-input requirements and output options of GSFLOW files. The last section presents an example simulation of the Sagehen Creek watershed in the eastern Sierra Nevada near Truckee, California. Appendix 1 provides detailed input instructions for the files that are required for a GSFLOW simulation. Users of GSFLOW are encouraged to review documentation for PRMS (Leavesley and others, 1983) and MODFLOW-2005 (Harbaugh, 2005).

Design of GSFLOW

GSFLOW simulates flow within and among three regions. The first region is bounded on top by the plant canopy and on the bottom by the lower limit of the soil zone; the second region consists of all streams and lakes; and the third region is the subsurface zone beneath the soil zone. PRMS is used to simulate hydrologic responses in the first region and MODFLOW-2005 is used to simulate hydrologic processes in the second and third regions. The following design principles guided development of the model:

- Use existing PRMS modules and MODFLOW-2005 packages where possible;

- Use a flexible and adaptive modular design that incorporates both PRMS and MODFLOW-2005 programming frameworks so that existing and new PRMS and MODFLOW-2005 simulation techniques can be added to GSFLOW in the future;

- Use general design procedures that can be used to integrate other simulation models into GSFLOW;

- Allow simulations using only PRMS or MODFLOW-2005 within the integrated model for the purpose of initial calibration of model parameters prior to a comprehensive calibration using the integrated model;

- Solve equations governing interdependent surface-water and ground-water flow using iterative solution techniques;

- Compute model-wide and detailed (for example, soil-zone flow and storage) water balances in both time and space;

- Allow flexibility in the spatial discretization of the hydrologic response units used for PRMS and the finite-difference grid used for MODFLOW-2005; and

- Allow model boundaries to be defined using standard specified-head, specified-flow, and head-dependent boundary conditions to account for inflows to and outflows from the modeled region.

The remainder of this section provides an overview of the PRMS and MODFLOW models, as well as a description of how the two models were integrated into a single model. Some of the components of PRMS and MODFLOW were modified during the integration process; these are described in the following sections. However, because GSFLOW can be used to run PRMS-only and MODFLOW-only simulations (that is, non-integrated simulations), descriptions of those components of PRMS and MODFLOW that are required for PRMS- and MODFLOW-only simulations also are included in this report.

Description of PRMS

PRMS is a modular deterministic, distributed-parameter, physical-process watershed model used to simulate and evaluate the effects of various combinations of precipitation, climate, and land use on watershed response. Response to normal and extreme rainfall and snowmelt can be simulated to evaluate changes in water-balance relations, streamflow regimes, soil-water relations, and ground-water recharge. Each hydrologic component used to model the generation of streamflow is represented within PRMS by a process algorithm that is based on a physical law or an empirical relation with measured or estimated characteristics. A description of how a hydrologic system is conceptualized and discretized in PRMS is included to provide a basic understanding of how a model can be developed to represent a particular hydrologic system. However, the focus is on the capabilities of PRMS that are integrated into GSFLOW; the reader is referred to Leavesley and others (1983), Leavesley and Stannard (1995), and Leavesley and others (2005) for a complete description.

Representation of Watershed Hydrologic Processes

A watershed is defined as the area of land that drains into a stream above a given location (Chow and others, 1988, p. 7). PRMS simulates the hydrologic processes of a watershed using a series of reservoirs that represent a volume of finite or infinite capacity. Water is collected and stored in each reservoir for simulation of flow, evapotranspiration, and sublimation. Flow to the drainage network, which consists of stream-channel and detention-reservoir (or simple-lake) segments, is simulated by surface runoff, interflow, and ground-water discharge (fig. 2).

Climate data consisting of measured or estimated precipitation, air temperature, and solar radiation are the driving factors used to compute evaporation, transpiration, sublimation, snowmelt, surface runoff, and infiltration in a PRMS simulation (Leavesley and others, 1983, p. 12-18). The form of precipitation (rain, snow, or mixture of both) is determined from temperature data or can be specified as input data. Precipitation can be intercepted by and evaporated from the plant canopy. Precipitation that is not intercepted by the plant canopy, which is referred to as throughfall, is distributed to the watershed land surface. Precipitation that reaches land surface can accumulate as part of the snowpack, be stored in impervious-zone reservoirs, infiltrate into the soil zone, be evaporated, or become surface runoff. Water and energy balances are computed for the snowpack to determine snow accumulation, snowmelt, or sublimation (fig. 3).

The subsurface is represented by a series of three reservoirs—the soil-zone, subsurface, and ground-water reservoirs (fig. 2). The soil-zone and subsurface reservoirs are used in PRMS to account for different fractions of water in pores within the matrix of the soil. Pores can range from relatively small intergranular space between grains of clay, silt, sand, and gravel to relatively large pores caused by cracks from seasonal shrinking and swelling of the soil, by holes from decaying plant matter such as roots and leaf litter, by holes from animal activity such as worms and gophers, and by cracks from landscape altering events such as earthquakes. The relatively larger pores commonly are called macropores (Selker and others, 1999). The two reservoirs include water in the pores that (1) can be removed by plants, (2) can drain downward to ground water or flow laterally downslope through relatively small intergranular space between grains caused by perching of water on top of a less permeable soil horizon, and (3) can flow laterally through macropores.

Water-saturation levels that are above the plant wilting threshold and below field capacity are simulated within the soil-zone reservoir. Antecedent water stored in the soil-zone reservoir controls the partitioning of available water (throughfall and snowmelt) at land surface to surface runoff and infiltration. Water stored in the soil-zone reservoir is referred to as capillary water and can be lost to evaporation and transpiration on the basis of plant type and cover, rooting depth, precipitation, solar radiation, and air temperature. The soil-zone reservoir is partitioned into two zones—the recharge zone and the lower zone. The recharge zone contains water up to a specified maximum water-saturation threshold that is available for evaporation and transpiration. The lower zone contains water when the water-saturation level in the soil-zone reservoir exceeds the maximum threshold. Lower-zone water is available only for transpiration. Water content in the soil-zone reservoir below the wilting threshold is assumed to be constant and is not included in water-balance computations. This water adheres strongly to the granular maxtrix and is unavailable to plants and relatively nonmoving (Bear, 1972, p. 3).

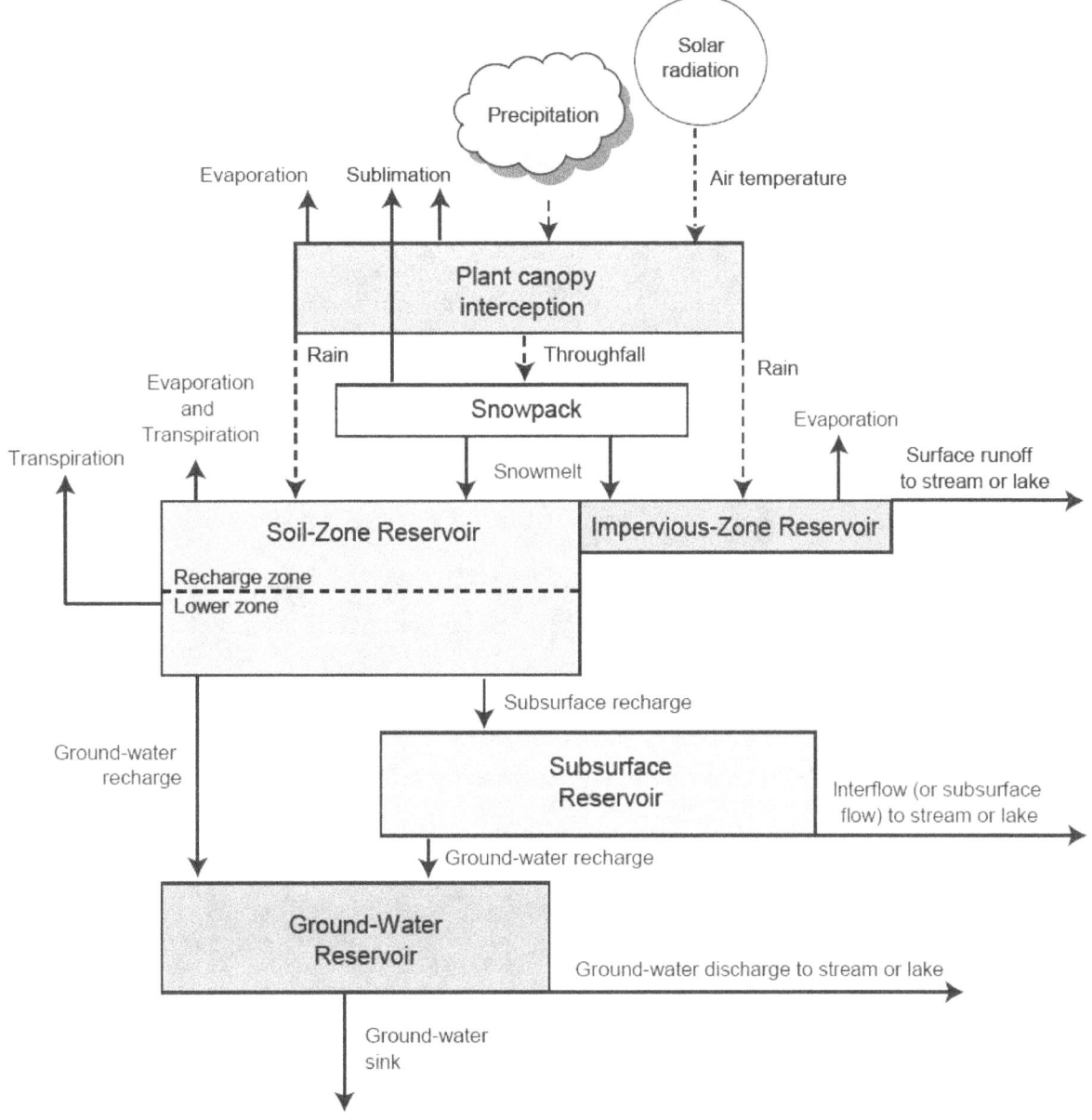

Figure 2.   Schematic diagram of a watershed and its climate inputs (precipitation, air temperature, and solar radiation) simulated by PRMS (modified from Leavesley and others, 1983).

Water that infiltrates into the soil-zone reservoir above field capacity is distributed to the subsurface and ground-water reservoirs (**fig. 2**). Water stored in the subsurface reservoir(s) is available for gravity drainage to a ground-water reservoir and for interflow to a stream or lake. Water stored in the ground-water reservoirs is available for ground-water discharge to a stream or lost to the ground-water sink.

A new Soil-Zone Module to simulate flow in the soil-zone and subsurface reservoirs was developed to facilitate integration between PRMS and MODFLOW and allow for flow through macropores and Dunnian runoff (Dunne and Black, 1970). The new module affects PRMS-only simulations as well as integrated simulations, and is described in the "**Soil Zone**" section of "**Computations of Flow.**" Flow beneath the soil zone is simulated using MODFLOW-2005 in a GSFLOW integrated simulation.

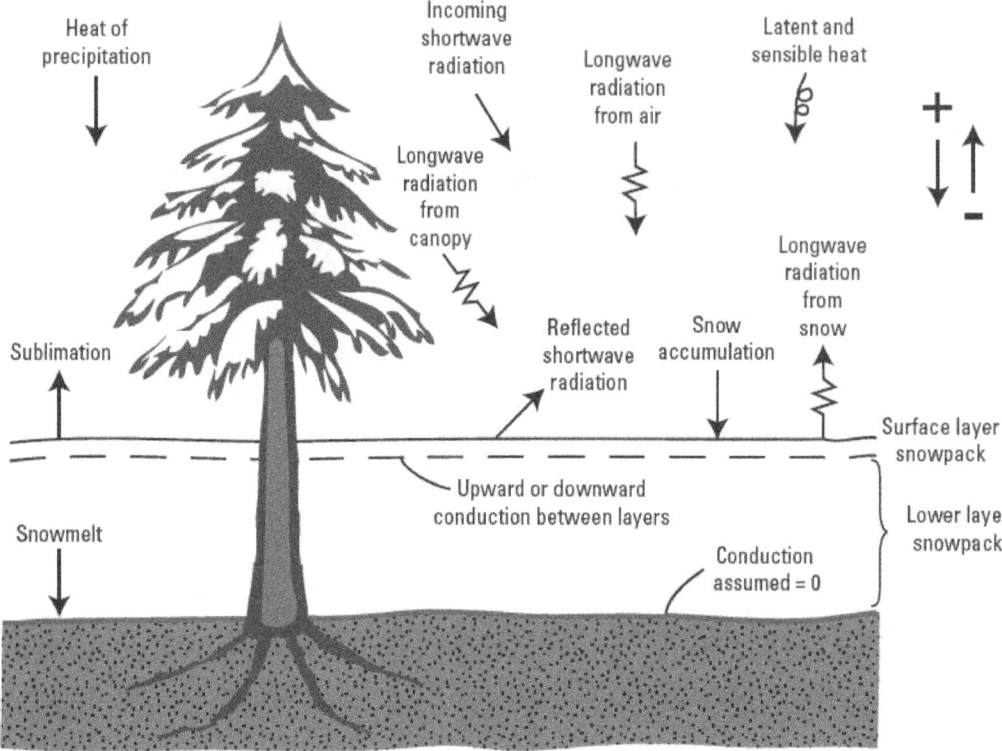

Figure 3. Components of the snowpack energy balance, accumulation, snowmelt, and sublimation (modified from Leavesley and others, 1983).

Watershed Discretization—Hydrologic Response Units

The area of a watershed is discretized into a network of hydrologic response units (HRUs). The discretization can be based on hydrologic and physical characteristics such as drainage boundaries, land-surface altitude, slope, and aspect; plant type and cover; land use; distribution of precipitation, temperature, and solar radiation; soil morphology and geology; and flow direction. Each HRU is assumed to be homogeneous with respect to these hydrologic and physical characteristics and to its hydrologic response. A water balance and an energy balance are computed daily for each HRU. Each HRU is identified by a numerical index. Assignment of the index to the HRU is arbitrary, but indices must be unique, consecutive, and start with 1.

The delineation of a watershed into HRUs can be automated with the aid of a Geographic Information System (GIS) analysis. Three delineation approaches typically are used (Viger and Leavesley, 2007): (1) a topological approach, which results in irregularly shaped polygons representing hill slopes and flow planes; (2) a grid-based approach, which results in regular rectangles; or (3) a noncontiguous approach, which results in many unique and irregularly shaped polygons determined by combinations of watershed characteristics. Two examples of HRU delineation of a hypothetical watershed are shown in **figure 4**.

Precipitation that falls on each HRU is routed through the soil-zone, subsurface, and ground-water reservoirs. Water is added to and subtracted from each HRU according to input data and model parameters. Each HRU includes separate reservoirs that represent water stored in the plant canopy, snowpack, impervious surfaces, and soil zone. A group of contiguous soil-zone reservoirs can be designated to add water to one subsurface reservoir. Similarly, contiguous subsurface reservoirs can be grouped to add water to one ground-water reservoir. Thus, there may be fewer subsurface and ground-water reservoirs than HRUs. Typically, a watershed is divided into many HRUs, several subsurface reservoirs, and one or two ground-water reservoirs. Each reservoir also is assumed to be homogeneous in hydrologic response and model parameterization. Water entering a reservoir is considered to be instantaneously mixed with water previously in the reservoir. The number of subsurface reservoirs must equal the number of HRUs for GSFLOW.

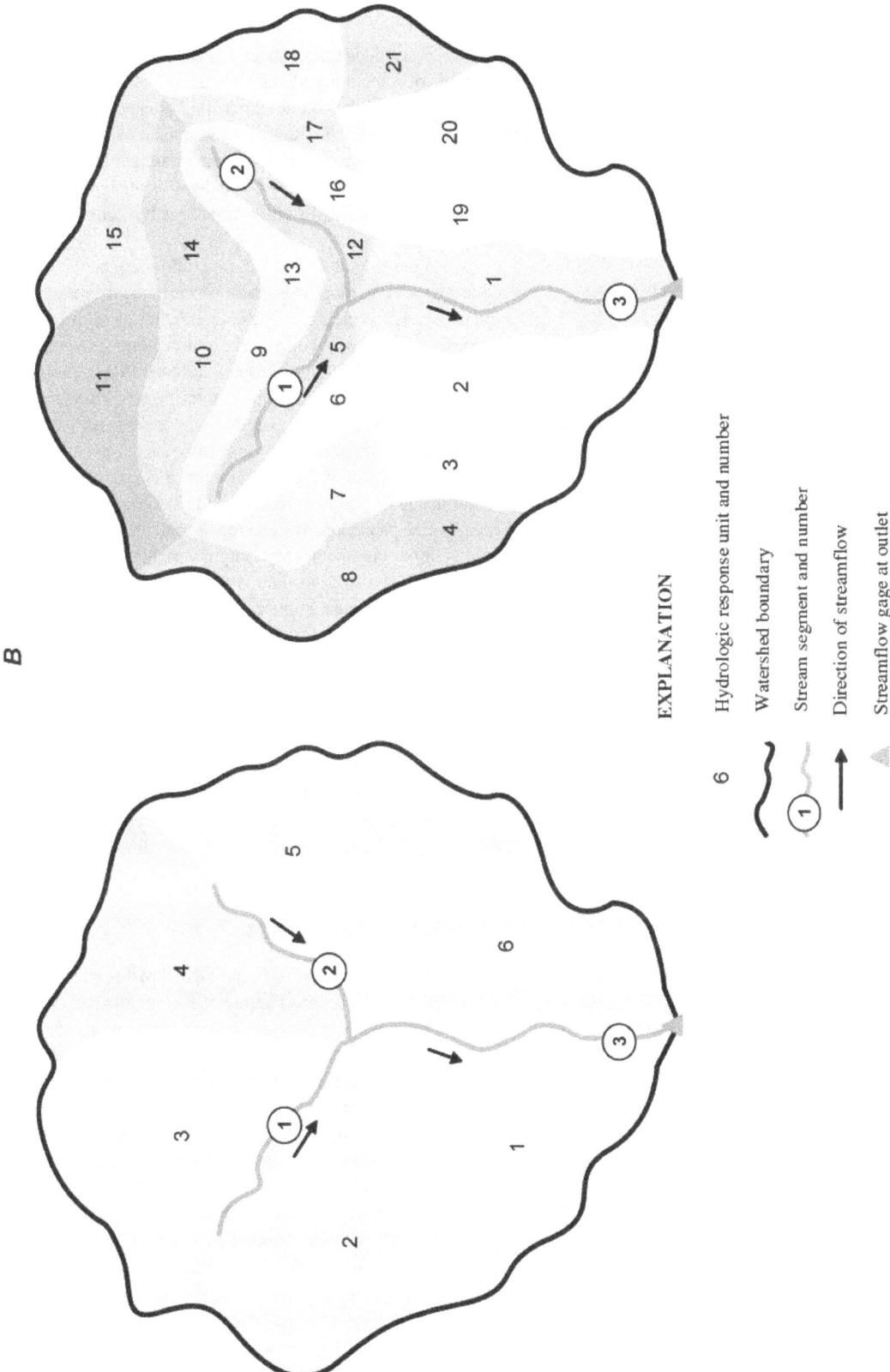

Figure 4. Delineation of hydrologic response units of a hypothetical watershed determined from (A) topology and (B) topology, climate, and vegetation.

Time Discretization—Time Steps

Simulation of watershed response to climate data is discretized into a series of time steps. Time steps can be specified equal to 1 day—referred to as the "daily mode"—or they can be specified equal to a time increment less than 1 day—referred to as the "storm mode" (Leavesley and others, 1983, p. 10). The length of each time step is determined by the smallest increment of time between consecutive measured input data values. Daily mode is used to simulate a watershed response to changes in daily precipitation, air temperature, and solar radiation. Storm mode is used to simulate a watershed response to individual storms on the basis of changes in precipitation, air temperature, and solar radiation that range from seconds to hours. Storm-mode simulations are not available in GSFLOW.

Channel-Flow and Detention Reservoir Routing

The drainage network within the watershed is characterized as a collection of stream-channel and detention reservoir (simple-lake) segments for daily simulations (PRMS "daily mode") and as a collection of stream-channel, detention-reservoir, and junction segments for storm simulations (PRMS "storm mode"). The PRMS allows for "storm mode" flow routing in the watershed drainage network to simulate storm events using a kinematic-wave approximation of single-direction, unsteady flow (Leavesley and others, 1983, p. 34-37). Outflow from as many as three upstream channel and detention-reservoir segments can be added as inflow to a downstream segment. Two routing methods are available to compute outflow from detention reservoirs: (1) a linear storage method computed using continuity; and (2) a modified-Puls method (U.S. Soil Conservation Service, 1971). Inflow to stream-channel and detention-reservoir segments includes upstream channel flow and surface runoff from all contiguous HRUs and flow from subsurface and ground-water reservoirs associated with a particular segment. Watershed outflow in "daily mode" includes flow from all contributing HRUs, detention-reservoir segments, and subsurface and ground-water reservoirs. Flow in the drainage network is simulated using MODFLOW-2005 in a GSFLOW integrated simulation.

Organization

PRMS is implemented as a set of modules in the Modular Modeling System (MMS) (Leavesley and others, 1996a and 1996b, http://wwwbrr.cr.usgs.gov/projects/SW_MoWS/). Modules are a group of subroutines written in the Fortran or C programming languages that simulate a particular model process. MMS was developed to: (1) support developing, testing, and evaluation of physical-process models as compatible sets of computer code; (2) facilitate integration of user-selected codes into operational physical-process models; (3) facilitate the coupling of models for application to complex, multidisciplinary problems; and (4) provide utility software for optimization, sensitivity, forecasting, visualization, and statistical analyses (Leavesley and others, 2005, p. 160).

MMS provides utility functions for models built within its framework, including: (1) file input and output; (2) time step looping and other model control; (3) function calls to pass data values between modules; (4) a graphical user interface; (5) tools for model calibration and uncertainty analysis; and (6) tools for statistical and visualization analysis. For GSFLOW, only the MMS utility functions for items 1, 2, and 3 are implemented.

MMS controls the PRMS computation sequence, including the time loop, using four procedures: declare, initialize, run, and cleanup. These procedures are similar to MODFLOW-2005 procedures. These procedures are executed through a call to the main subroutine of each PRMS module, which then calls the appropriate subroutine developed for the active MMS procedure. The declare procedure is executed first and is used to: (1) specify the valid value range, default value, units, and definition of input parameters; (2) specify the units and definition of output variables; and (3) allocate memory. The initialize procedure is executed second and is used to initialize input parameters and output variables and open any output files. The run procedure is executed each time step and is used to compute a particular simulation process. The cleanup procedure is executed after all time steps have been simulated and is used to release allocated memory and close files.

PRMS modules simulate a particular hydrologic component or computational method and are called within the MMS in the proper sequence. As such, PRMS modules serve similar purposes as packages in MODFLOW-2005. All PRMS modules include a declare and initialize procedure. Most PRMS modules have a run procedure and a few have a cleanup procedure.

Computational Sequence

Model execution begins in the PRMS main program, where the Control File and Parameter File are opened and read. Hydrologic processes and program options are determined from input data followed by declaration and initialization of all required parameters and variables (fig. 5). Next, the time-step loop begins and precipitation, air temperature, and optionally solar radiation values are read from the Data File for each measurement station and then distributed over the HRUs in the watershed. The computations of flows and storages follows the progression of water as it moves through the plant canopy, across the land surface, into the soil zone, and through the subsurface and ground-water reservoirs until reaching a stream (fig. 2). Results are saved to output files and a new time step is begun when the computation of flow and storage has been completed for a time step (fig. 5). The computation of flow and storage continues for each time step until the last time step is completed. After all time steps have been simulated, output files are closed and memory is deallocated.

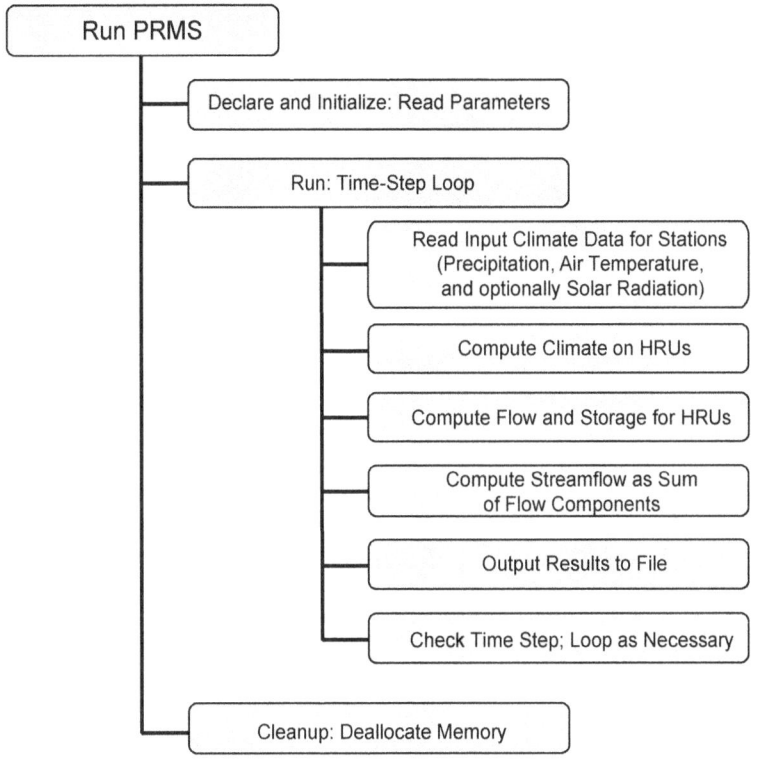

Figure 5. Computational sequence of PRMS used for simulating streamflow response in watersheds.

Description of MODFLOW-2005

MODFLOW-2005 (Harbaugh, 2005) is a new version of the finite-difference, three-dimensional ground-water flow model commonly called MODFLOW (McDonald and Harbaugh, 1988; Harbaugh and McDonald, 1996; Harbaugh and others, 2000). MODFLOW can be used to simulate steady-state or transient flow of constant density ground water through porous earth. A general description of how a ground-water flow system is conceptualized and discretized in all versions of MODFLOW is included to provide a basic understanding of how a model is developed to represent a particular ground-water system. However, the focus herein is on the capabilities of MODFLOW-2005 that are integrated into GSFLOW; the reader is referred to Harbaugh (2005) for a complete description.

Representation of Subsurface Processes

Regions of porous and permeable rocks where water flows and is stored below land surface are called aquifers (Meinzer, 1923, p. 30). The subsurface in most regions of the United States generally consists of many types of rocks or geologic formations, some of which are aquifers. Saturated low-permeability regions below land surface commonly are called confining units (Lohman and others, 1972; Sun, 1986). The subsurface consists of geologic materials that are either partially saturated (unsaturated zone) or fully saturated (saturated zone). The modeled region of the saturated zone may have multiple aquifers separated by confining units that collectively are called an aquifer system. MODFLOW-2005 allows for many types of processes, such as areal recharge across the water table, leakage to aquifers from streams and lakes, subsurface inflows either at fixed rates or at rates that vary as a function of head, discharge by evapotranspiration from phreatophytes, discharge from pumping wells, and discharge to streams and lakes (fig. 6). Confining units can be simulated explicitly as separate units or implicitly in a quasi-three-dimensional approach as averaged into the overall properties of the aquifer system. Wells can be simulated as removing water from or adding water to one or more aquifers. Although MODFLOW-2005 originally was designed to simulate only the saturated zone beneath the water table, the Unsaturated-Zone Flow Package (Niswonger and others, 2006a) for MODFLOW-2005 provides the capability to simulate one-dimensional flow through the unsaturated zone above the water table (fig. 6).

Aquifer-System Discretization—Finite-Difference Cells

An aquifer system is discretized in MODFLOW-2005 with a finite-difference grid that represents the computational cells used to calculate ground-water heads and flows. The finite-difference grid should be aligned with the major axes of hydraulic conductivity. Finite-difference cells are numbered according to the indices i, j, and k, where i is the row index, j is the column index, and k is the layer index (Harbaugh, 2005, p. 2-2). Horizontal discretization is defined as a rectangular grid of rows and columns. Rows are aligned parallel to the x axis. Increments of the row index i correspond to decreases in the y direction. Columns are aligned parallel to the y axis. Increments of column index j correspond to increases in the x direction. Vertical discretization is defined by layers. Layers of cells are aligned parallel to the horizontal plane. Increments of layer index k correspond to increases in the z direction (fig. 7). Thus, the top layer of the model corresponds with $k = 1$; the upper-left cell in each layer corresponds with row 1 ($i = 1$) and column 1 ($j = 1$). Layers can have uniform thickness or can have variable thickness among cells in a layer.

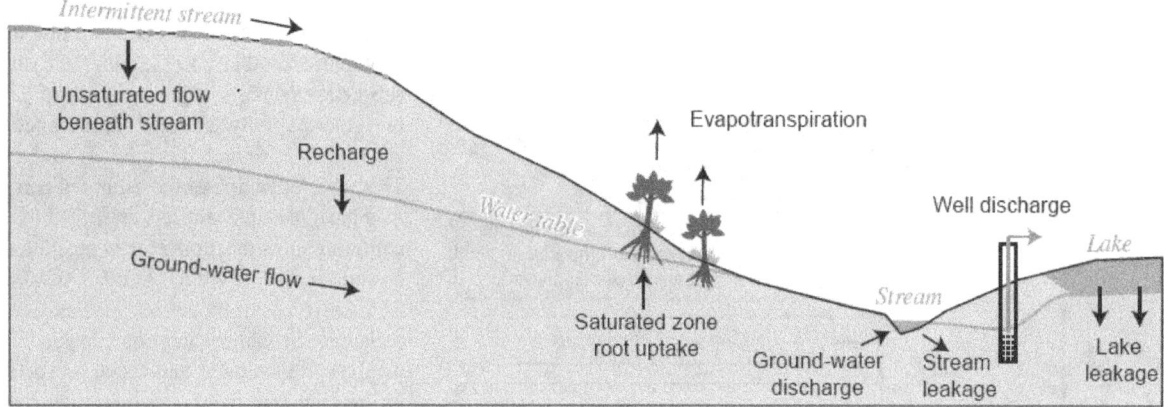

Figure 6.   Flow through a hypothetical aquifer system that can be simulated using MODFLOW-2005.

Some error is introduced into the solution when the grid is not oriented with the principal axes of hydraulic conductivity, when a cell represents a combination of aquifers and confining units, or when cells in a layer have variable thicknesses (Harbaugh, 2005, p. 2-15).

Head is computed at the center of each cell called a "node." The Block-Centered Flow Package (Harbaugh, 2005, p. 5-15), the Layer-Property Flow Package (Harbaugh, 2005, p. 5-17), or the Hydrogeologic-Unit Flow Package (Anderman and Hill, 2000; Anderman and others, 2002) can be used to calculate conductance terms for flow between cells and storage terms within cells. The Horizontal-Flow Barrier Package (Hsieh and Freckleton, 1993) can be used with the Block-Centered Flow, Layer-Property Flow, and Hydrogeologic-Unit Flow Packages to simulate the effects of ground-water flow barriers such as a fault.

Cells used to simulate boundary conditions are grouped into two categories—specified-head (or constant-head for certain conditions in which the specified head does not change with time) and no-flow cells (Harbaugh, 2005, p. 2-13). A head is specified in the data input for specified-head cells and the head value does not change unless it is changed in the data input for subsequent time or stress periods. No-flow cells are cells where flow into or out of a cell is not permitted. Ground-water head is not calculated for no-flow cells; for this reason, they are referred to as no-flow cells. The remaining cells are referred to as variable-head cells because ground-water head is computed in these cells and the computed heads can vary with time (Harbaugh, 2005, p. 2-13). No special designation is needed along the top or bottom of the finite-difference grid or where an edge of the aquifer system is coincident with the outside edge of the grid (fig. 7). Any cell within the finite-difference grid can be designated as no-flow cells in order to approximate no-flow boundaries of aquifer systems. No-flow cells can be surrounded by variable-head cells to represent internal regions in the finite-difference grid where flow does not occur.

Time-variant inflow and(or) outflow stress rates (boundary conditions) can be assigned to variable-head cells using MODFLOW stress packages such as the Well Package, where the stress rates only can change at the beginning of a stress period. Another type of possible boundary condition applied to variable-head cells is the head-dependent flow boundary condition. Interaction of ground water with streams and lakes is an example of a head-dependent flow boundary condition, and are covered in detail in section "**Discretization of Streams and Lakes**." In addition to boundary conditions applied to variable-head cells representing the area of interest, flow can be simulated into and(or) out of the finite-difference grid by using the General-Head Boundary or the Flow and Head Boundary Packages.

Discretization of Streams and Lakes

Because streams and lakes are a primary linkage between the PRMS and MODFLOW-2005 codes, a more detailed description of these MODFLOW-2005 packages is provided here. In the GSFLOW implementation of MODFLOW-2005, ground-water interactions with streams are simulated with the Streamflow-Routing Package (Prudic and others, 2004; Niswonger and Prudic, 2005). This package determines ground-water interactions along streams using a head-dependent boundary condition in which the head or stage along the stream is determined from the depth of flow in the channel. Streamflow can be routed along a network of channels that may include rivers, streams, canals, or ditches, and are referred to collectively as streams. The Streamflow-Routing Package originally was developed for MODFLOW-2000 (Prudic and others, 2004) and was revised by Niswonger and Prudic (2005) to simulate unsaturated flow beneath streams. The package was subsequently revised for MODFLOW-2005 and then for GSFLOW to simulate kinematic-wave routing.

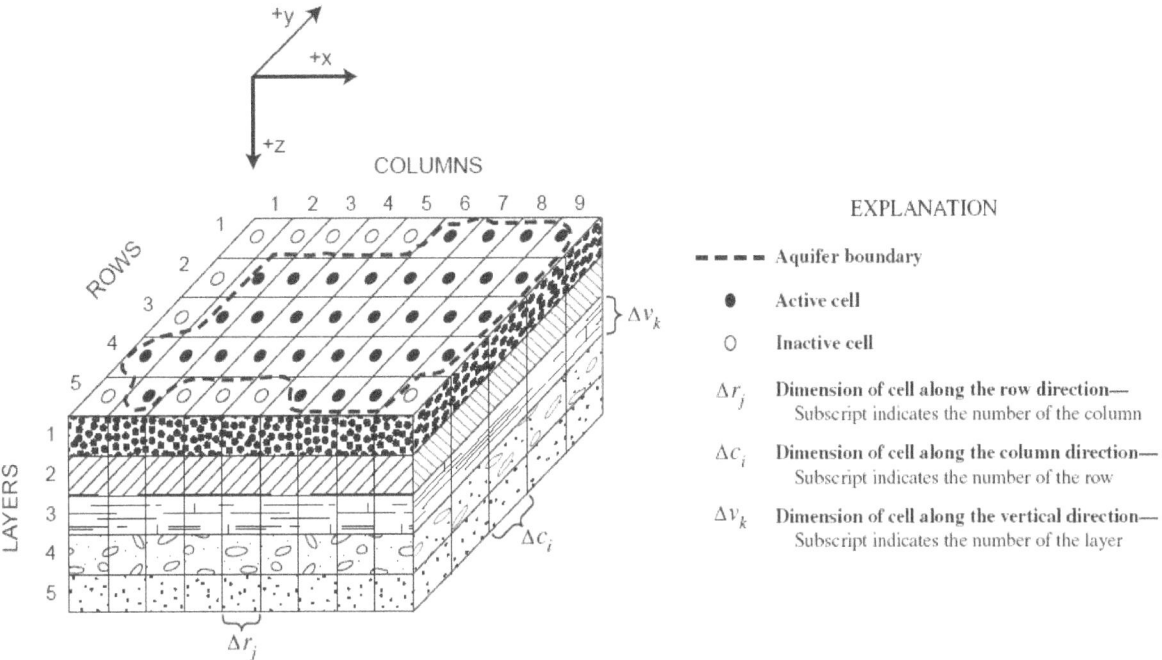

Figure 7. A discretized hypothetical aquifer system (modified from Harbaugh, 2005).

Streams are divided into reaches and segments (Prudic and others, 2004) (**fig. 8**). A reach is defined as a section of a stream that is associated with a particular finite-difference cell. More than one stream reach can be assigned to a particular finite-difference cell, but only one finite-difference cell can be assigned to a single reach. Reaches are grouped into segments that represent lengths of the stream between connections with another stream or tributary, a lake, or a watershed boundary. User-specified inflows to a stream that are external to inflows calculated by the model are added to the stream at the upstream end of a segment. Specified outflow at the upstream or downstream end of a segment can be used to divert water from a stream to a pipeline or lined canal; the water that is diverted in this way is removed from the modeled area without ground-water interaction. The quantity of water diverted is limited to the available flow entering the reach where a diversion is specified. Diverted water may be added back into another segment as a specified inflow. Four different options are available for diverting flow from the farthest downstream reach of a stream segment; these options are described in **appendix 1** (input instructions for MODFLOW-2005). Streambed properties can vary among reaches within a segment but are considered uniform within the reach (**appendix 1**).

Stream segments typically are numbered in downstream order starting at the upstream segment and ending at the farthest downstream segment (**fig. 8**) (Prudic and others, 2004). There is flexibility in how segments can

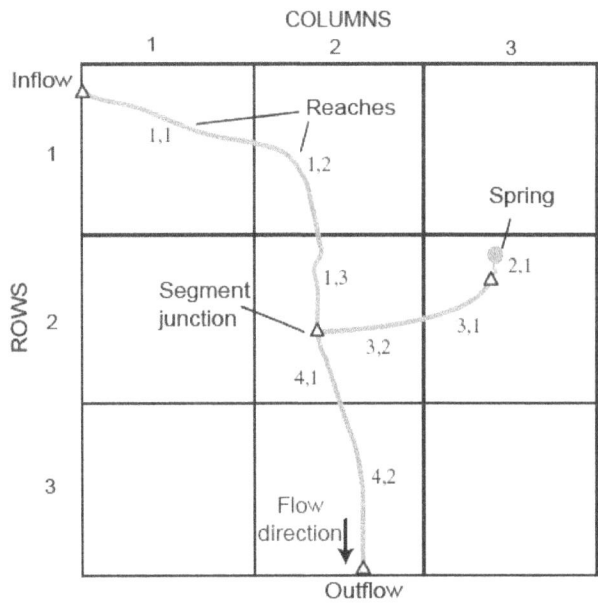

EXPLANATION

1,3   Segment number and reach number

Figure 8. Simple stream network having four segments and eight reaches in a finite-difference model grid consisting of three rows and three columns (modified from Prudic and others, 2004).

be numbered, and an upstream tributary segment can have a lower number than the downstream segment receiving its flow. However, within a segment, reach numbers must increase in downstream order. The direction of streamflow remains the same throughout the simulation. Multiple segments may be joined as tributaries of a downstream segment and outflow from all segments entering a confluence are summed and added to the downstream segment. Only a single segment may emanate from a confluence.

A simple stream network is illustrated in figure 8 with a rectangular finite-difference grid having 1 layer, 3 rows, and 3 columns. Each finite-difference cell is designated first by the layer number and followed by the row and column numbers, respectively. Thus, cell–1,1,2 is the cell in layer 1, row 1, and column 2. Segment 1 in this example has three reaches and inflow is specified for the upstream end of the first reach. Reach 1 of segment 1 is in cell–1,1,1; reach 2 is in cell–1,1,2; and reach 3 is in cell–1,2,2. Segment 2 has one reach in cell–1,2,3 and is used to represent ground-water discharge from a spring into a stream. Segment 3 has two reaches and receives inflow from segment 2. Reach 1 of segment 3 is in cell–1,2,3 and reach 2 is in cell–1,2,2. Segment 4 has two reaches and receives inflow from the confluence of segments 1 and 3 (called a segment junction in fig. 8). Reach 1 of segment 4 is in cell–1,2,2 and reach 2 is in cell–1,3,2. When multiple stream reaches are assigned to a finite-difference cell such as cell–1,2,2 in figure 8, the simulated ground-water head in the cell is used to calculate the flow of water between the cell and each reach.

Ground-water interactions with lakes can be simulated with the Lake Package (Merritt and Konikow, 2000). The package determines ground-water interactions with lakes using a head-dependent boundary condition in which the head or stage in a lake is determined from the difference between inflows to and outflows from a lake. The package originally was developed for MODFLOW-2000 (Harbaugh and others, 2000) and has been revised and converted for use with MODFLOW-2005 and GSFLOW. The Lake Package also was modified to include the capability to simulate unsaturated flow beneath lakes.

Lakes are represented as a volume within a group of designated finite-difference cells, referred to here as lake cells. Lake cells are different than other cells as they are not directly considered in the computation of ground-water heads in MODFLOW-2005. Rather, a lake stage is computed by the Lake Package and applied to all cells that comprise the lake. This lake stage is used to calculate ground-water – lake interaction. Multiple lakes can be delineated in the finite-difference grid. Lakes are numbered in any order from one to the total number of simulated lakes. Lakes can be grouped to allow them to coalesce during high lake stages and separate during low lake stages. Surface inflows to a lake can be from precipitation, surface runoff, point sources, and streams. Surface outflows from a lake can be from evaporation, point withdrawals, and streams. Multiple streams can enter and leave a lake from different locations.

A simple stream-lake network associated with a finite-difference grid of an aquifer having 3 layers, 7 rows, and 7 columns is illustrated in figure 9. A total of 147 finite-difference cells are used to define the rectangular aquifer and lake. Each finite-difference cell is designated first by the layer number and followed by the row and column numbers, respectively. Cell–1,1,2 is the cell in layer 1 at row 1, and column 2. The example includes two stream segments and one lake. Stream segment 1 has two reaches. Reach 1 of segment 1 is in cell–1,1,3 and reach 2 is in cell–1,2,4. Inflow to stream segment 1 is specified at the beginning of reach 1. Outflow from reach 2 in segment 1 enters the lake at cell–1,3,4. Stream segment 2 also has two reaches. Reach 1 of segment 2 begins in cell–1,6,5 and reach 2 is in cell–1,7,6. The lake occupies nine cells in layer 1, three cells in layer 2, and zero cells in layer 3. Vertical discretization of the lake is shown along row 4 (fig. 9B). Outflow from the lake to stream segment 2 is added as inflow to reach 1. Outflow from segment 2 exits the modeled area.

Time Discretization—Stress Periods and Time Steps

Time also is discretized into increments because the calculation of ground-water heads can be a function of time as well as space (Harbaugh, 2005, p. 2-3). Time increments are allowed to vary during a simulation and are divided into two levels (Harbaugh, 2005, p. 4-4). The first level is called a stress period. Stress periods correspond to changes in the specified boundary conditions such as fluctuations in the rate of pumping from a cell. Stress periods also can be used to change user-specified inflow and outflow to streams and lakes, such as those resulting from diversions. A simulation must have at least one stress period and stress-period lengths can vary. Stress periods can be either steady state or transient (Harbaugh, 2005, p. 3-6). A steady-state stress period neglects changes in storage, whereas a transient stress period considers the effects of changes in storage in the calculation of ground-water heads. Multiple steady-state stress periods can be interspersed with transient stress periods when GSFLOW is run in MODFLOW only mode; however, when run in integrated mode, the stress periods must all be transient after the initial stress period.

The second level of time discretization is called a time step. Time steps are used to divide a change in stress into finer time increments that may be needed for solution convergence or to accurately solve the distribution of ground-water heads. In addition, time steps also can be specified for printing model output at specific times. The solution accuracy can be tested by increasing the number of time steps for a stress period until ground-water heads at the end of the stress period no longer change with increasing time steps. A steady-state stress period has only one time step. Transient stress periods within GSFLOW must have time steps that are divided into uniform daily time increments.

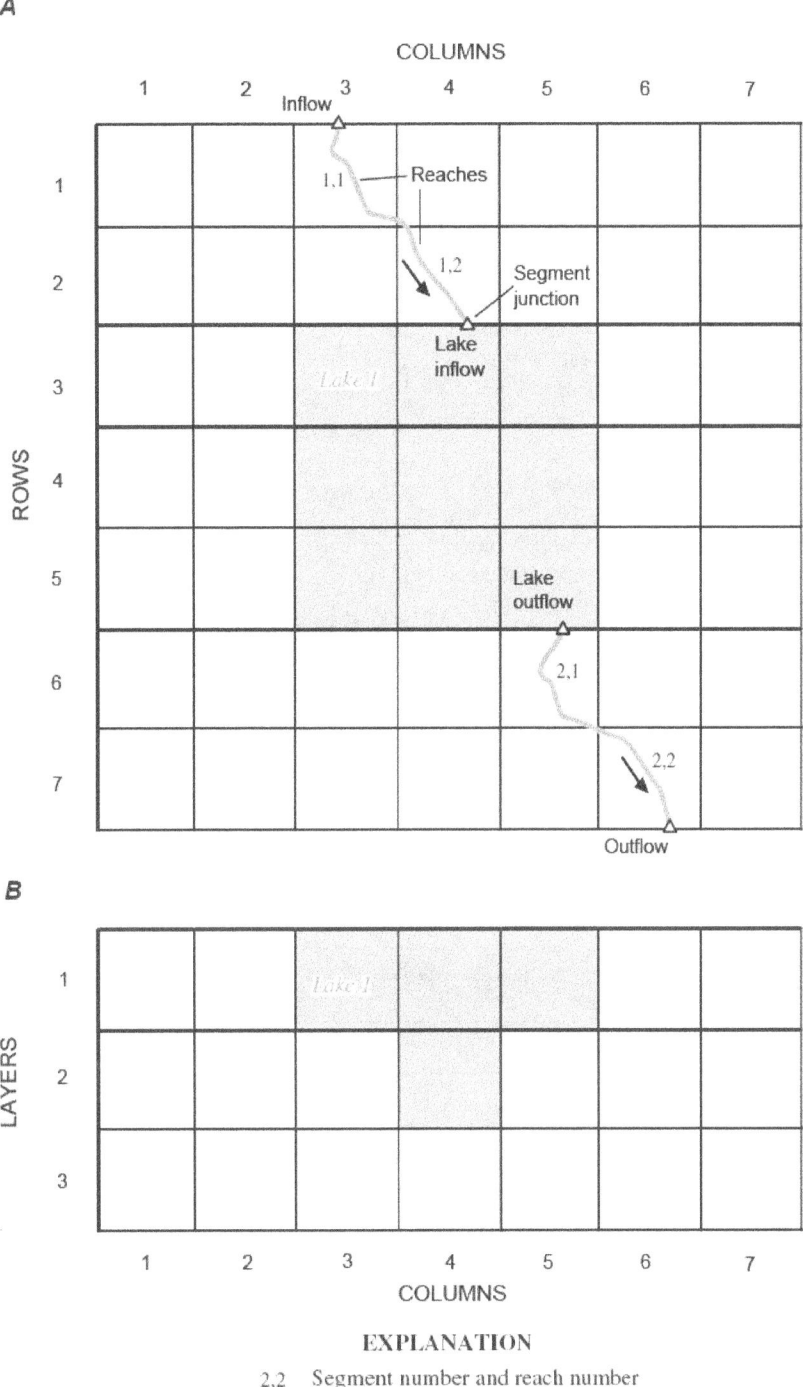

Figure 9. Simple stream and lake network having one lake, two stream segments, and four stream reaches in a finite-difference model grid consisting of seven rows and columns and three layers: (*A*) plan-view of active finite difference grid, (*B*) vertical section along model row 4.

Organization

MODFLOW-2005 is organized into processes and packages (Harbaugh, 2005, p. 1-2). Each process and package contributes to the overall capabilities of MODFLOW-2005. A process is defined as a section of the computer code that solves a major equation or set of related equations. Thus, the section of the computer program that solves the ground-water flow equation is called the Ground-Water Flow Process. GSFLOW consists of the Ground-Water Flow (Harbaugh, 2005) and Observation Processes (Hill and others, 2000). The Observation Process is used to compare model-calculated heads and flows to measured values of heads and flows. Processes that work with MODFLOW but are not currently (2007) supported by GSFLOW include: the Sensitivity and Parameter Estimation Processes of MODFLOW-2000 (Hill and others, 2000); the Ground-Water Transport Process (Konikow and others, 1996); the Local Grid Refinement Process (Mehl and Hill, 2006); the Ground-Water Management Process (Ahlfeld and others, 2005); and the Farm Process (Schmid and others, 2006). However, GSFLOW was designed with the intention to support these or other processes as they become needed to solve future problems.

Processes are divided into packages. A package is defined as the part of a process that pertains to a single aspect of the model's operations. Packages that are used to simulate time-varying, head-dependent or specified-flow boundary conditions are called stress packages, whereas packages used to solve the system of simultaneous finite-difference equations are called solver packages. Packages consist of a group of subroutines that are accessed from the MODFLOW-2005 main program. A detailed description of the organization of MODFLOW-2005 is presented by Harbaugh (2005, Chapter 1).

Computational Sequence

A MODFLOW-2005 simulation begins by reading input data that do not change with stress periods. The number of cells in the grid for the problem to be simulated is determined as well as the stress-package options and solution method. Subsequently, memory is allocated for all aspects of the simulation. Following memory allocation, the finite-difference grid, stress periods and time steps, and starting ground-water heads and hydraulic properties are read for each finite-difference cell (fig. 10).

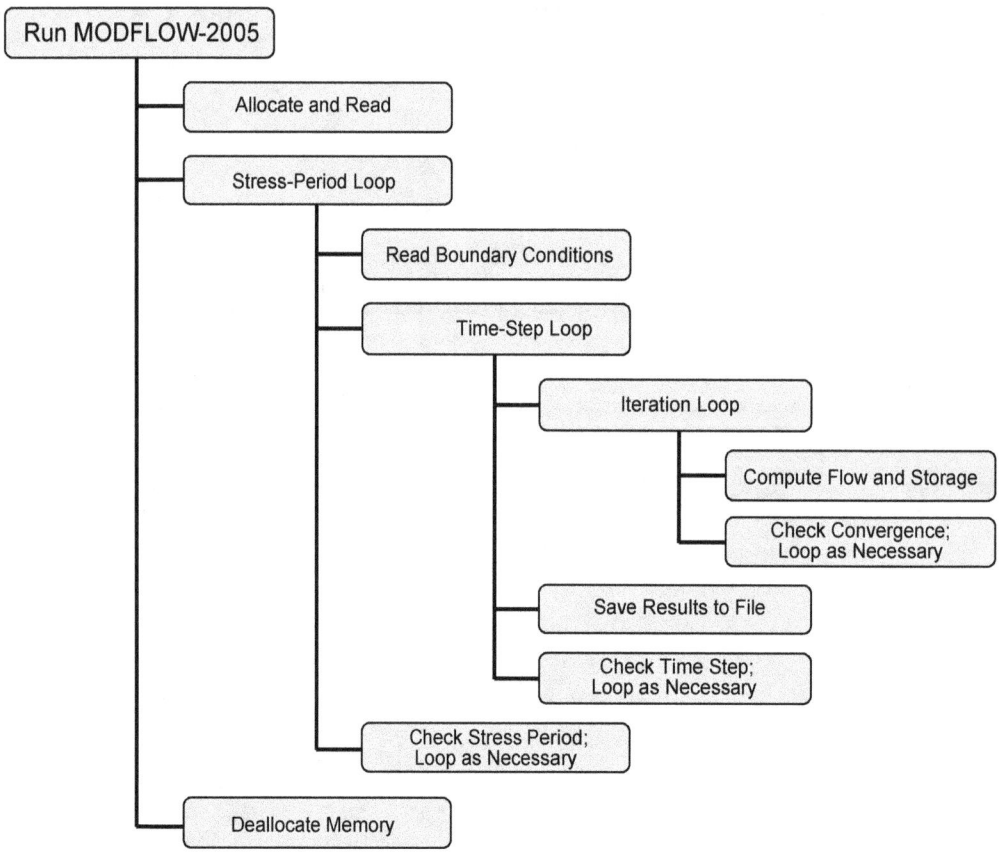

Figure 10. Computational sequence of MODFLOW-2005 used for simulating ground-water flow through aquifer systems (modified from Harbaugh, 2005).

The computational sequence follows three basic loops (Harbaugh, 2005, p. 3-1). The first loop is the stress-period loop, during which specified stress data remain constant. Inside the stress-period loop is the time-step loop where ground-water heads and flows to and from head-dependent boundaries are computed at the end of each time step. Iterative solution methods are used to solve for the heads for each time step in the iteration loop (Harbaugh, 2005, p. 2-10).

Flows, ground-water heads, and a summary water budget are saved or printed at the end of a time step when requested. The program continues through each time step and stress period until the end of the simulation. However, the program may terminate during a time step if the difference in heads between iterations does not decrease below a specified error tolerance within a specified number of iterations; this type of program termination is referred to as convergence failure. Convergence failure may be caused by various conditions in the model that can usually be mitigated by increasing the number of time steps (for MODFLOW-only simulations) or by changing solver input parameters or other data input.

The choice of daily time steps for GSFLOW was made considering limitations of time-step length on model convergence. Experience has shown that daily time steps are sufficiently small for MODFLOW to converge to a solution for most simulations, and GSFLOW applications should not require reduction in time-step length less than 1 day.

Initial ground-water heads are required for the first stress period of a simulation regardless of whether the stress period is designated as steady state or transient. Initial heads are needed to calculate heads at the beginning of the simulation and to check for no-flow cells where the initial head is beneath the bottom of the cell. Initial heads do not affect the steady-state solution other than determining if a cell is a no-flow cell or a variable-head cell. Initial heads are critically important for transient solutions, however, because they set the initial storage conditions available for the subsequent transient run. A steady-state stress period can be used to determine the initial head distribution for a subsequent transient stress period. Initial heads that are generated from a steady-state stress period help reduce unrealistic changes in heads at the beginning of a transient stress period and provide a defensible and representative basis for model calibration.

Integration of PRMS and MODFLOW-2005

Flow is exchanged among the three regions of GSFLOW on the basis of interdependent equations that calculate flow and storage of water throughout the simulated hydrologic system (fig. 11). The first region includes the plant canopy, snowpack, impervious storage, and soil zone.

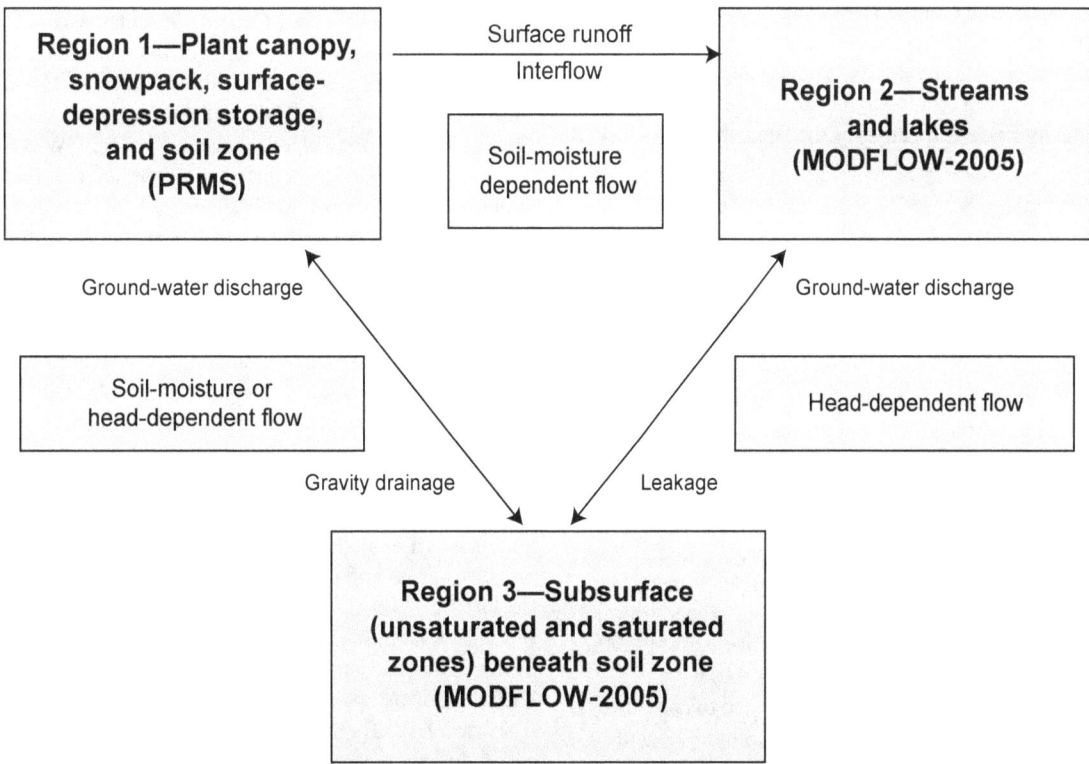

Figure 11.   Schematic diagram of the exchange of flow among the three regions in GSFLOW. The dependency on soil moisture and head in the computation of flow among the regions also is shown.

The second region consists of streams and lakes. The subsurface or third region is beneath regions 1 and 2, which consists of the unsaturated and saturated zones. The first region is simulated with PRMS modules, while the second and third regions are simulated with MODFLOW-2005 packages. A description of how GSFLOW simulates water flow within all three regions is provided in section "Computations of Flow."

Specified inputs of precipitation and temperature and specified inputs or model-estimated potential solar radiation are distributed to each HRU to compute energy budgets, flow, and storage within region 1. A portion of the water entering region 1 infiltrates into the soil zone, where it is evaporated and transpired back to the atmosphere, flows to streams and lakes (region 2), and (or) drains to the deeper unsaturated and saturated zones (region 3).

The rate at which water flows from the soil zone to streams and lakes is dependent on: (1) the rate at which water is added to the land surface by snowmelt and rain, (2) the rate of infiltration into the soil zone, and (3) the antecedent soil-zone storage. Water that flows from the soil zone to the unsaturated and saturated zones (region 3) is called gravity drainage. Gravity drainage is dependent on the vertical hydraulic conductivity of the unsaturated zone and the volume of water stored in the soil zone. Additionally, gravity drainage ceases as the water table rises into the soil zone. Water also can flow from the saturated zone into the soil zone as ground-water discharge; the rate of discharge is dependent on the hydraulic conductivity and ground-water head relative to the altitude of the soil-zone base. Flow between the unsaturated and saturated zones to streams and lakes is dependent on the ground-water head in relation to the stream- or lake-surface altitude, the hydraulic properties of the streambed and lakebed sediments, and the hydraulic properties of the unsaturated and saturated zones.

PRMS Modules and MODFLOW-2005 Packages Implemented in GSFLOW

A total of 25 PRMS modules and 16 MODFLOW-2005 packages are implemented in GSFLOW, (table 1). PRMS modules used in GSFLOW were reformatted to a consistent coding style and updated using Fortran 90 (American National Standards Institute, 1992) language elements. Some PRMS modules were revised during this conversion to include more accurate constants and to correct existing programming errors. The functionality of the revised modules is identical to those in the original PRMS documentation except where noted in table 1 and described in section "Changes to PRMS." Additionally, some MODFLOW-2005 packages were modified to allow linkages to PRMS flows and to correct existing programming errors. Several PRMS modules and MODFLOW-2005 packages are not implemented in the initial version of GSFLOW (table 2). Developers that intend to add new modules and packages to GSFLOW are cautioned to evaluate how these additions could affect surface runoff and

interflow to streams and lakes or how they could affect the link between the soil zone in PRMS with the unsaturated and saturated zones in MODFLOW-2005.

GSFLOW allows for three simulation modes—integrated, PRMS-only, and MODFLOW-only. All but two PRMS modules can be used in the integrated mode; the two modules that can not be used in the integrated mode are the Ground-Water Flow and Streamflow Modules (gwflow_casc_prms and strmflow_prms, table 1). All MODFLOW-2005 packages listed in table 1 can be used for simulation using either the integrated or MODFLOW-only modes. PRMS modules, MODFLOW-2005 packages, and GSFLOW modules that have been modified or created also are denoted in table 1 and modules and packages are listed in their computational order. The capability of having PRMS-only and MODFLOW-only simulations in GSFLOW allows incremental model setup that provides flexibility in calibration. For example, independent calibration of the example test problem during the development of GSFLOW allowed for a better understanding of the sensitivity of the models to the various model parameters, and resulted in an easier calibration of the integrated model.

Changes to PRMS

Changes to PRMS modules include: (1) a new HRU-based distribution method for solar radiation that may improve the calculations of evapotranspiration and the energy budget of the snowpack; (2) a new Cascade Module designed to simulate more complex flow paths in watersheds; (3) a new Soil-Zone Module designed to link with MODFLOW-2005; and (4) a new type of HRU designed to link precipitation and evaporation with the Lake Package in MODFLOW-2005.

The new HRU-based distribution method for solar radiation was developed to distribute daily solar radiation on the basis of the average slope and aspect and the latitude of the centroid of each HRU. The new method computes potential solar radiation for each Julian day of the year for each HRU. This differs from the original distribution method in which each HRU was assigned to a radiation plane determined by slope, aspect, and latitude. Daily potential solar radiation for each radiation plane also was estimated by linear interpolation of 13 values that represent the annual cycle of potential solar radiation (Leavesley and others, 1983, p. 14-15). Changes to the computation and distribution of solar radiation were implemented in modules soltab_hru_prms, ddsolrad_hru_prms, ccsolrad_hru_prms, and potet_hamon_hru_prms (table 1).

The new Cascade Module (cascade_prms; table 1) is designed to route surface runoff and interflow from upslope HRUs to downslope HRUs. Surface runoff and interflow from each HRU for the daily mode of PRMS originally was added directly to streamflow (fig. 2 and Leavesley and others, 1983, p. 25 and 31). The new method allows surface runoff and interflow to satisfy soil-zone storage of downslope HRUs before being added as streamflow or inflow to a lake.

Table 1. Description of PRMS and GSFLOW modules and MODFLOW-2005 packages implemented in GSFLOW, listed in computational order.

[**Module or package name:** User selects only one of the modules or packages in each group indicated by a number from 1 to 7. **Model mode:** G is integrated simulation, P is PRMS-only simulation, M is MODFLOW-only simulation. HRU, hydrologic response unit]

Hydrologic or computational process	Module or package	Description	New or modified	Model mode
		PRMS modules		
Basin	basin_prms	Defines shared watershed-wide and HRU physical parameters and variables		G,P
Cascade	cascade_prms	Determines computational order of the HRUs and ground-water reservoirs for routing flow downslope	new	G,P
Observed data	obs_prms	Reads and stores observed data from all specified measurement stations		G,P
Solar table	soltab_hru_prms	Computes potential solar radiation for each HRU; modification of soltab_prms	modified	G,P
Temperature (1)	temp_1sta_prms	Distributes maximum and minimum temperatures to each HRU using temperature data measured at one station and an estimated monthly lapse rate		G,P
	temp_laps_prms	Distributes maximum and minimum temperatures to each HRU using temperature data measured at a base station and lapse station with differing altitudes		G,P
	xyz_dist	Distributes precipitation and maximum and minimum temperatures to each HRU using a multiple linear regression of measured data from a group of measurement stations or from atmospheric model results. Selection requires that module also be used for precipitation		G,P
	temp_dist2_prms	Distributes temperature from one or more stations to each HRU by an inverse distance weighting scheme		G,P
Precipitation (2)	precip_prms	Distributes precipitation from one or more stations to each HRU and determines the form of precipitation using monthly correction factors to account for differences in altitude, spatial variation, topography, and measurement gage efficiency		G,P
	precip_laps_prms	Distributes precipitation from one or more stations to each HRU and determines form of precipitation using monthly lapse rate adjustments to account for the differences in altitude of each HRU relative to each measurement station		G,P
	xyz_dist	Distributes precipitation and maximum and minimum temperatures to each HRU using a multiple linear regression of measured data from a group of measurement stations or from atmospheric model results. Selection requires that module also be used for temperature		G,P
	precip_dist2_prms	Distributes precipitation from one or more stations to each HRU by an inverse distance weighting scheme		G,P
Solar radiation (3)	ddsolrad_hru_prms	Distributes solar radiation to each HRU and estimates missing solar radiation data using a maximum temperature per degree-day relation; modification of ddsolrad_prms	modified	G,P
	ccsolrad_hru_prms	Distributes solar radiation to each HRU and estimates missing solar radiation data using a relation between solar radiation and cloud cover; modification of ccsolrad_prms	modified	G,P
Potential evapotranspiration (4)	potet_jh_prms	Determines whether current time period is one of active transpiration, and computes the potential evapotranspiration for each HRU using the Jensen-Haise formulation (Jensen and others, 1969)		G,P
	potet_hamon_hru_prms	Determines whether current time period is one of active transpiration, and computes the potential evapotranspiration for each HRU using the Hamon formulation (Hamon, 1961); modification of potet_hamon_prms	modified	G,P
	potet_epan_prms	Determines whether current time period is one of active transpiration and computes the potential evapotranspiration for each HRU using pan-evaporation data		G,P

Table 1.   Description of PRMS and GSFLOW modules and MODFLOW-2005 packages implemented in GSFLOW, listed in computational order.—Continued

[**Module or package name:** User selects only one of the modules or packages in each group indicated by a number from 1 to 7. **Model mode:** G is integrated simulation, P is PRMS-only simulation, M is MODFLOW-only simulation; HRU, hydrologic response unit]

Hydrologic or computational process	Module or package	Description	New or modified	Model mode
PRMS modules—Continued				
Interception	intcp_prms	Computes volume of intercepted precipitation, evaporation from intercepted precipitation, and throughfall that reaches the soil or snowpack		G,P
Snow	snowcomp_prms	Initiates development of a snowpack and simulates snow accumulation and depletion processes using an energy-budget approach		G,P
Surface runoff and infiltration (5)	srunoff_smidx_casc	Computes surface runoff and infiltration for each HRU using a non-linear variable-source-area method allowing for cascading flow; modification of srunoff_smidx_prms	modified	G,P
	srunoff_carea_casc	Computes surface runoff and infiltration for each HRU using a linear variable-source-area method allowing for cascading flow; modification of srunoff_carea_prms	modified	G,P
Soil zone	soilzone_gsflow	Computes inflows to and outflows from soil zone of each HRU and includes inflows from infiltration, ground-water, and upslope HRUs, and outflows to gravity drainage, interflow, and surface runoff to down-slope HRUs. Inflow from ground water is not allowed when using the PRMS-only mode; merge of smbal_prms and ssflow_prms with enhancements	new	G,P
Ground water	gwflow_casc_prms	Sums inflow to and outflow from PRMS ground-water reservoirs; outflow can be routed to downslope ground-water reservoirs, stream segments, and sink modification of gwflow_prms	modified	P
Streamflow	strmflow_prms	Computes daily streamflow as the sum of surface runoff, interflow, detention reservoir interflow, and ground-water flow		P
Summary	hru_sum_prms	Computes daily, monthly, yearly, and total flow summaries of volumes and flows for each HRU		G,P
	basin_sum_prms	Sums values for daily, monthly, yearly, and total flow summaries of volumes and flows for all HRUs		G,P
MODFLOW Packages				
Basic	BAS	Handles a number of basic administrative tasks		G,M
Block centered flow (6)	BCF	Calculates conductance coefficients for ground-water flow equations using a block-centered flow package		G,M
Layer property flow (6)	LPF	Calculates conductance coefficients for ground-water flow equations using a layer-property flow package		G,M
Hydrostratigraphic unit flow (6)	HUF	Calculates effective hydraulic properties for model layers using hydrostratigraphic units		G,M
Horizontal flow barrier	HFB	Simulates flow barriers by reducing horizontal conductance		G,M
Well	WEL	Adds terms to flow equation to represent wells		G,M
Multi-node well	MNW	Adds terms to flow equation for wells that extract or inject water in more than one cell		G,M
General head boundary	GHB	Adds terms to flow equation to represent general head-dependent boundaries		G,M
Constand head boundary	CHD	Adds terms to flow equation to represent constant-head boundaries		G,M
Flow and head boundary	FHB	Adds terms to flow equation to represent flow and head boundaries		G,M
Streamflow routing	SFR	Adds terms to flow equation to represent ground-water and stream interactions; modification of SFR2	modified	G,M
Lake	LAK	Adds terms to flow equation to represent ground-water and lake interactions; modification of LAK3	modified	G,M
Unsaturated zone flow	UZF	Adds terms to flow equation to represent recharge from the unsaturated zone, evapotranspiration, and ground-water discharge to land surface	modified	G,M

Table 1. Description of PRMS and GSFLOW modules and MODFLOW-2005 packages implemented in GSFLOW, listed in computational order.—Continued

[**Module or package name:** User selects only one of the modules or packages in each group indicated by a number from 1 to 7. **Model mode:** G is integrated simulation, P is PRMS-only simulation, M is MODFLOW-only simulation; HRU, hydrologic response unit]

Hydrologic or computational process	Module or package	Description	New or modified	Model mode
MODFLOW Packages—Continued				
Gage	GAG	Prints time series gage output for selected stream reaches and lakes; modification of GAG5	modified	G,M
Solver (7)	SIP	Solves simultaneous equations resulting from finite-difference approximations using the strongly implicit procedure		G,M
	PCG	Solves simultaneous equations resulting from finite-difference approximations using a preconditioned conjugate-gradient procedure	modified	G,M
	DE4	Solves simultaneous equations resulting from finite-difference approximations using a direct solution procedure		G,M
	GMG	Solves simultaneous equations resulting from finite-difference approximations using a geometric multigrid solution of the preconditioned conjugate-gradient procedure		G,M
GSFLOW modules				
Computation Order	gsflow_prms	Controls model mode and computational sequence order of PRMS and GSFLOW modules for each time step; modification of the PRMS call_modules.c subroutine	modified	G,P,M
	gsflow_modflow	Controls sequence order for reading time-dependent input data and controls the computational sequence of calculations in the time-step and iteration loops for MODFLOW-2005 packages and PRMS and GSFLOW modules dependent on flows to and from MODFLOW; modification of the MODFLOW-2005 main program	modified	G,M
Conversion	gsflow_setconv	Defines variables used for converting between PRMS and MODFLOW-2005 units	new	G
Integration	gsflow_prms2mf	Distributes gravity drainage from gravity reservoirs in soil zone of PRMS to finite-difference cells in MODFLOW-2005; also distributes surface runoff and interflow from HRUs to stream segments and lakes in MODFLOW-2005	new	G
	gsflow_mf2prms	Distributes ground-water discharge from finite-difference cells in MODFLOW-2005 to gravity reservoirs in soil zone of PRMS	new	G
Summary	gsflow_budget	Computes watershed budget for GSFLOW and adjusts final storage in gravity reservoirs using flows to and from finite-difference cells at end of each time step	new	G
	gsflow_sum	Computes detailed water budgets for all flow components at end of each time step	new	G

Table 2. PRMS modules and MODFLOW-2005 packages not implemented in GSFLOW.

[HRU, hydrologic response unit]

Hydrologic or computational process	Module or package	Description
PRMS modules		
Solar radiation	ccsolrad_prms	Distributes solar radiation as determined for radiation planes to each HRU and estimates missing solar radiation data using a relation between solar radiation and cloud cover
	ddsolrad_prms	Distributes solar radiation as determined for radiation planes to each HRU and estimates missing solar radiation data using a maximum temperature per degree-day relation
Storm infiltration	grnampt_infil_prms	Compute infiltration for each HRU during storm events using a modified Green and Ampt infiltration approach (Green and Ampt, 1911)
Storm surface runoff	krout_ofpl_prms	Computes surface runoff for each overland flow plane during storm events using kinematic routing and sediment detachment and transport using a rill-interrill concept
Surface runoff and infiltration	srunoff_carea_prms	Computes surface runoff and infiltration for each HRU using a non-linear variable-source-area method
	srunoff_smidx_prms	Computes surface runoff and infiltration for each HRU using a linear variable-source-area method
Soil moisture accounting	smbal_prms	Computes soil-moisture accounting for each HRU including addition of infiltration, computation of actual evapotranspiration, and seepage to subsurface and ground-water reservoirs
Subsurface flow	ssflow_prms	Adds inflow to subsurface reservoirs and computes outflow to ground-water reservoirs and to streamflow
Ground water	gwflow_prms	Sums inflow to ground-water reservoirs and computes outflow to streamflow and to a ground-water sink if specified
Storm channel flow	krout_chan_prms	Computes flow for each stream segment during storm events using kinematic routing, reservoir routing, and sediment transport
Streamflow	strmflow_st_prms	Computes daily streamflow as the sum of surface, subsurface, and ground-water flow contributions, storm runoff totals for storm periods, and daily detention reservoir routing
MODFLOW-2005 Packages		
Recharge	RCH	Adds terms to ground-water flow equation to represent areal recharge to ground-water system
Evapotranspiration	EVT	Adds terms to ground-water flow equation to represent head-dependent evapotranspiration from ground-water system
Segmented evapotranspiration	ETS	Adds terms to ground-water flow equation to represent segmented head-dependent evapotranspiration from ground-water system
Interbed storage	IBS	Adds terms to ground-water flow equation to represent inelastic compaction of fine-grained sediments
Subsidence	SUB	Simulates aquifer-system compaction and land subsidence
Tile drain	DRT	Adds terms to ground-water flow equation to represent ground-water discharge to drains while accounting for irrigation return flows
River	RIV	Adds terms to ground-water flow equation to represent rivers to represent head-dependent flow between a surface water body and a ground-water system.
Drain	DRN	Adds terms to ground-water flow equation to represent ground-water discharge to drains
Reservoir	RES	Adds terms to ground-water flow equation to represent leakage from reservoirs

The new Soil-Zone Module (table 1) was developed to link the soil zone in PRMS to the subsurface zone, streams, and lake simulated by MODFLOW-2005 (fig. 12). The new module is a replacement of the original PRMS Soil-Moisture Accounting and Subsurface-Flow Modules (Leavesley and others, 1983, p. 19 and 31); thus, the original modules are not implemented in GSFLOW (table 2). The old Soil-Moisture Accounting Module was used to compute flows into and out of the soil-zone reservoir, and the Subsurface-Flow Module was used to compute flows into and out of the subsurface reservoir (fig. 12A).

The new Soil-Zone Module replaces the finite-volume used to account for water content below field capacity in the soil-zone reservoir and the infinite-volume used for the subsurface reservoir with three finite-volume reservoirs. These finite-volume reservoirs are the capillary reservoir, preferential-flow reservoir, and gravity reservoir (fig. 12B). The finite-volume reservoirs were developed to allow for saturation excess in the soil zone to become surface runoff (Dunne and Black, 1970; Freeze, 1972). Saturation excess is referred to here as Dunnian runoff. Dunnian runoff

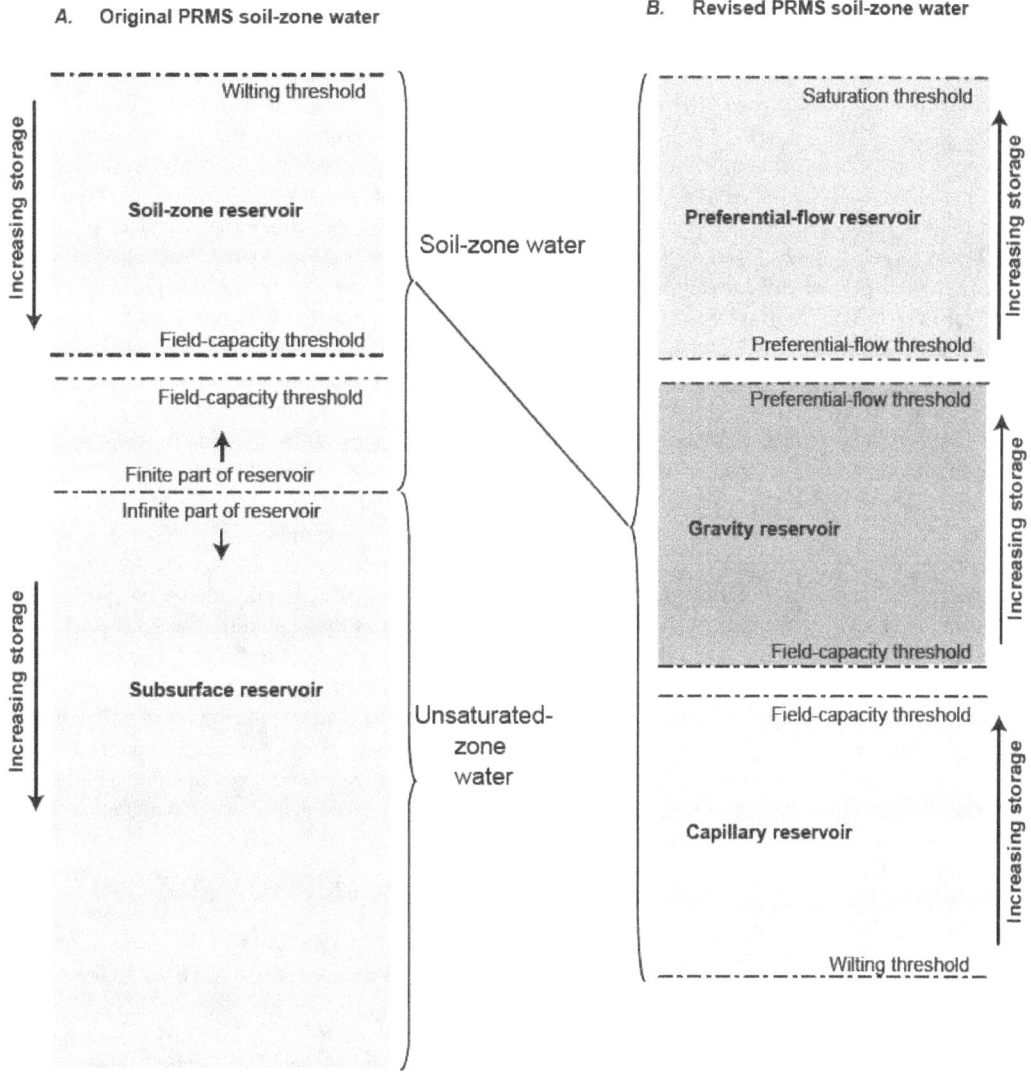

Figure 12. Differences in conceptualization of soil-zone water simulated by PRMS, where boxes are storage reservoirs representing pore-space volumes for a given volume of soil: (A) original conceptualization of a finite-volume soil-zone reservoir and an infinite-volume subsurface reservoir; (B) revised conceptualization of three finite-volume reservoirs: preferential flow, gravity, and capillary.

differs from Hortonian runoff, which is generated when the precipitation rate exceeds the infiltration rate of the soil that may not be saturated (Horton, 1933). Hortonian runoff is computed in both PRMS Surface Runoff Modules (srunoff_smidx_casc or srunoff_carea_casc; table 1). Capillary, preferential-flow, and gravity reservoirs are shown as separate boxes within figure 12 because they represent separate storage reservoirs for computational purposes in the model; however, in reality, they are all contained within the same physical space. Another reason for separating them into separate stacked boxes is that their order in the vertical sequence corresponds to the order in which they are filled and drained, from the bottom up.

The capillary reservoir is the same as the original soil-zone reservoir and represents water held in the soil by capillary forces between the wilting and field-capacity thresholds. Water is removed from the reservoir by evaporation and transpiration.

The gravity reservoir replaces the subsurface reservoir and represents water in the soil zone between field-capacity and saturation thresholds that is not subject to the preferential-flow threshold. This reservoir was developed to provide gravity drainage from the soil zone to the unsaturated zone simulated by MODFLOW-2005. The gravity reservoir also is capable of receiving ground-water discharge into the soil zone whenever the ground-water head in a connected finite-difference cell is greater than the soil-zone base (fig. 12B). Gravity drainage is added to the ground-water reservoir in PRMS instead of to finite-difference cells when using the PRMS-only mode in GSFLOW. The ground-water reservoir in PRMS only can discharge water to a stream, a downslope ground-water reservoir, and/or a ground-water sink. Thus, the gravity reservoir is not capable of receiving ground-water discharge when using the PRMS-only mode. Water in the gravity reservoir also is available for downslope flow within the soil zone. The downslope flow is referred to herein as slow interflow.

The preferential-flow reservoir is a new reservoir that is not part of the original concept of the soil zone. This reservoir represents soil water between field capacity and saturation that is available for fast interflow through relatively large openings in the soil of each HRU (fig. 12B). The capacity of this reservoir is depicted in figure 12B as the area between the preferential-flow and saturation thresholds. Both fast interflow and slow interflow are routed to downslope HRUs, streams, and lakes at the end of a time step.

The original conceptualization of the subsurface reservoir allowed for fewer reservoirs than HRUs (fig. 13). This conceptualization has been changed with the revision of the soil zone. At least one gravity reservoir must correspond to each HRU. Datasets created for PRMS using fewer subsurface reservoirs than HRUs must be revised before importing the dataset into GSFLOW by replacing the subsurface reservoirs with gravity reservoirs. Some of the original parameters used for the subsurface reservoirs are used for the gravity reservoirs to maintain compatibility with existing PRMS model

applications. Several new parameters are required for both the gravity and preferential-flow reservoirs and must be created or added to the imported datasets. Descriptions of the parameters needed for PRMS in GSFLOW are in appendix 1.

PRMS-only simulations can include fewer ground-water reservoirs than HRUs (fig. 13). The revised Ground-Water Flow Module (gwflow_casc_prms, table 1), computes ground-water flow to streams and allows for cascading flow among the ground-water reservoirs. The original PRMS Streamflow Module (strmflow_prms, table 1) is unchanged and associated parameters must be specified in the PRMS Parameter File. This PRMS module, like the Ground-Water Flow Module, is limited to PRMS-only simulations (table 1).

A designation for HRUs was added as a parameter to PRMS (parameter `hru_type`) to distinguish between land-based, lake, and inactive HRUs (fig. 14, appendix 1). Land-based HRUs are assigned an integer value of 1 and lake HRUs, which encompass the maximum surface area of a lake, are assigned an integer value of 2. Only precipitation, potential solar radiation, and lake evaporation are computed for lake HRUs. Other processes such as plant canopy interception, surface runoff, and interflows through the soil zone are assumed zero and snowpack is computed as a liquid equivalent. HRUs that are inactive in the simulation are assigned an integer value of zero and no computations are performed.

Several PRMS modules are not implemented in GSFLOW (table 2). The PRMS modules ccsolrad_prms, ddsolrad_prms, srunoff_carea_prms, srunoff_smidx_prms, and gwflow_prms were replaced by modified versions for GSFLOW. The Soil-Zone Module (table 1) was developed for the purpose of linking PRMS with MODFLOW-2005. The new module replaces PRMS Soil-Moisture Accounting and Subsurface-Flow Modules (smbal_prms and ssflow_prms, table 2). PRMS modules grnampt_infil_prms, krout_chan_prms, krout_ofpl_prms, and strmflow_st_prms are not available because PRMS storm mode is not implemented in this version of GSFLOW. Details on the implementation of the changes to PRMS that affect the rate and timing of surface runoff and interflow to streams and lakes, and interactions with ground water are described in section, "Computations of Flow."

Changes to MODFLOW-2005

Changes to MODFLOW-2005 needed for GSFLOW include refinements to existing MODFLOW-2005 packages. The Unsaturated-Zone Flow Package (Niswonger and others, 2006a) was developed for GSFLOW to simulate ground-water recharge and discharge to land surface by accounting for vertical flow through an unsaturated zone. However, the Unsaturated-Zone Flow Package also was designed to be used in MODFLOW-only simulations and was released as an independent MODFLOW-2005 package prior to the release of GSFLOW (Niswonger and others, 2006a). Modifications to existing packages included revisions to the Streamflow-Routing Package (Niswonger and Prudic, 2006) so that

A.

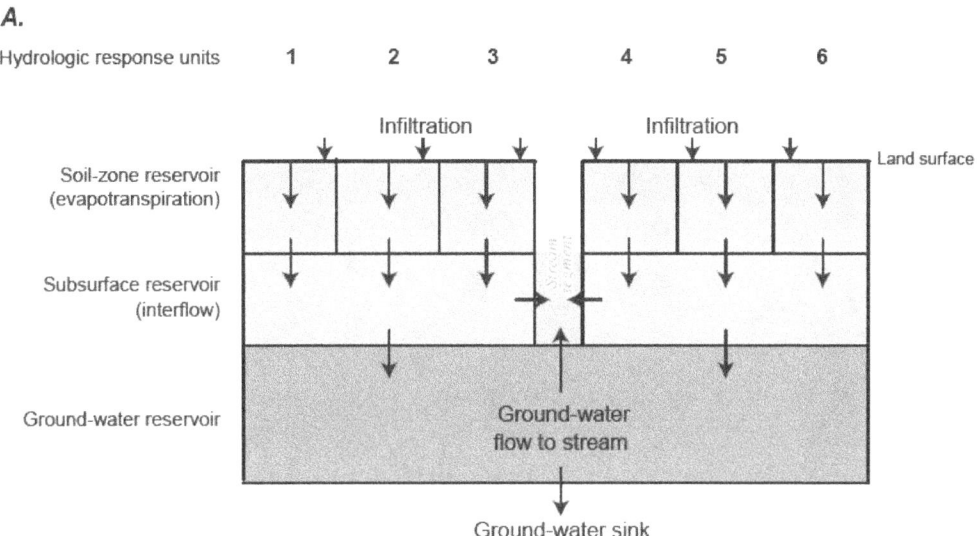

B.

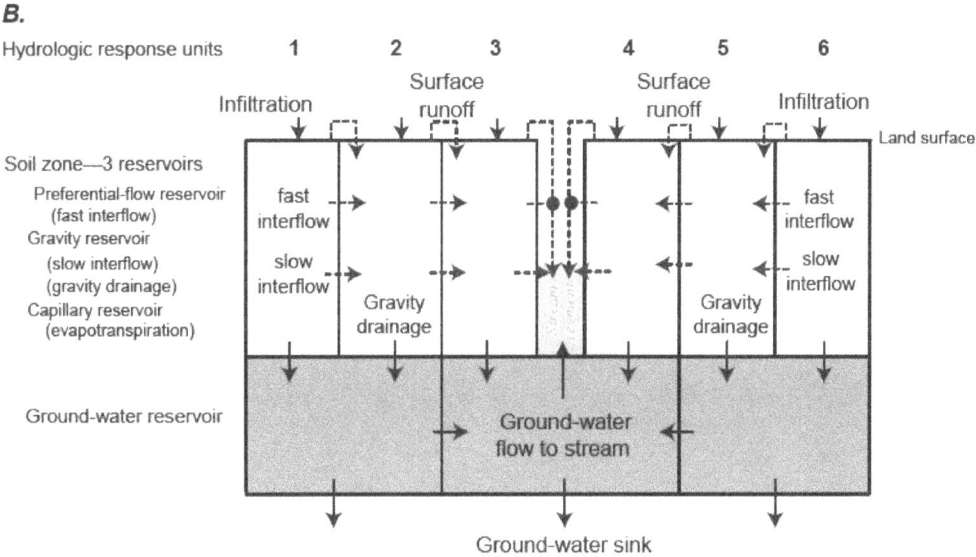

Figure 13.   Conceptual changes to the soil-zone, shallow-subsurface, and ground-water reservoirs for a PRMS-only simulation: (*A*) original conceptualization of PRMS (Leavesley and others, 1983), and (*B*) revised conceptualization that includes cascading flow of surface runoff and interflow from upslope hydrologic response units to downslope hydrologic response units.

streams can receive surface runoff and interflow from HRUs and for kinematic-wave routing of streamflow down channels; revisions also were made to the Lake Package (Merritt and Konikow, 2000) to link a MODFLOW designated lake with a lake HRU, to refill intermittent lakes from surface–water inflow, and to simulate unsaturated flow beneath lakes.

The Unsaturated-Zone Flow Package is used in GSFLOW to simulate flow through the unsaturated zone, ground-water recharge, and ground-water discharge to land surface (fig. 15). It replaces the standard Recharge and Evapotranspiration Packages of MODFLOW-2005, and these packages are not included in GSFLOW. Three modifications were made to the original version of the Unsaturated-Zone Flow Package released with MODFLOW-2005. These changes included adding the ability to simulate unsaturated flow beneath lakes and changing the formulation used for transitioning between conditions of ground-water recharge and ground-water discharge within a finite-difference cell.

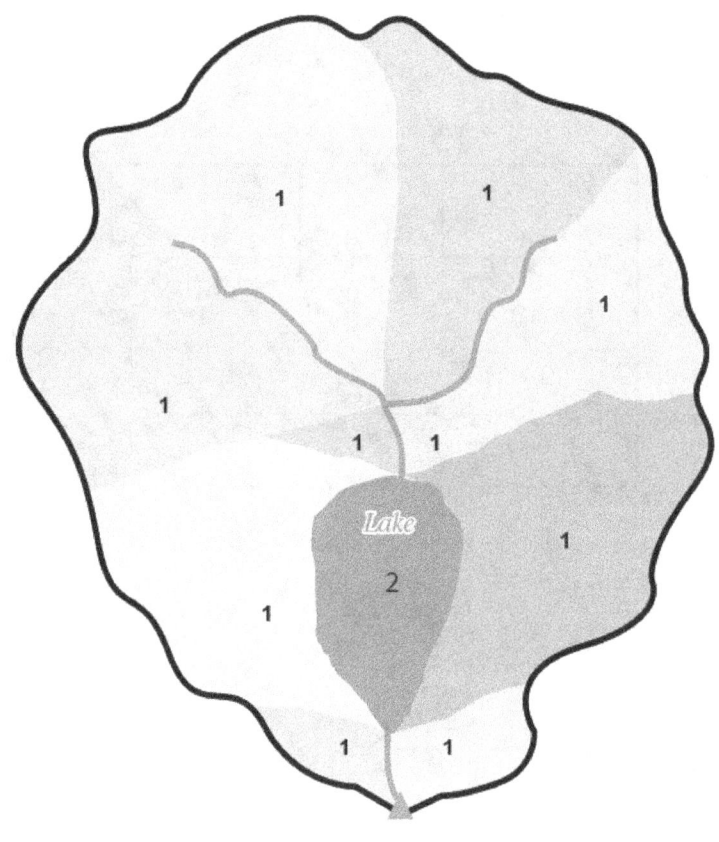

Figure 14. Delineation of land-based and lake hydrologic response units of a hypothetical watershed.

The latter modification was made to improve the efficiency of MODFLOW convergence. Both of these modifications are described in more detail in section "**Computations of Flow**."

The Streamflow-Routing Package is used in GSFLOW to route water in channels, to calculate streambed leakage, and to receive surface runoff and interflow from PRMS. The capability of routing transient streamflow down a network of channels was added to the Streamflow-Routing Package using a kinematic-wave routing method. Transient routing was added to consider simulations in which changes in channel storage are important on a daily time step, such as in long rivers.

The Streamflow-Routing Package also was revised to receive flows from PRMS as surface runoff and interflow from adjacent HRUs. Surface runoff and interflow are added to stream reaches by connecting HRUs to stream segments. The volume of runoff and interflow are distributed to each stream reach in a segment on the basis of the fraction of HRU associated with a stream reach. The River and Drain Packages in MODFLOW-2005 are not implemented in GSFLOW because they do not have an accounting procedure for surface flows and because these head-dependent boundary types can be simulated with the Streamflow-Routing Package.

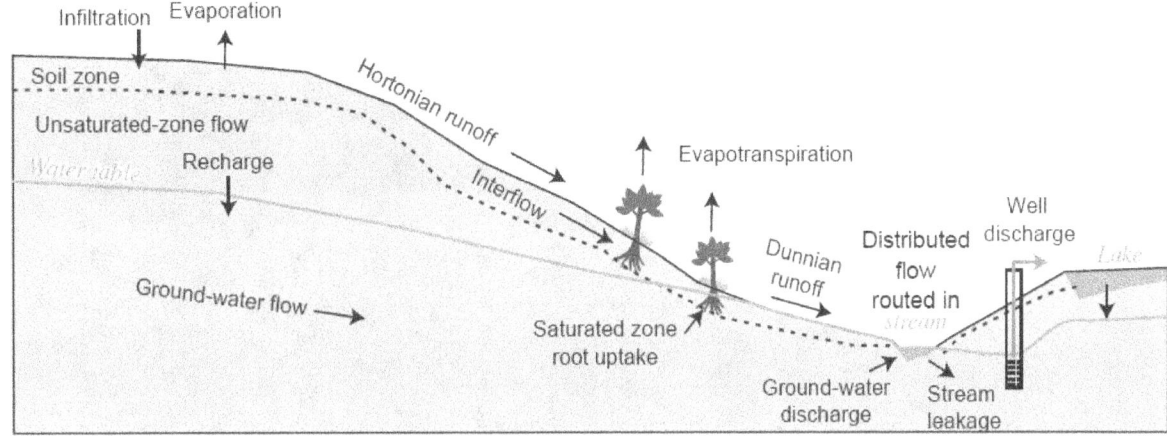

EXPLANATION

- - - - -   Soil-zone base

Figure 15. Changes to MODFLOW-2005 implemented in GSFLOW include: unsaturated zone is connected to PRMS at the soil-zone base; unsaturated flow is simulated beneath lakes; distributed flow is routed in streams.

The Lake Package (Merritt and Konikow, 2000) is used in GSFLOW to simulate (1) lake storage, stream inflows and outflows, and lakebed leakage, and (2) to receive precipitation, interflow, and surface flow from PRMS lake HRUs. PRMS also calculates the evapotranspiration rate from the surface of lakes that is removed from available lake storage. Other changes to the Lake Package include a new method for simulating dry lake conditions that occur while there are inflows to and outflows from the lake. This revision can simulate refilling of a dry lake caused by increases in surface inflows. The Lake Package published by Merritt and Konikow (2000) only allowed an empty lake to refill when the average ground-water head in cells beneath the lake was greater than the altitude of the lake bottom. The final revision was the linking of the Unsaturated-Zone Flow Package to the Lake Package for the purpose of simulating unsaturated flow beneath lakes. Unsaturated flow beneath lakes can occur when the ground-water head is beneath the lakebed.

Three additional packages available in MODFLOW-2005 were not included in GSFLOW. The Reservoir Package (Fenske and others, 1996) was not implemented because its capabilities are available in the Lake Package. Two other packages were not implemented in GSFLOW because they have not been properly linked with the Unsaturated-Zone Flow Package. These packages are the Interbed Storage Package (Leake and Prudic, 1991), and the Subsidence Package (Hoffmann and others, 2003). Details on the implementation of changes to MODFLOW-2005 that affect the rate and timing of ground-water flow and interactions with surface water are described in section "**Computations of Flow.**"

Connecting Hydrologic Response Units to Finite-Difference Cells

An important component to the coupling of the PRMS and MODFLOW models is the process used to spatially link the HRUs used by PRMS with the finite-difference cells used by MODFLOW. This is done through the generation of gravity reservoirs, which are used to transfer water between HRUs and finite-difference cells (**fig. 16**). Because HRUs and finite-difference cells can have different spatial extents, the spatial extent of each gravity reservoir is defined by the intersection of the HRUs and finite-difference cells (**fig. 16**). A unique identification number is assigned to each gravity reservoir, starting with 1 and ending with the total number of gravity reservoirs. HRUs and finite-difference cells can have more than one gravity reservoir within their areas. In **figure 16**, for example, there are 10 HRUs (including the lake HRU), 65 active MODFLOW finite-difference cells, and 98 gravity reservoirs. Any gravity reservoirs associated with lake HRUs are ignored by the model. The minimum number of gravity reservoirs in a GSFLOW simulation results when an HRU is specified for each finite-difference cell (the plan-view area of each HRU and finite-difference cell corresponds); in that case, the total number of gravity reservoirs would equal the number of HRUs.

Four topologic parameters need to be specified to relate each gravity reservoir to their corresponding HRU and finite-difference cell (**table 3**). Gravity drainage from each gravity reservoir is added to the connected finite-difference cell in GSFLOW module gsflow_prms2mf (**table 1**).

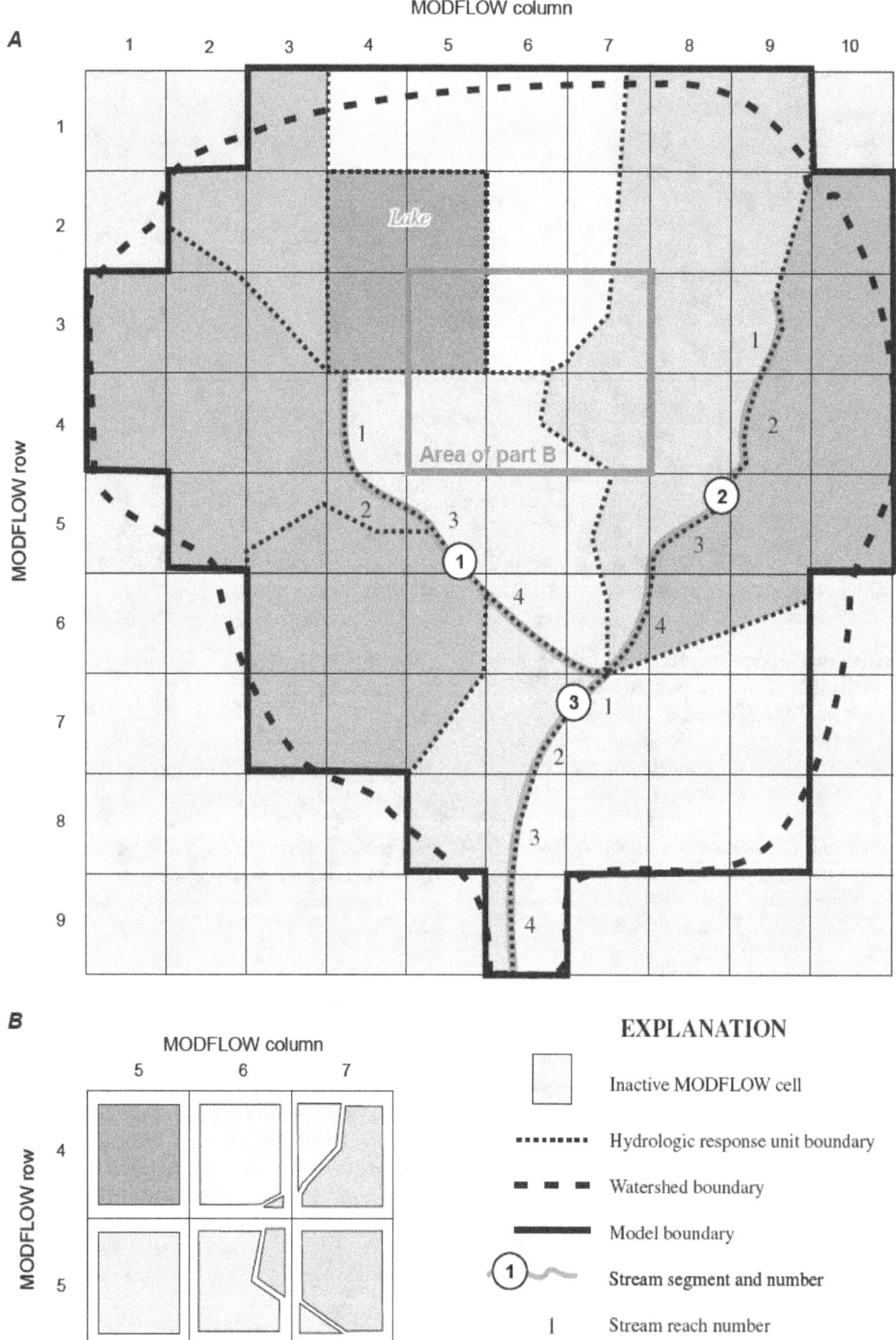

Figure 16. (*A*) example distribution of gravity reservoirs in the soil zone of hydrologic response units used to connect hydrologic response units in PRMS with finite-difference cells in MODFLOW. (*B*) includes 4 hydrologic response units, 6 finite-difference cells, and 9 gravity reservoirs. Lake hydrologic response units do not have gravity reservoirs because the soil zone is absent.

Similarly, ground-water discharge from a finite-difference cell is added to the connected gravity reservoir using GSFLOW module gsflow_mf2prms (table 1). Slow interflow out of each gravity reservoir is aggregated by HRU. Interflow can be routed as cascading flow from upslope HRUs to downslope HRUs, terminating in a stream segment or lake.

The total area of all gravity reservoirs in an HRU must equal the total area of the HRU in order to conserve mass (that is, the sum of PRMS parameter gvr_hru_pct for each HRU must equal 1). Similarly, the total area of all gravity reservoirs connected to a finite-difference cell must equal the top area of the cell (that is, the sum of GSFLOW parameter gvr_cell_pct for each finite-difference cell must equal 1). These parameters are described in appendix 1. Standard analyses using GIS may not produce parameters of sufficient accuracy and, consequently, the percentage of each gravity reservoir area in an HRU or connected to a finite-difference cell may require additional processing. The GSFLOW model will issue warning messages when the absolute errors of the area-weighted sums exceed 0.000001.

Connecting Hydrologic Response Units to Streams and Lakes

HRUs are connected to streams for the purpose of routing surface runoff and interflow to stream segments and reaches defined in the Streamflow-Routing Package of MODFLOW-2005. Five topologic parameters need to be specified as part of the model input to relate stream segments and reaches to corresponding HRUs that are contiguous to them (table 4). These parameters can be determined using a GIS analysis by identifying all unique intersections among HRUs, stream segments, and stream reaches. Multiple HRUs can contribute surface runoff and interflow to a stream segment. Parameter hru_segment is used to route flow from HRUs to segments when the option to cascade flow is not used. Surface runoff and interflow from one HRU can be distributed to multiple stream segments when an HRU cascades. Surface runoff and interflow to lakes are routed from land-based HRUs to lake HRUs (fig. 14) on the basis of the topological parameters for cascading flow among HRUs. Refer to section "Cascading Flow Procedure".

Table 3. Topological parameters used to relate areas of gravity reservoirs in the soil zone to hydrologic response units in PRMS and to areas of finite-difference cells in MODFLOW-2005.

[HRU, hydrologic response unit; nhrucell, total number of intersections between HRUs and active finite-difference cells]

Parameter name	Array dimension	Description
gvr_hru_id	nhrucell	HRU identification number corresponding to a gravity reservoir.
gvr_cell_id	nhrucell	Finite-difference cell identification number corresponding to a gravity reservoir. Finite-difference cell identification number is a unique number determined by the row, column, and layer number of the cell. The unique cell number starts with one for a cell in row 1, column 1, and layer 1. The unique cell number is incremented by one for each column along each row until the last column in the last row of layer 1. The unique number is then continued using the same method through the last layer. The last unique cell identification number is equal to the product of the number of columns, rows, and layers in the finite-difference grid.
gvr_hru_pct	nhrucell	Decimal fraction of HRU occupied by gravity reservoir.
gvr_cell_pct	nhrucell	Decimal fraction of finite-difference cell occupied by gravity reservoir.

Table 4. PRMS parameters for module gsflow_prms2mf that relate hydrologic response units in PRMS to stream segments and reaches in the Streamflow-Routing Package of MODFLOW-2005.

[HRU, hydrologic response unit; nhru, total number of HRUs; nreach, total number of stream reaches; and nsegment, total number of stream segments]

Parameter name	Array dimension	Description
hru_segment	nhru	Stream segment number corresponding to each HRU. When an HRU cascades to another HRU or stream segment, hru_segment is set to zero in module gsflow_prms2mf.
local_reachid	nreach	Stream reach identification number within each stream segment listed in sequence from the first to last stream segment.
numreach_segment	nsegment	Number of stream reaches in each stream segment.
reach_segment	nreach	Stream segment number of each stream reach.
segment_pct_area	nreach	Decimal fraction of HRU area contributing to a stream reach.

PRMS-calculated inflows to a stream segment or a lake from contiguous HRUs include Hortonian and Dunnian runoff, slow interflow through the gravity reservoirs, and fast interflow through the preferential-flow reservoirs. The total contribution of surface runoff and interflow to a stream segment or lake is the area-weighted sum from all contiguous HRUs. The total contribution of inflow to a stream segment is partitioned to individual stream reaches on the basis of the decimal fraction of the HRU associated with each stream reach (table 4). Refer to section "Computations of Flow" for explanations of how runoff and interflow are calculated for routing to HRUs, streams, or lakes.

Time Discretization—Stress Periods, Time Steps, and Iteration

GSFLOW operates on a daily time step, and volumetric flow rates are exchanged between PRMS and MODFLOW during each daily time step. MODFLOW-2005 variable-length stress periods are used to specify changes in stress or boundary conditions that are not calculated by PRMS, such as recharge and discharge rates specified in the Well Package input file. However, boundary conditions applied to MODFLOW-2005 that are calculated by PRMS are applied on a daily basis, such as gravity drainage through the base of the soil zone. In order to ensure synchronization between MODFLOW-2005 variable-length stress periods and GSFLOW daily time steps, GSFLOW requires that variable-length stress periods also use daily time steps. That is, although the number or time unit of variable-length stress periods in GSFLOW is not restricted, the time increment of each stress period must be an even multiple of days. For example, the time increment of a stress period in the input for the Discretization File (appendix 1) could be described in units of seconds, but the value must result in a whole number of days when divided by 86,400 seconds. Moreover, the capability in MODFLOW-2005 to increase the length of time steps during a stress period is not included in GSFLOW. The multiplier that typically is used to increase the time step during a stress period for MODFLOW-2005 simulations is automatically set to 1.0 in GSFLOW, and the specified time of each stress period is checked for compliance with the requirement of 1-day time steps. The program will print an error message before stopping when a stress period cannot be evenly divided into uniform time steps of 1 day.

Only the first stress period in the Discretization File of MODFLOW-2005 can be designated as steady state for integrated simulations. The steady-state designation is ignored for all other stress periods. No computations pertaining to PRMS are executed for an initial steady-state stress period. Consequently, gravity drainage beneath the soil zone, as well as any other inflows and outflows to streams, lakes, and ground water are required input for a steady-state simulation. Gravity drainage beneath the soil zone is specified as an infiltration rate in the Unsaturated-Zone Flow Package File

for a steady-state stress period. Similarly any specified stream, lake, and ground-water inflows and outflows are specified in the respective MODFLOW package input file(s). Results from the steady-state stress period are automatically used as initial conditions for a steady-state time step of the subsequent stress period in the integrated model. Although PRMS does not have a steady-state capability for determining initial storages in the various reservoirs, an initial volume per unit area can be specified for each PRMS reservoir prior to beginning a transient simulation. Alternatively, a "spin up" period can be included at the beginning of the simulation.

Organization

PRMS and MODFLOW-2005 were integrated without much alteration to the underlying codes because both models have similar programming frameworks. PRMS and MODFLOW-2005 remain separate in the integrated model by continuing to handle their respective operations, such as reading input files, determining simulation options, computing solutions, and printing output results to files. Additionally, much of each model's computational sequence remains the same within the integrated model, other than modifications to make some PRMS and MODFLOW-2005 computations sequential within the iteration loop.

GSFLOW is organized using two Computation-Control Modules—one for PRMS (gsflow_prms; table 1) and another for MODFLOW-2005 (gsflow_modflow; table 1). The module gsflow_prms is used to call PRMS and GSFLOW modules in the proper sequence, which varies for each simulation mode (integrated, PRMS-only, and MODFLOW-only).

GSFLOW time-step incrementing is the same as that used for the PRMS daily mode. Time is incremented through the total number of days in the specified simulation time period. Rather than having separate stress-period and time-step loops as in MODFLOW-2005, the conditions represented by the variable-length stress periods commonly used in MODFLOW-2005 are changed on the appropriate day of the GSFLOW simulation. Counters representing the stress-period and time-step numbers used for MODFLOW-2005 computations are incremented appropriately within the module gsflow_modflow according to the specified number of time steps within a stress period.

All MODFLOW-2005 procedures and the PRMS Soil-Zone Module (table 1) are processed in the module gsflow_modflow. The contents and coding style of module gsflow_modflow are similar to those in the MODFLOW-2005 main program, the iteration loop is retained but the MODFLOW time-step and stress-period loops have been removed and are controlled through module gsflow_prms. This module determines the current MODFLOW stress period and calls the procedures specified in the MODFLOW Name File in the proper sequence. The Soil-Zone Module is called from within the MODFLOW formulate and budget procedures. Computations of the volume of water transferred between the Soil-Zone Module and the Unsaturated-Zone Flow, Lake, and Streamflow-Routing Packages are made during each cycle through the iteration loop. All other PRMS computations are made before the iteration loop as described in section "Computations of Flow."

Additionally, five new modules were developed for integrating PRMS and MODFLOW-2005. Module gsflow_setconv computes unit conversions between PRMS and MODFLOW-2005. Modules gsflow_prms2mf and gsflow_mf2prms integrate the spatial units and transfer dependent variables and volumetric flow rates between PRMS and MODFLOW-2005. Module gsflow_budget is used to compute and write an overall water budget for the simulated system and module gsflow_sum is used to produce detailed water budgets for all flow components. These new modules were written in Fortran 90.

GSFLOW simulations require input for each active PRMS module and MODFLOW-2005 package. Input for all PRMS modules is contained within the PRMS Parameter File and PRMS Data File, whereas there is at least one individual input file for each MODFLOW-2005 package. New parameters were added to the PRMS Parameter File to specify data needed for new and revised modules. These data include parameters that connect HRUs to finite-difference cells, lakes, and streams. A complete description of the data-input requirements for GSFLOW is included in appendix 1.

Additional source-code files related to the Modular Modeling System and the MODFLOW-2005 Ground-Water Flow and Observation Processes also are required to build the GSFLOW executable (table 5).

Computational Sequence

The computational sequence of GSFLOW is illustrated with a flow chart shown in figure 17 and descriptions listed in table 6. The sequence of these computations depends on the GSFLOW simulation mode. If a simulation does not include MODFLOW-2005, then MODFLOW-2005 packages are skipped, and the computational sequence has a single daily time-step loop (fig. 5) and follows the sequence described in section "Description of PRMS." Similarly, if a simulation does not include PRMS, then PRMS modules are skipped, and the sequence of computations includes the standard MODFLOW-2005 stress period, time step, and iteration loops (fig. 10) and follows the sequence described in section "Description of MODFLOW-2005." An integrated simulation executes both PRMS modules and MODFLOW-2005 packages on a daily time step that includes variable-length stress periods and the iteration loop. The iteration loop solves the interdependent equations within GSFLOW, such as ground-water head dependent flows between PRMS and MODFLOW-2005.

Table 5. GSFLOW main program, Modular Modeling System utility functions used with PRMS, and additional MODFLOW-2005 packages and files used in GSFLOW.

Computer program file	Programming language	Description
GSFLOW main program		
gsflow_main.c	C	GSFLOW program main, which reads the Control File and initiates execution.
Modular Modeling System utility functions used with PRMS		
mms_util.c	C	Utility functions that primarily handle verification, storage and retrieval of input parameters and output variables for PRMS and GSFLOW modules.
defs.h, globals.h, mms.h, nodes.h, protos.h, structs.h	C	Include files that define variables and values used by the Modular Modeling System.
Packages and file used with MODFLOW for the ground-water flow process		
openspec.inc	Fortran	Include file that defines values for OPEN-statements.
u2ddbl.f	Fortran	Package of two-dimensional array reading utility subroutines.
utl7.f	Fortran	Package of array reading utility subroutines that includes reading of text and optional conversions to numeric values.
Packages used with MODFLOW for the observations process		
obs2bas7.f	Fortran	Package of subroutines that handle observations for the Basic Package.
obs2chd7.f	Fortran	Package of subroutines that handle observations for the Constant-Head Package.
obs2ghb7.f	Fortran	Package of subroutines that handle observations for the General-Head Boundary Package.

Program control begins using Modular Modeling System (MMS) utility functions to read the GSFLOW Control File, PRMS Parameter File, and PRMS Data File. Next, the MMS declare procedure is executed in which PRMS data consistency is checked, memory requirements are allocated, the simulation mode is determined, and variables are initialized and read (steps 1 and 2; fig. 17 and table 6). PRMS computations required for each time step are made by accessing PRMS modules within module gsflow_prms (steps 3 and 7). Program control also enters the module gsflow_modflow for each cycle through the time-step loop (steps 3–16). If specified by the user, integrated GSFLOW simulations that begin with a steady-state stress period execute MODFLOW-2005 procedures and exclude PRMS computations from the initial steady-state stress period (step 5), and data related to boundary conditions are read for some MODFLOW-2005 packages at the beginning of any additional stress periods not designated as an initial steady state (step 6).

Table 6. Description of computational sequence for GSFLOW.

[HRU, hydrologic response unit]

Sequence No.	Computation
1	**Declare**—Run PRMS and GSFLOW declare procedures.
2	**Initialize, allocate, and read**—Run PRMS and GSFLOW initialize procedures, MODFLOW-2005 allocate and read and prepare subroutines, and an optional MODFLOW-2005 steady-state simulation.
3	**Daily time-step loop**—Read measured daily values for PRMS, read data for active MODFLOW-2005 packages when new stress period begins, and run MODFLOW-2005, advance time subroutines.
4	**Check for new MODFLOW-2005 stress period**—If time step loop is at 1 and initial stress period is steady state then do step 5, otherwise if new stress period then do step 6.
5	**Steady-state simulation**—Add source and sink terms for all packages and Run MODFLOW-only simulation including MODFLOW-2005 formulate and budget processes. Go to step 3 and increment daily time-step loop.
6	**Read and set new MODFLOW-2005 boundary conditions**—Add source and sink terms for all packages except Unsaturated-Zone Flow, Streamflow-Routing, and Lake Packages.
7	**Compute land-surface and soil-zone hydrologic processes in PRMS**—Distribute precipitation, compute canopy interception and evaporation, determine snowpack accumulation and snowmelt, impervious surface evaporation and storage, infiltration, and Hortonian surface runoff (Horton, 1933) for each HRU.
8	**Begin iteration loop**—Sequence through coupled PRMS and MODFLOW-2005 components.
9	**Compute soil-zone flow and storage in PRMS**—Compute evapotranspiration, interflow and Dunnian runoff (Dunne and Black, 1970) to downslope HRUs and stream reaches, and gravity drainage to the unsaturated zone.
10	**Transfer dependent variables and volumetric flow rates to MODFLOW-2005**—Transfer flow-dependent variables and volumetric flow rates needed for computing unsaturated and saturated flow, streamflow routing, and lake stages and volumes in MODFLOW-2005.
11	**Begin formulate procedure in MODFLOW-2005**—Add source and sink terms for Unsaturated-Zone Flow, Streamflow-Routing and Lake Packages.
12	**Approximate a solution to the ground-water flow equation**—Use one of the MODFLOW-2005 solver packages to calculate ground-water heads and head-dependent flows.
13	**Transfer dependent variables and volumetric flow rates to PRMS**—Transfer flow-dependent variables and volumetric flow rates needed to compute soil-zone flow and storage in PRMS from the Unsaturated-Zone Flow Package in MODFLOW-2005.
14	**Check for convergence**—Iterate (repeat steps 9–13) until solution has converged to a specified closure criterion for changes in ground-water head and volumetric flow rate in MODFLOW-2005 and a specified closure criterion for changes in storage in soil zone of PRMS.
15	**Compute budgets**—Run MODFLOW-2005 and GSFLOW budget procedures.
16	**Write results**—Run GSFLOW, PRMS, and MODFLOW-2005 output procedures.
17	**Check for end of simulation**—Repeat time loop (steps 3–14) until end of simulation period.
18	**End of simulation**—Close files and clear computer memory.

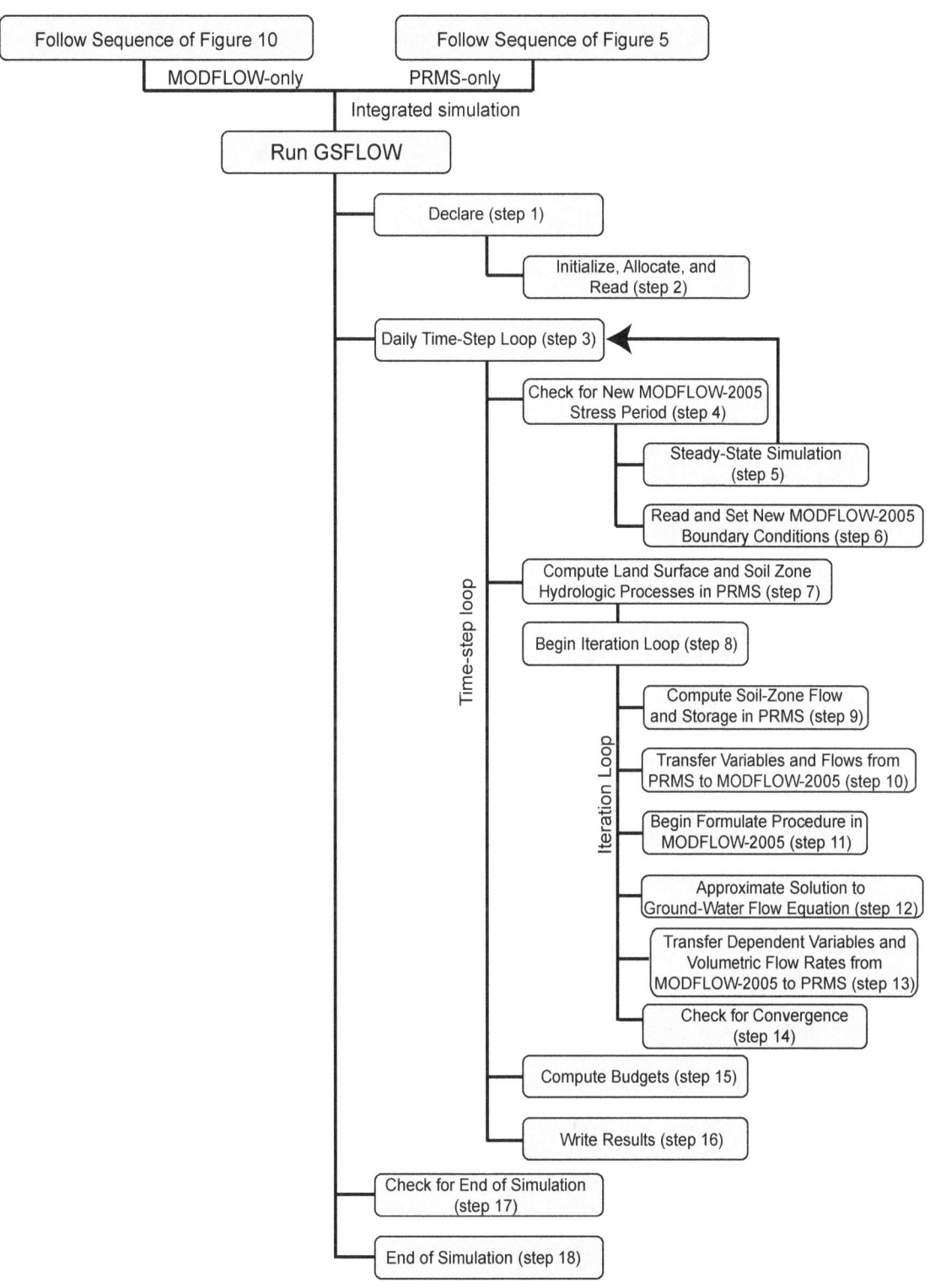

Figure 17. Overall computational sequence used for an integrated PRMS and MODFLOW-2005 simulation in GSFLOW. (See table 6 for further explanation of sequence steps.)

Module gsflow_modflow also includes the iteration loop in which computations are made for the soil, streams, lakes, unsaturated, and saturated zones, and for transferring dependent variables and volumetric flow rates between PRMS modules and MODFLOW-2005 packages (steps 8–14). Ground-water heads are calculated and model convergence is checked at the end of iterations (step 14). Subroutines associated with the budget procedure (step 15) are executed at the end of a steady-state stress period (if specified) and prior to the beginning of each new time step. Results and budgets computed for all simulation modes can be saved at the end of every time step (step 14). Volumetric flow and storage is computed each time step (steps 3–17) until the specified ending time is reached (step 17). Once the simulation has reached the end of the simulation, all files are closed and memory is released (step 18).

Climate-data distribution and land-surface hydrologic processes are computed by PRMS at the beginning of each time step (step 7). After these processes are computed, the coupled equations in GSFLOW (steps 8–14) are solved in the iteration loop. The computational sequence within the iteration loop is important for linking PRMS to MODFLOW-2005, and follows a required order for solving the coupled equations representing the soil zone, streams, and lakes, and the unsaturated and saturated zones. Soil-zone calculations are made prior to computing flow through streams, lakes, and the unsaturated and saturated zones (step 9). Flow among the soil zone and unsaturated and saturated zones is dependent on ground-water heads and storage in soil-zone reservoirs from the previous time step, for the first iteration, and the previous iteration subsequently. Initial ground-water heads can be obtained from the optional initial steady-state stress period simulation or specified by the user. Computations for the MODFLOW-2005 formulate procedure that involve head-dependent boundaries (step 11) are made after the soil-zone computations are made, followed by computations to solve for ground-water heads (step 12). The dependent variables and volumetric flows computed by MODFLOW are then transferred to PRMS (step 13) for revising flow and storage in the soil-zone reservoirs during the subsequent iteration (step 9).

A solution is found after the storage in the soil zone, depths in streams and lakes, and ground-water heads change and volumetric flows change less than the specified tolerances between iterations (step 14). Separate convergence tolerances are used for the soil zone, stream stages, lakes stages, and ground-water heads. Non-convergence occurs if any of these tolerances are not achieved within the specified maximum number of iterations. A warning message is printed if the model does not converge for a time step; however, GSFLOW continues to the next time step using the non-converged solution. Similarly, warning messages are printed to the MODFLOW-2005 Listing File if stream or lake stages do not converge for a time step.

Conversion of Units Between PRMS and MODFLOW-2005

Volumes in PRMS are expressed as acre-inch and volumetric flow rates are expressed in acre-inch per day. For example, if the total soil-water equivalent depth in the soil zone of an HRU is 3.0 inches at the beginning time step, then 3.0 inches of water are available over the HRU for evapotranspiration, gravity drainage, and interflow. If the area of an HRU is 100 acres, then the volume of water in the soil zone is 300 acre-inch or 1,089,000 cubic feet or 30,837 cubic meters. Volumes in MODFLOW-2005 are expressed in units of cubic length and volumetric flow rates are expressed in units of cubic length per time. Time can be in units of seconds, minutes, days, or years; length can be specified in units of inches, feet, centimeters, or meters. Time and length units are specified by MODFLOW parameters `ITMUNI` and `LENUNI`, respectively, which are specified in the Discretization File of the Basic Package (Harbaugh, 2005; p. 8-11 and appendix 1). Historically, `ITMUNI` and `LENUNI` were not critical to MODFLOW calculations as the input was expected to be of consistent units. These parameters take on a new importance in GSFLOW in that they control how units are converted between the integrated models.

Volumetric flow rates in PRMS are converted to units specified in MODFLOW-2005 when the rates are transferred from PRMS to MODFLOW-2005. Conversely, volumetric flow rates in the specified units in MODFLOW-2005 are converted to PRMS units when the rates are transferred from MODFLOW-2005 to PRMS. A set of variables is determined in the GSFLOW module gsflow_setconv (table 1) to calculate the particular conversion between PRMS and MODFLOW-2005 units for a simulation. The variables are summarized in table 7. Due to the empirical nature of some algorithms in PRMS, units other than acre-inch or acre-inch per day are used in some PRMS modules (Leavesley and others, 1983). These different units do not affect the conversion of dependent variables and volumetric flow rates between PRMS and MODFLOW-2005.

Table 7. Conversion of volumes, volumetric flow rates, length, and velocity for variables in modules gsflow_prms2mf and gsflow_mf2prms in GSFLOW.

Convert from	Multiply by GSFLOW variable	To obtain
Volume		
acre-inch in PRMS cubic length in MODFLOW	`Acre_inch_to_Mfl3` `Mfl3_to_ft3`	cubic length in MODFLOW cubic feet in PRMS
Volumetric flow rate		

cubic length per time step in MODFLOW
cubic feet per second in PRMS
acre-inch per time step
	`Mfl3t_to_cfs`
	`Sfr_conv`
	`Acre_inch_to_cfs`
		cubic feet per second in PRMS
		cubic length per time step in MODFLOW
		cubic feet per second in PRMS

Length		
length in MODFLOW	`Mfl_to_inch`	inch in PRMS
Velocity		
inch per day in PRMS	`Inch_to_mfl_t`	length per time step in MODFLOW

Cascading-Flow Procedure

The cascading-flow procedure developed for routing surface runoff and interflow among HRUs and to streams and lakes uses a directed, acyclic-flow network (Ford and Fulkerson, 1956). Flow paths start at the highest upslope HRUs and continue through downslope HRUs until reaching a stream segment or lake. Cascading flow may occur along many different paths and satisfies soil-zone storage of downslope HRUs before being added as inflow to stream segments or lakes. Surface runoff and interflow from one upslope HRU may provide inflow to as many downslope HRUs, stream segments, and lakes as is required to conceptualize drainage in the watershed.

The cascading-flow procedure for routing surface runoff was added to two Surface-Runoff Modules (srunoff_smidx_casc and srunoff_carea_casc, table 1). Cascading interflow was included in the new Soil-Zone Module. Additionally, the cascading-flow procedure was added to the Ground-Water Flow Module (gwflow_casc_prms, table 1) used for the PRMS-only mode of simulation to route upslope PRMS ground-water reservoir discharge to downslope ground-water reservoirs until reaching a stream segment.

PRMS parameters are used to define the flows between one reservoir and another and the fraction of area in the upslope HRU that contributes flow to downslope HRUs (table 8 and appendix 1). Input parameters can be derived on the basis of flow accumulation and direction analysis of a Digital Elevation Model (U.S. Geological Survey, 2000) of the watershed typically used in the delineation of HRUs (fig. 18A), or, if HRUs are discretized using a finite-difference grid, flow paths among the finite-difference cells can be determined from their land-surface altitudes (fig. 18B). The cascading parameters must account for all HRUs and each path must terminate in a stream segment or lake and not include loops or circles within the path.

The model checks the cascading parameters and prints error and warning messages prior to starting the time-step loop of either an integrated or PRMS-only simulation. An error message stops model execution, whereas a warning message results in an adjustment of parameter values. The following describes the error and warning messages as related to cascading flow routing from HRUs. Similar error and warning messages are printed for cascade paths involving PRMS ground-water reservoirs in a PRMS-only simulation. Error messages are printed when: (1) a lake HRU is specified to cascade to another HRU, as lake HRUs can only cascade to a stream segment; (2) the sum of the contributing area from an HRU to downslope HRUs of all paths originating from an HRU is substantially different than the area of the HRU; (3) an HRU(s) is not included in the cascading pattern; and (4) a circular path is detected. Warning messages are printed when: (1) a cascade path is found with a contributing area from the upslope HRU that is less than PRMS parameter `cascade_tol`—the area is evenly distributed to other paths originating from the HRU; and (2) when two or more flow paths that are specified having the same source and destination—the paths are combined and computed as a single path.

PRMS parameter `cascade_tol` (table 8) can be used to ignore any downslope connections when the contributing area of the upslope HRU is a negligible fraction of total area of the upslope HRU (below 0.01 for example). PRMS parameter `cascade_flg` can be set to a value of 1 to add all flow from an upslope HRU to one downslope HRU. Each downslope HRU is automatically determined on the basis of which downslope HRU has the greatest fraction of contributing area in the upslope HRU as specified by PRMS parameter `hru_down_pct`. PRMS parameter `hru_strmseg_down_id` is used to define the connection of HRUs to stream segments (fig. 18). Stream segments are identified in the Streamflow-Routing Package of MODFLOW-2005. A separate set of PRMS parameters is used to define the direction of flow among ground-water reservoirs, which allows for different patterns of cascading flow among ground-water reservoirs than those for surface runoff and interflow (table 8).

Computations of Flow

This section describes all of the computations that are made by GSFLOW, beginning with climate inputs of temperature, precipitation, and solar radiation, and ending with ground water and its interactions with streams and lakes. Additional details on the computations of flow are described in the documentation for PRMS (Leavesley and others, 1983; Leavesley and others, 1996a) and MODFLOW-2005 (Harbaugh, 2005).

Descriptions of parameters and variables used in equations for computations of flow include units for each parameter and variable. Units that are specified in a generic format—such as "cubic length"—indicate that the units are defined by data input. Units that are specified in a particular format—such as "acre-inch"—indicate that the units in GSFLOW are predetermined. The type of units used for

Table 8. Parameters for PRMS Cascade Module (cascade_prms) used to define connections for routing flow from upslope to downslope hydrologic response units and stream segments and among ground-water reservoirs.

[HRU, hydrologic response unit; `ncascade`, total number of connections among HRUs and stream segments; `ncascdgw`, total number of connections among ground-water reservoirs]

Parameter name	Array dimension	Description
Parameters for routing surface runoff and interflow among HRUs and to stream segments[1]		
cascade_tol	1	Minimum contributing area of an upslope HRU considered for flow to a downslope HRU.
cascade_flg	1	Flag used to force outflow from an upslope HRU to be added as inflow to one downslope HRU[3].
hru_up_id	ncascade	Identification number of upslope HRU.
hru_down_id	ncascade	Identification number of downslope HRU.
hru_strmseg_down_id	ncascade	Identification number of stream segment that receives inflow from an upslope HRU.
hru_pct_up	ncascade	Fraction of outflow from upslope HRU to be added as inflow to a downslope HRU or stream segment[2].
Parameters for routing flow among ground-water reservoirs in PRMS-only simulations[1]		
gw_up_id	ncascdgw	Identification number of upslope ground-water reservoir.
gw_down_id	ncascdgw	Identification number of downslope ground-water reservoir.
gw_strmseg_down_id	ncascdgw	Identification number of stream segment in PRMS that receives cascading flow from a ground-water reservoir.
gw_pct_up	ncascdgw	Fraction of outflow from a ground-water reservoir to be added as inflow to a ground-water reservoir or stream segment[4].

[1] Routing of surface runoff and interflow in the integrated simulation mode is to stream segments defined in the Streamflow-Routing Package of MODFLOW-2005, or, when using the PRMS-only mode, to detainment reservoirs defined in the PRMS module strmflow_prms.

[2] Fraction is the ratio of contributing area to total area of upslope HRU.

[3] A zero allows for multiple routing of inflows and outflows among HRUs and a value of one forces routing from an upslope HRU to one downslope HRU that has the greatest fraction of contributing area in the upslope HRU.

[4] Fraction is the ratio of contributing area to total area of the ground-water reservoir that produces cascading flow.

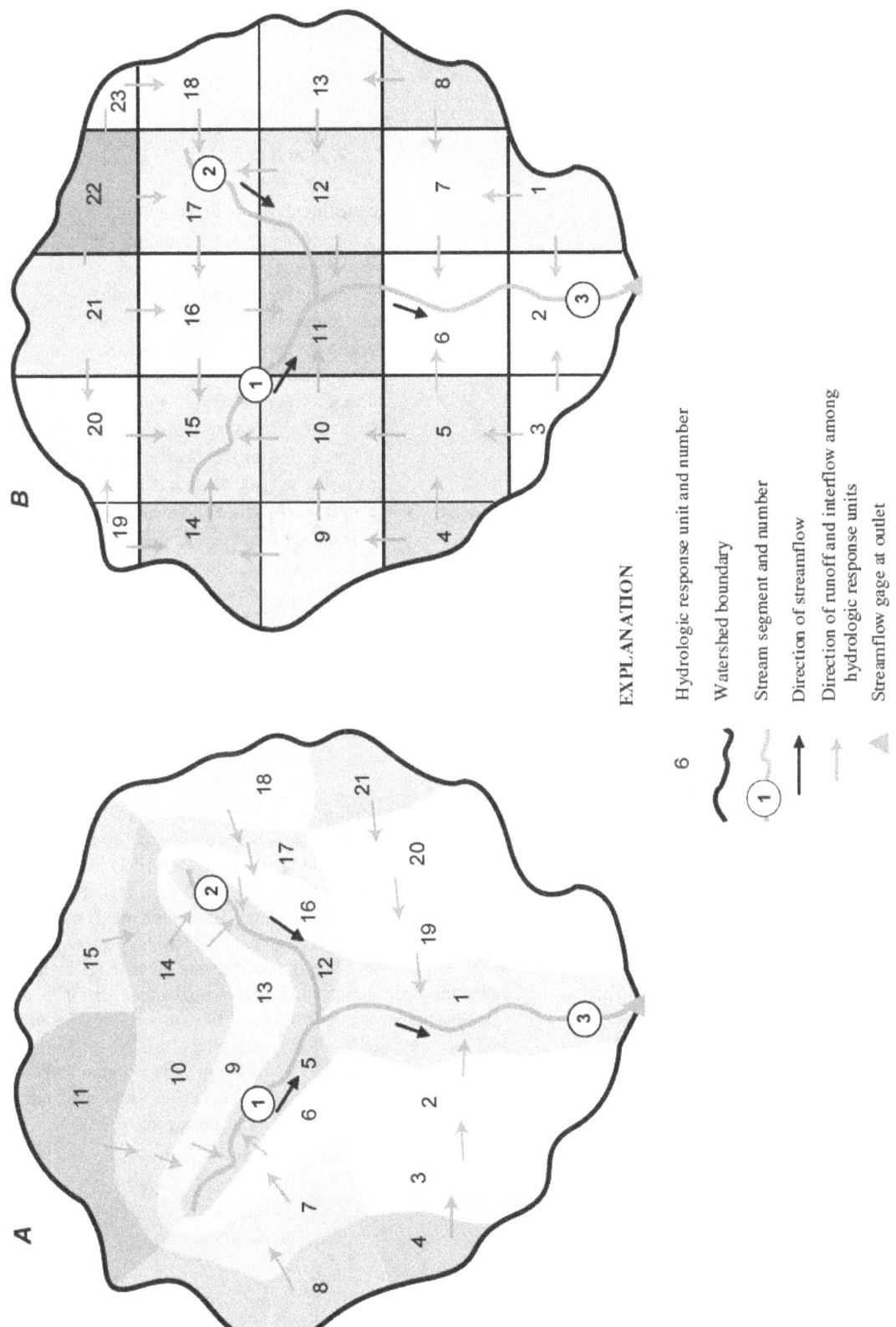

EXPLANATION

Hydrologic response unit and number

Watershed boundary

Stream segment and number

Direction of streamflow

Direction of runoff and interflow among
hydrologic response units

Streamflow gage at outlet

6

1

Figure 18. Cascading flow of surface runoff and interflow among hydrologic response units and streams delineated from (A) topology, climate, and vegetation; and (B) a finite-difference grid.

MODFLOW-2005 variables are specified in consistent units of length and time as defined by variables `LENUNI` for length and `ITMUNI` for time in the Discretization File. Most PRMS parameters and variables must be in units specified in the data input instructions (appendix 1). The Definition of Symbols (appendix 2) lists all equation variables in alphabetical order. Parameters represented by equation variables are identified in the variable definitions in `courier` font.

References to other publications are used if complete descriptions of all equations used to represent a hydrologic process in GSFLOW are not stated. Derivations of equations only are presented when they are not available in other publications or if they better explain the hydrologic processes represented by the model.

Distribution of Temperature, Precipitation, and Solar Radiation

Model inputs are daily precipitation, maximum and minimum air temperature, and, optionally, solar radiation. Air temperature and solar radiation are used in the computation of evaporation, transpiration, sublimation, and snowmelt. These point data are extrapolated to each HRU using a set of adjustment coefficients developed from regional climate data. The coefficients typically include the effects of altitude, slope, aspect, and distance to one or more measurement sites (Linsley and others, 1975).

Temperature

Maximum and minimum daily air temperatures are extrapolated to each HRU by one of four user-specified options that depend on the number of air-temperature measurement stations to be used in estimating the air temperature of each HRU. The first option is used when data from only one station are used to assign air temperature to each HRU. The second through fourth options are used when air temperature assigned to each HRU is based on data from more than one station. Temperature units can be specified as either degrees Fahrenheit or Celsius, and the altitude of the measurement station to which the temperature corresponds can be specified as either feet or meters. Lapse rate coefficients that are input to the model to represent changes in temperature with land-surface altitude must be expressed in units that are consistent with the specified temperature and altitude units.

The first option (module temp_1sta_prms) is used when data from one or more measurement stations are available. The station associated with each HRU is specified by PRMS parameter `hru_tsta` (appendix 1). The maximum and minimum daily HRU temperatures are computed from the measured station temperatures and a monthly lapse rate according to:

$$T_{HRU}^{m} = T_{sta}^{m} - b_{month} \left(\frac{Z_{HRU} - Z_{tsta}}{1000} \right) - taf_{HRU} \, , \qquad (1)$$

where

T_{HRU}^{m} is the maximum (or minimum) daily temperature at each HRU for time step m, in degrees Fahrenheit or Celsius;

T_{sta}^{m} is the measured maximum (or minimum) daily temperature at the station for time step m, in degrees Fahrenheit or Celsius;

b_{month} is the monthly maximum (or minimum) daily temperature lapse rate representing the change in maximum (or minimum) air temperature per 1,000 feet or meters of altitude change for each month, January to December—parameter `tmax_lapse` (or `tmin_lapse`), in degrees Fahrenheit or Celsius per 1,000 length units;

Z_{HRU} is the mean land-surface altitude of the HRU —parameter `hru_elev`, in length;

Z_{tsta} is the altitude of the air temperature measurement station—parameter `tsta_elev`, in length; and

taf_{HRU} is the maximum (or minimum) daily HRU temperature adjustment factor—parameter `tmax_adj` (or `tmin_adj`), which is estimated on the basis of slope and aspect, in degrees Fahrenheit or Celsius.

The second option (module temp_laps_prms) is used when data from at least two stations at different altitudes are used to estimate air temperature of each HRU. Each HRU is assigned air temperatures on the basis of computed lapse rates from a pair of stations. The station that is most representative of the temperature at the HRU is designated as the base station. The station that is not the base station is called the lapse station and is used with the base station to calculate the lapse rate. The two stations associated with each HRU are specified by PRMS parameters `hru_tsta` and `hru_tlaps` (appendix 1). Maximum and minimum daily HRU temperatures are computed by a temperature lapse rate from the two associated stations according to:

$$T_{HRU}^m = T_{base}^m + \left(T_{lapse}^m - T_{base}^m\right)\left(\frac{Z_{HRU} - Z_{base}}{Z_{lapse} - Z_{base}}\right) - taf_{HRU} \ , \qquad (2)$$

where

T_{base}^m is the measured maximum (or minimum) daily temperature at the base station assigned to an HRU for time step m, in degrees Fahrenheit or Celsius;

T_{lapse}^m is the measured maximum (or minimum) daily temperature at the lapse station assigned to an HRU during time step m, in degrees Fahrenheit or Celsius;

Z_{lapse} is the altitude of the lapse station, in length; and

Z_{base} is the altitude of the base station, in length.

The third option (module xyz_dist) uses a three-dimensional multiple-linear regression based on longitude (x), latitude (y), and altitude (z) to distribute temperature from two or more stations (Hay and others, 2000; Hay and Clark, 2003). The independent variables used in the regression are normalized by subtracting the mean and dividing by the standard deviation to remove the effects of units, magnitude, and inconsistency in specification of the origin. The resulting equation is:

$$T_{HRU}^m = c_{T1} + c_{T2} - taf_{HRU} \ , \qquad (3a)$$

$$c_{T1} = bt_{X,month}\,\overline{X}_{HRU} + bt_{Y,month}\,\overline{Y}_{HRU} + bt_{Z,month}\,\overline{Z}_{HRU} \ , \text{ and} \qquad (3b)$$

$$c_{T2} = \overline{T}_{stas}^m - \left(bt_{X,month}\,\overline{X}_{sta} + bt_{Y,month}\,\overline{Y}_{sta} + bt_{Z,month}\,\overline{Z}_{sta}\right) \ , \qquad (3c)$$

where

$bt_{X,month}$, $bt_{Y,month}$, and $bt_{Z,month}$ are the maximum (or minimum) air temperature regression coefficients for longitude, latitude, and altitude, respectively by month, starting with January—parameter `max_lapse` (or `min_lapse`), in degrees Fahrenheit or Celsius;

$\overline{X}_{HRU}$ is the normalized longitude of the HRU, dimensionless;

$\overline{Y}_{HRU}$ is the normalized latitude of the HRU, dimensionless;

$\overline{Z}_{HRU}$ is the normalized altitude of the HRU, dimensionless;

$\overline{T}_{stas}^m$ is the mean measured maximum or (minimum) daily temperature of all stations for time step m, in degrees Fahrenheit or Celsius; and

$\overline{X}_{sta}$, $\overline{Y}_{sta}$, and $\overline{Z}_{sta}$ are the mean normalized, longitude, latitude, and altitude of all stations respectively, dimensionless.

The fourth option (module temp_dist2_prms) weights measured daily air temperatures from two or more stations by the inverse of the square of the distance between the centroid of an HRU and each station location (Dean and Snyder, 1977; Bauer and Vaccaro, 1987; Vaccaro, 2007). Extrapolated values also are adjusted by daily maximum and minimum lapse rates that are calculated using the daily values from all measurement stations. Air temperature data and adjustment parameters must be in the same units. Daily basin-average lapse rates for maximum and minimum temperature are first computed by:

$$b_{lapse}^m = \sum_{i=1}^{ntstas-1} X_i^2 \left(\left(T_i^m - T_{i+1}^m \right) / \left(Z_{tsta,i} - Z_{tsta,i+1} \right) \right) / ntstas, \quad (4)$$

where

b_{lapse}^m is the daily basin-average maximum (or minimum) temperature lapse rate for time step m, in degrees Fahrenheit or Celsius per length;

T_i^m and T_{i+1}^m are the measured maximum (or minimum) temperature at measurement stations i and $i+1$ for time step m, in degrees Fahrenheit or Celsius;

$Z_{tsta,i}$ and $Z_{tsta,i+1}$ are the altitudes of measurement stations i and $i+1$—parameter tsta_elev, in length; and

$ntstas$ is the number of air temperature measurement stations, dimensionless.

The daily maximum and minimum temperatures are computed on the basis of data from two or more stations according to:

$$T_{HRU}^m = \frac{\sum_{i=1}^{ntstas} \left[\left(T_i^m - \left(b_{lapse}^m \frac{Z_{HRU} - Z_i}{1,000} \right) \right) \left(Ldist_{HRU,i} \right)^2 \right]}{\sum_{i=1}^{ntstas} \left(Ldist_{HRU,i} \right)^2} - taf_{HRU} \quad (5a)$$

and

$$Ldist_{HRU,i} = 1 / \left(\sqrt{ \left(Y_{HRU} - Ytemp_i \right)^2 + \left(X_{HRU} - Xtemp_i \right)^2 } \right), \quad (5b)$$

where

$Ldist_i$ is the inverse distance between the HRU centroid and measurement station i, in length;

X_{HRU} is the longitude of the HRU centroid— parameter hru_xlong, in length;

$Xtemp_i$ is the longitude of each air temperature measurement station—parameter tsta_xlong, in length;

Y_{HRU} is the latitude of the HRU centroid- parameter hru_ylat, in length; and

$Ytemp_i$ is the latitude of each air temperature measurement station—parameter tsta_ylat, in length.

Precipitation

Precipitation, which can be measured at one or more stations, is extrapolated to each HRU by one of four user-specified options that depend on the number of stations available. The altitude of the precipitation measurements can be specified in units of either feet or meters.

The form of the precipitation (rain, snow, or a mixture of both) is important to the simulation of snow accumulation, snowmelt, infiltration, and runoff. Precipitation form on each HRU can be specified or it can be estimated from the HRU maximum and minimum daily air temperatures. Precipitation is considered all snow on an HRU when the maximum daily air temperature is less than or equal to parameter tmax_allsnow. Precipitation is all rain when either the minimum daily air temperature is greater than or equal to parameter tmax_allsnow, or the maximum daily air temperature is greater than or equal to parameter `tmax_allrain`. When neither condition for all snow nor rain is met, then precipitation on an HRU is considered a mixture, and the rain is assumed to occur first. The fraction of the total precipitation occurring as rain is computed by:

$$Frain_{HRU}^{m} = \left(\frac{Tmx_{HRU}^{m} - Tmxsnow_{HRU}}{Tmx_{HRU}^{m} - Tmn_{HRU}^{m}} \right) pmixaf_{month} , \quad (6)$$

where

$Frain_{HRU}^{m}$ is the decimal fraction of total precipitation occurring as rain on an HRU for time step m, dimensionless;

Tmx_{HRU}^{m} is the maximum air temperature assigned to the HRU for time step m, in degrees Fahrenheit or Celsius;

Tmn_{HRU}^{m} is the minimum air temperature assigned to the HRU for time step m, in degrees Fahrenheit or Celsius;

$Tmxsnow_{HRU}$ is the monthly maximum air temperature at which precipitation is all snow for the HRU—parameter `tmax_allsnow`, in degrees Fahrenheit or Celsius; and

$pmixaf_{month}$ is the monthly rain adjustment factor for a mixed precipitation event (usually 1.0)—parameter `adjmix_rain`, dimensionless.

The form of precipitation also may be explicitly specified through use of the variable `form_data` in the PRMS Data File (appendix 1).

One or more precipitation stations can be used to estimate precipitation throughout the model domain. The first option requires assigning a precipitation station to an HRU (module precip_prms). The station assigned to an HRU is specified by PRMS parameter `hru_psta` (appendix 1). A monthly correction factor, which accounts for elevation, spatial variation, topography, gage location, deficiencies in gage catch due to the effects of wind and other factors, is used to estimate the daily precipitation at the HRU according to:

$$P_{HRU}^{m} = P_{sta}^{m} CF_{HRU} , \quad (7)$$

where

P_{HRU}^{m} is the precipitation at the HRU during time step m, in inches;

P_{sta}^{m} is the measured precipitation at the station during time step m, in inches; and

CF_{HRU} is the monthly correction factor as a decimal fraction used to adjust rain (or snow) at the HRU—parameter `rain_adj` (or `snow_adj`), dimensionless.

The second option (module precip_laps_prms) also uses equation 7, except that the correction factor is calculated for each HRU on the basis of computed lapse rates from two precipitation stations. The station that is most representative of precipitation on an HRU is designated as the base station. The station that is not the base station is called the lapse station and is used with the base station to calculate the lapse rate. The two stations assigned to each HRU are specified by PRMS parameters `hru_psta` and `hru_plaps` (appendix 1). The monthly correction factor can be calculated as:

$$CF_{HRU} = 1.0 + paf_{base} \left[\frac{\left(\frac{\overline{P}_{lapse} - \overline{P}_{base}}{Z_{lapse} - Z_{base}} \right)(Z_{HRU} - Z_{base})}{\overline{P}_{base}} \right] , \quad (8)$$

where

paf_{base} is the mean monthly factor used to adjust rain (or snow) lapse rate (usually 1.0)—parameter `padj_rn` (or `padj_sn`), dimensionless;

$\overline{P}_{base}$ is the mean monthly precipitation at the base station—parameter `pmn_mo`, in inches per day; and

$\overline{P}_{lapse}$ is the mean monthly precipitation at the lapse station—parameter `pmn_mo`, in inches per day.

Z_{HRU} is the mean land-surface altitude of HRU—parameter `hru_elev`, in length;

Z_{base} and Z_{lapse} are the land-surface altitudes of the base and lapse stations, respectively—parameter `psta_elev`, in length.

The third option uses a three-dimensional multiple-linear regression to distribute precipitation from two or more stations (Hay and others, 2000; Hay and Clark, 2003) and is computed by module xyz_dist. Computations are similar to the multiple-linear regression for the calculation of temperature distribution given in equations 3a-3c:

$$P_{HRU}^m = c_{P1} + c_{P2} - CF_{xyz} , \qquad (9a)$$

$$c_{P1} = bp_{X,month} \overline{X}_{HRU} + bp_{Y,month} \overline{Y}_{HRU} + bp_{Z,month} \overline{Z}_{HRU}, \quad \text{and} \quad (9b)$$

$$c_{P2} = \overline{P}_{stas}^m - (bp_{X,month} \overline{X}_{sta} + bp_{Y,month} \overline{Y}_{sta} + bp_{Z,month} \overline{Z}_{sta}) , \quad (9c)$$

where

CF_{xyz} is the monthly correction factor as a decimal fraction used to adjust rain (or snow) values— parameter adjust_rain (or adjust_snow), dimensionless;

$bp_{X,month}, bp_{Y,month},$
and $bp_{Z,month}$ are the precipitation regression coefficients for longitude, latitude, and altitude, respectively by month, starting with January— parameter ppt_lapse, in inches; and

$\overline{P}_{stas}^m$ is the mean measured precipitation of all stations during time step m, in inches.

The fourth option (module precip_dist2_prms) weights measured precipitation from two or more stations by the inverse of the square of the distance between the centroid of an HRU and each station location (Dean and Snyder, 1977, Bauer and Vaccaro, 1987, Vaccaro, 2007). Estimates of daily precipitation on an HRU are adjusted by the ratio of the mean-monthly precipitation on an HRU to the mean monthly precipitation at each precipitation station. Precipitation data and the adjustment parameters must be in the same units. Daily precipitation on an HRU is computed according to:

$$P_{HRU}^m = \frac{\sum_{i=1}^{npstas} \left[\left(CF_{HRU,rain} P_i^m + CF_{HRU,snow} P_i^m \right) \left(Ldist_{HRU,i} \right)^2 \right]}{\sum_{i=1}^{npstas} \left(Ldist_{HRU,i} \right)^2} \quad (10a)$$

$$Ldist_{HRU,i} = 1 / \left(\sqrt{\left(Y_{HRU} - Yppt_i \right)^2 + \left(X_{HRU} - Xppt_i \right)^2} \right) , \quad (10b)$$

$$CF_{HRU,rain} = \frac{\overline{Rain}_{HRU,month}}{\overline{P}_{sta,month}} , \text{ for monthly fraction of} \quad (10c)$$

precipitation that is rain, and

$$CF_{HRU,snow} = \frac{\overline{Snow}_{HRU,month}}{\overline{P}_{sta,month}} , \text{ for monthly fraction of} \quad (10d)$$

precipitation that is snow,

where

$nptas$ is the number of precipitation stations, dimensionless;

$Xppt_i$ is the longitude of each precipitation measurement station—parameter psta_xlong, in length;

$Yppt_i$ is the latitude of each precipitation measurement station—parameter psta_ylat, in length;

$\overline{Rain}_{HRU,month}$ is the mean monthly rain on each HRU that can be obtained from National Weather Service's spatial distribution of mean annual precipitation for the 1971–2000 climate normal period— parameter rain_mon, in length;

$\overline{Snow}_{HRU,month}$ is the mean monthly snow on each HRU that can be obtained from National Weather Service's spatial distribution of mean annual precipitation for the 1971–2000 climate normal period— parameter snow_mon, in length;

$\overline{P}_{sta,month}$ is the mean monthly precipitation at each measurement station— parameter psta_mon, in length;

$CF_{HRU,rain}$ is the monthly rain correction factor as a decimal fraction of precipitation at the measurement station, dimensionless; and

$CF_{HRU,snow}$ is the monthly snow correction factor as a decimal fraction of precipitation at the measurement station, dimensionless.

Solar Radiation

Tables consisting of daily estimates of the potential (clear sky) short-wave solar radiation for each HRU are computed on the basis of hours between sunrise and sunset for each Julian day of the year in module soltab_hru_prms. The potential short-wave solar radiation also is computed for each Julian day of the year for a horizontal plane at the surface of the centroid of the modeled domain. The computations of the solar tables were modified for GSFLOW to use double precision and more accurate constants, such as decimal days per year, rotational degrees per day, eccentricity of the Earth's orbit, and the number pi and constants based on pi such as radians and radians per year.

Daily estimates of obliquity are computed from (Meeus, 1999):

$$E^m = 1 - [EC * \cos(jd - 3) * rad] , \qquad (11)$$

where

E^m is the obliquity of the Sun's ecliptic for time step m, in angular degrees;

EC is the eccentricity of the Earth's orbit (~ 0.01671), in radians;

jd is the Julian day number (3 is subtracted as the solar year begins on December 29), in days; and

rad is the revolution speed of the Earth (~ 0.0172), in radians per day.

Daily estimates of solar declination are computed from (Meeus, 1999):

$$
\begin{aligned}
DM^m = {} & 0.006918 - 0.399912 * \cos(E_{rt}) + 0.070257 \\
& * \sin(E_{rt}) - 0.006758 * \cos(E_{rt}) \\
& + 0.000907 * \sin(2 * E_{rt}) - 0.002697 \\
& * \cos(3 * E_{rt}) + 0.00148 * \sin(3 * E_{rt}) ,
\end{aligned} \qquad (12)
$$

where

DM^m is the solar declination for time step m, in angular degrees; and

$E_{rt} = rad * (jd - 1)$.

Sunset and sunrise times are computed for each HRU with a three-step procedure. First, the sunset and sunrise times are computed for a horizontal plane at the centroid of the HRU. Next, a horizontal surface on the terrestrial spheroid is found which is parallel to the slope and aspect of the surface of the HRU. This is called the equivalent-slope surface (Lee,

1963). Sunset and sunrise times also are computed for this surface. Finally, sunset time on the sloped surface of the HRU is taken as the earliest of the computed horizontal surface sunset times of the HRU and equivalent-slope surface. Likewise, sunrise time is taken as the latest of the computed horizontal surface sunrise times. Daylight length for each HRU is computed from these sunset and sunrise times.

The angle between the local meridian and the sunset (or sunrise) meridian, referred to as the hour angle of sunset (or sunrise), for the horizontal surface of both the HRU and equivalent-slope surface is calculated according to (Swift, 1976):

$$ss^m = \cos^{-1}[-\tan(lat)\tan(DM^m)] , \qquad (13a)$$

$$sr^m = -ss^m, \text{ and} \qquad (13b)$$

$$sh_{HRU}^m = \frac{\left(ss_{HRU}^m - sr_{HRU}^m\right)24}{2\pi} , \qquad (13c)$$

where

ss^m is the hour angle of sunset, measured from the local meridian of a horizontal surface (HRU or equivalent-slope surface) for time step m, in radians;

sr^m is the hour angle of sunrise, measured from the local meridian of a horizontal surface (HRU or equivalent-slope surface) for time stem m, in radians;

ss_{HRU}^m is the hour angle of sunset on the sloped surface of the HRU for time step m, in radians;

sr_{HRU}^m is the hour angle of sunrise on the sloped surface of the HRU for time step m, in radians;

sh_{HRU}^m is the daylight length on the HRU for time step m, in hours;

π is the constant pi (~ 3.1415926535898), dimensionless; and

lat is the latitude of the horizontal surface (basin centroid—parameter basin_lat, HRU centroid—parameter hru_lat or equivalent-slope surface), positive values are in the northern hemisphere and negative values are in the southern hemisphere, in radians.

Daily estimates of potential solar radiation for each HRU are calculated as described in Frank and Lee (1966), and Swift (1976):

$$Rsp_{HRU}^{m} = sc^{m}(c1_{PSR} + c2_{PSR}) \,, \tag{14a}$$

$$c1_{PSR} = \sin(DM^{m})\sin(lat'_{HRU})sh_{HRU}^{m} \,, \text{ and} \tag{14b}$$

$$c2_{PSR} = \frac{\cos(DM^{m})\cos(lat'_{HRU})\left[\sin(ss_{HRU}^{m} + long'_{HRU}) - \sin(sr_{HRU}^{m} + long'_{HRU})\right]24}{2\pi} \,, \tag{14c}$$

where

Rsp_{HRU}^{m} is the potential solar radiation on the HRU during time step m, in calories per square centimeter per day;

lat'_{HRU} is the latitude of the equivalent-slope surface of the HRU, in radians;

$long'_{HRU}$ is the longitude offset between the equivalent-slope surface and the HRU, in radians; and

sc^{m} is the 60-minute period solar constant for time step m, in calories per square centimeter per hour.

Computed daily shortwave radiation data for each HRU is estimated using one of two methods. The first is a modification of the degree-day method described by Leaf and Brink (1973) and is computed by module ddsolrad_hru_prms. This method was developed for the Rocky Mountain region of the United States. It is most applicable to this region where predominantly clear skies prevail on days without precipitation. The computed shortwave radiation is calculated according to:

$$Rah_{HRU}^{m} = ap_{HRU}^{m} Rsp_{HRU}^{m} \,, \text{ and} \tag{15a}$$

$$Rasw_{HRU}^{m} = \frac{Rah_{HRU}^{m}}{\cos(\tan^{-1}(slope_{HRU}))} pptadj_{HRU} \,, \tag{15b}$$

where

Rah_{HRU}^{m} is the measured or computed as the horizontal plane shortwave radiation on the HRU during time step m, in calories per square centimeter per day;

ap_{HRU}^{m} is the degree-day based ratio of actual to potential shortwave radiation for the HRU during time step m, dimensionless;

$Rasw_{HRU}^{m}$ is the computed shortwave radiation on the HRU during time step m, in calories per square centimeter per day;

$slope_{HRU}$ is the slope of the HRU—parameter hru_slope, dimensionless; and

$pptadj_{HRU}$ is the precipitation-day adjustment factor to solar radiation—parameters radj_wppt (for a winter day) and radj_sppt (for a summer day), dimensionless.

The second procedure uses a relation between sky cover and daily range in air temperature demonstrated by Tangborn (1978), and a relation between solar radiation and sky cover developed by Thompson (1976) and is computed by module ccsolrad_hru_prms. This procedure is applicable to more humid regions where extensive periods of cloud cover occur with and without precipitation. Shortwave radiation applicable for humid regions is calculated as:

$$CC_{HRU}^m = crn_{month} \left(Tmx_{HRU}^m - Tmn_{HRU}^m \right) + crb_{month} \quad \text{(16a)}$$

$$Rah_{HRU}^m = \left[B + (1.0 - B)(1.0 - CC_{HRU}^m)^{cre} \right] Rsp_{HRU}^m , \text{ and } \quad \text{(16b)}$$

$$Rasw_{HRU}^m = \frac{Rah_{HRU}^m}{\cos(\arctan(slope_{HRU}))} , \quad \text{(17)}$$

where

B is the constant used in the cloud – cover to solar – radiation relation, a value can be obtained from Thompson (1976, fig. 1)—parameter `crad_coef`, dimensionless;

CC_{HRU}^m is the decimal fraction of cloud cover on the HRU during time step m, dimensionless;

cre is the exponent used in the cloud – cover to solar – radiation relation, a value of 0.61 is suggested by Thompson (1976)—parameter `crad_exp`, dimensionless;

crn_{month} is the slope in the regression equation that relates cloud cover to daily minimum and maximum air temperature by month, starting with January—parameter `ccov_slope`, dimensionless; and

crb_{month} is the intercept in the regression equation that relates cloud cover to daily minimum and maximum air temperature by month, starting with January—parameter `ccov_intcp`, dimensionless.

Potential Evapotranspiration

Potential evapotranspiration is computed for each HRU by one of three user-specified options. The first option (module potet_hamon_hru_prms) is the empirical Hamon formulation, in which PET is computed as a function of daily mean air temperature and possible hours of sunshine according to (Hamon, 1961; Murray, 1967; and Federer and Lash, 1978):

$$PET_{HRU}^m = HC_{HRU} \left(\frac{sh_{HRU}^m}{12} \right)^2 \rho_{HRU} , \text{ and } \quad \text{(18)}$$

$$\rho_{HRU} = 216.7 \frac{6.108 e^{17.26939 \frac{\overline{T}_{HRU}}{\overline{T}_{HRU} + 273.3}}}{\overline{T}_{HRU} + 273.3} , \quad \text{(19)}$$

where

PET_{HRU}^m is the potential evapotranspiration for the HRU during time step m, in inches;

HC_{HRU} is the Hamon monthly air-temperature coefficient—parameter `hamon_coef`, in inch-cubic meter per gram;

ρ_{HRU} is the saturated water-vapor density (absolute humidity), in grams per cubic meter;

e is the exponential function constant (~ 2.7182818), dimensionless; and

$\overline{T}_{HRU}$ is the mean daily temperature on the HRU, in degrees Celsius.

The second option (module potet_jh_prms) is the modified Jensen-Haise formulation (Jensen and others, 1969), in which potential evapotranspiration is computed as a function of air temperature, solar radiation, and two coefficients that can be estimated using regional air temperature, altitude, vapor pressure, and plant cover according to:

$$PET_{HRU}^m = JH_{month} \left(\overline{T}_{HRU} - JH_{HRU} \right) \frac{Rasw_{HRU}^m}{2.54 \lambda_{HRU}} , \text{ and } \quad \text{(20a)}$$

$$\lambda_{HRU} = 597.3 - (0.5653 \overline{T}_{HRU}) , \quad \text{(20b)}$$

where

JH_{month} is the monthly Jensen-Haise air-temperature coefficient—parameter `jh_coef`, in degrees Fahrenheit;

2.54 is the conversion from inch to centimeters;

JH_{HRU} is the Jensen-Haise air-temperature coefficient for the HRU—parameter `jh_coef_hru`, in degrees Fahrenheit; and

λ_{HRU} is the latent heat of vaporization on the HRU for time step m, in calories per gram.

The third option (module potet_pan_prms) is used when pan evaporation data from one or more measurement stations are available. The station associated with each HRU is specified by PRMS parameter hru_pansta (appendix 1). Potential evapotranspiration is computed from the measured pan evaporation and a monthly coefficient according to:

$$PET_{HRU}^m = Pancoef_{month} Panevap_{sta}^m , \qquad (21)$$

where

$Pancoef_{month}$ is the monthly pan-evaporation coefficient—parameter epan_coef, dimensionless; and

$Panevap_{sta}^m$ is the pan evaporation for the corresponding measurement station for time step m, in length.

Canopy Interception

Interception of precipitation by the plant canopy is computed during a time step as a function of plant-cover density and the storage available on the predominant plant-cover type in each HRU (Leavesley and others, 1983) using module intcp_prms. Throughfall precipitation, which is precipitation that is not intercepted by the plant canopy, is computed as:

$$Spca_{HRU}^m = (Spcmx_{HRU} - Spc_{HRU}^m)(A_{HRU}\rho'_{HRU})$$

$$Ptf_{HRU}^m = P_{HRU}^m - \frac{Spca_{HRU}^m}{A_{HRU}\rho'_{HRU}} \quad \text{when } P_{HRU}^m > \frac{Spca_{HRU}^m}{A_{HRU}\rho'_{HRU}}$$

$$Ptf_{HRU}^m = 0.0 \quad \text{when } P_{HRU}^m \le \frac{Spca_{HRU}^m}{A_{HRU}\rho'_{HRU}} , \qquad (22)$$

where

$Spca_{HRU}^m$ is the available storage in the plant canopy of the HRU during time step m, in acre-inch;

$Spcmx_{HRU}$ is the maximum storage in the plant canopy for snow, summer rain, and winter rain on each HRU— parameter snow_intcp (snow), srain_intcp (summer rain), wrain_intcp (winter rain), in inches;

Spc_{HRU}^m is the storage in the plant canopy (summer or winter) on the HRU during time step m, in acre-inch;

A_{HRU} is the area of the HRU—parameter hru_area, in acres;

ρ'_{HRU} is the plant canopy density as a decimal fraction of the HRU area— parameter covden_sum (summer) covden_win (winter), dimensionless; and

Ptf_{HRU}^m is the precipitation throughfall on the HRU during time step m, in inches.

The precipitation that reaches the ground during time step m is referred to as net precipitation, and is the sum of throughfall and precipitation on the HRU not covered by plants. Net precipitation is calculated according to:

$$Pnet_{HRU}^m = P_{HRU}^m(1.0 - \rho'_{HRU}) + (Ptf_{HRU}^m\rho'_{HRU}) , \qquad (23)$$

where

$Pnet_{HRU}^m$ is the precipitaiton that reaches the ground during time step m, in inches.

Plant-cover density varies by season and type. The types of plant cover that can be specified are bare (no cover), grass, shrubs, and trees. Intercepted rain is assumed to evaporate at a free-water surface rate. Intercepted snow is assumed to sublimate at a rate that is expressed as a decimal fraction of the potential evapotranspiration.

Snowpack

PRMS simulates the initiation, accumulation, and depletion of a snowpack on each HRU (fig. 3; Leavesley and others, 1983; Leavesley and others, 2005) using module snowcomp_prms. A snowpack is modeled as a two-layered system that is maintained and modified on both a water equivalent basis and as a dynamic heat reservoir, as depicted in figure 3 (Obled and Rosse, 1977). The surface layer consists of the upper 3–5 cm of the snowpack, and the lower layer is the remaining snowpack. Only equations that describe the water and energy balance of the snowpack are described herein. Refer to Leavesley and others (1983) for a comprehensive discussion of the PRMS snowpack calculations.

A snowpack develops when precipitation occurs as snow. If a snowpack already exists, the precipitation received during that time step is added to the snowpack and the new snowpack depth is calculated. Snowpack depth is calculated by solving the following ordinary differential equation with a finite-difference approximation of the time derivative (Riley and others, 1973):

$$\frac{dDs_{HRU}^m}{dt} + cs_{HRU}Ds_{HRU}^m = \frac{Ps_{HRU}^m}{\rho s} + \left(\frac{cs_{HRU}}{\rho sm}Dse_{HRU}^m\right),\qquad(24)$$

where

Ds_{HRU}^m is the precipitaiton that reaches the ground during time step m, in inches.

cs_{HRU} is the snowpack settlement-time constant —parameter `settle_const`, in per day;

Ps_{HRU}^m is the net snowfall rate, as a liquid water equivalent, on the HRU during time step m, in inches per day;

ρs is the density of new-fallen snow, as a decimal fraction—parameter `den_init`, dimensionless;

ρsm is the average maximum snowpack density, as a decimal fraction of the liquid water equivalent—parameter `den_max`, dimensionless; and

Dse_{HRU}^m is the snowpack liquid water equivalent depth for the HRU at time step m, in inches.

The snowpack energy balance is computed for the day over two 12-hour intervals (designated day and night) and aggregated to a daily time step. Heat transfer between the surface layer and the lower layer of the snowpack occurs by conduction when the temperature of the surface layer is less than freezing (0° Celsius), as discussed by Leavesley and others (1983, p. 39-41), and calculated as:

$$Ht_{HRU}^m = 2\rho snow_{HRU}^m C_{ice}\sqrt{\frac{ke_{HRU}^m \Delta t_{snow}}{\pi\rho snow_{HRU}^m C_{ice}}}\left(T_{surf}^m - T_{lower}^m\right),\qquad(25)$$

where

Ht_{HRU}^m is the daily heat transferred from the surface layer to the lower layer of the snowpack during time step m, in calories per square centimeter;

$\rho snow_{HRU}^m$ is the density of the snow pack, in grams per cubic centimeter;

C_{ice} is the specific heat of ice, in calories per gram-degree Celsius;

ke_{HRU}^m is the effective thermal conductivity of the snowpack, in calories per second-gram-degree Celsius;

Δt_{snow} is the snow computation time step, in 43,200 seconds (half-day interval);

T_{surf}^m is the temperature of the snowpack surface layer during time step m, in degrees Celsius; and

T_{lower}^m is the temperature of the snowpack lower layer during time step m, in degrees Celsius.

The effective thermal conductivity of the snowpack (ke_{HRU}^m) is estimated by (Anderson, 1968, p. 22):

$$ke_{HRU}^m = 0.0077\left(\rho_{HRU}^{''m}\right)^2.\qquad(26)$$

The volumetric density fraction of the snowpack used in equation 26 is computed as (Leavesley and other, 1983, p. 41):

$$\rho_{HRU}^{''m} = \frac{Dse_{HRU}^m}{Ds_{HRU}^m},\qquad(27)$$

where

$\rho_{HRU}^{''m}$ is the volumetric density fraction of the snowpack for the HRU during time step m, dimensionless.

The computation of the energy components available to melt snow when both layers of the snowpack reach isothermal conditions at 0° Celsius are described by Male and Gray (1981); the energy available to melt snow is calculated according to:

$$Hm_{HRU}^m = Hs_{HRU}^m + Hl_{HRU}^m + Hc_{HRU}^m + He_{HRU}^m \\ + Hg_{HRU}^m + Hp_{HRU}^m + Hq_{HRU}^m , \qquad (28)$$

where

Hm_{HRU}^m is the energy available for snowmelt during time step m, in calories;

Hs_{HRU}^m is the energy gained due to shortwave radiation during time step m, in calories;

Hl_{HRU}^m is the energy gained due to longwave radiation during time step m, in calories;

Hc_{HRU}^m is the convective or sensible heat at air-snow interface during time step m, in calories;

He_{HRU}^m is the latent heat (sublimation and condensation) at the air-snow interface during time step m, in calories;

Hg_{HRU}^m is the heat gained from the ground during time step m, in calories;

Hp_{HRU}^m is the heat gained from precipitation during time step m, in calories; and

Hq_{HRU}^m is the heat required for internal state change during time step m, in calories.

Further details of variables used in equation 28, assumptions made for the PRMS formulation, and calculation are presented by Leavesley and others (1983, p. 42-45).

The snowmelt on an HRU is computed by:

$$Vsm_{HRU}^m = \frac{Hm_{HRU}^m}{HF} Asc_{HRU}^m , \qquad (29)$$

where

Vsm_{HRU}^m is the volume of snowpack melted during time step m, in acre-inch;

HF is the specific latent heat of fusion to melt one inch of water-equivalent ice at 0° Celsius, in 203.2 calories per inch; and

Asc_{HRU}^m is the snow-covered area of the HRU determined by snow-cover areal depletion curve (Anderson, 1973), in acres.

Sublimation from the snow surface and evaporation from snow on the plant canopy are assumed to occur only when there is no transpiration from plants. The daily loss from the snowpack is computed as a decimal fraction of the potential evapotranspiration defined by input parameter potet_sublim (see appendix 1):

$$sub_{HRU}^m = dfsub_{HRU} PET_{HRU}^m - qce_{HRU}^m \rho_{HRU}''^m , \qquad (30)$$

where

sub_{HRU}^m is the sublimation from the HRU during time step m, in inches;

$dfsub_{HRU}$ is the decimal fraction of potential evapotranspiration that is sublimated from snow surface—parameter potet_sublim, dimensionless; and

qce_{HRU}^m is the evaporation loss from interception storage for the HRU during time step m, in inches.

Impervious Storage, Hortonian Runoff, and Infiltration

Snowmelt and net precipitation that reach the soil surface during a time step is partitioned to the pervious and impervious parts of each HRU. Surface runoff due to infiltration excess, hereafter referred to as Hortonian runoff, and infiltration are computed on the pervious parts of each HRU; whereas storage, evaporation, and Hortonian runoff are computed on the impervious parts of each HRU.

Impervious Storage and Evaporation

If rain throughfall and snowmelt satisfy available retention storage on the impervious parts of the HRU, Hortonian runoff is generated. Hortonian runoff from impervious parts of each HRU is calculated from continuity according to:

$$C_{imper}^m = D_{imper}^{m-1} - Dimx_{imper} + Pnet_{HRU}^m$$
$$+ \frac{Vsm_{HRU}^m}{A_{HRU}} + ROhup_{HRU}^m$$

$$ROh_{imper}^m = C_{imper}^m \quad \text{When} \quad C_{imper}^m > 0$$

$$ROh_{imper}^m = 0 \quad \text{When} \quad C_{imper}^m \leq 0,$$

(31)

where

C_{imper}^m is the water available for Hortonian runoff from the impervious part per unit area of the HRU during time step m, in inches;

ROh_{imper}^m is the Hortonian runoff from the impervious part of the HRU per unit area during time step m, in inches;

D_{imper}^{m-1} is the impervious storage, as calculated by equation 33, for the last iteration of time step $m-1$, as volume per unit area for the HRU, in inches;

$Dimx_{imper}$ is the maximum retention storage for HRU impervious area, in inches; and

$ROhup_{HRU}^m$ is the sum of Hortonian runoff from all upslope contributing HRUs as a volume per unit area of the HRU for time step m, in inches.

Evaporation from impervious parts of HRUs is computed for each time step by;

$$C1_{imper}^m = PET_{HRU}^m - sub_{HRU}^m$$

$$C2_{imper}^m = D_{imper}^{m-1} + Pnet_{HRU}^m + \frac{Vsm_{HRU}^m}{A_{HRU}} + ROhup_{HRU}^m$$
$$- ROh_{imper}^m$$

$$Evap_{imper}^m = C2_{imper}^m \left(1 - \frac{Asc_{HRU}^m}{A_{HRU}}\right) \text{When} \ C1_{imper}^m \geq C2_{imper}^m \quad (32)$$

$$Evap_{imper}^m = C1_{imper}^m \left(1 - \frac{Asc_{HRU}^m}{A_{HRU}}\right) \text{When} \ C1_{imper}^m < C2_{imper}^m \ ,$$

where

$Evap_{imper}^m$ is the evaporation from the impervious part of the HRU for time step m, in inches.

Storage on the impervious parts of the HRU is calculated according to:

$$D_{imper}^m = D_{imper}^{m-1} + Pnet_{HRU}^m + \frac{Vsm_{HRU}^m}{A_{HRU}}$$
$$+ ROhup_{HRU}^m - ROh_{imper}^m - Evap_{imper}^m \ .$$

(33)

Hortonian Runoff and Infiltration

Hortonian runoff for the pervious part of each HRU is computed using a contributing-area concept (Dickinson and Whiteley, 1970; Hewlett and Nutter, 1970). Hortonian runoff from pervious parts of a HRU is related to the area where the throughfall and snowmelt exceed the soil infiltration rate. This is represented by either a linear (module srunoff_carea_casc) or nonlinear function (module srunoff_smidx_casc) of antecedent soil-moisture content. The linear form of computing the contributing area for pervious runoff can be written as:

$$Fperv_{HRU}^m = Fmn_{HRU} + \left(Fmx_{HRU} - Fmn_{HRU}\right)\left(\frac{Dup_{HRU}^m}{Dupmx_{HRU}}\right), \tag{34a}$$

whereas the nonlinear form can be written as:

$$Smidx_{HRU}^m = D_{CPR}^{m-1} + 0.5 Pnet_{HRU}^m$$

$$C3_{HRU}^m = Smc_{HRU} * 10^{\left(Smex_{HRU} \, Smidx_{HRU}^m\right)}$$

$$Fperv_{HRU}^m = C3_{HRU}^m \quad \text{when} \quad C3_{HRU}^m \leq Fmx_{HRU} \tag{34b}$$

$$Fperv_{HRU}^m = Fmx_{HRU} \quad \text{when} \quad C3_{HRU}^m > Fmx_{HRU} ,$$

where

$Fperv_{HRU}^m$ is the surface-runoff-contributing area of the pervious parts in the HRU for time step m, as a decimal fraction of HRU area, dimensionless;

Fmn_{HRU} is the minimum possible area contributing to surface runoff, as a decimal fraction of HRU area—parameter carea_min, dimensionless;

Fmx_{HRU} is the maximum possible area contributing to surface runoff, as a decimal fraction of HRU area—parameter carea_max, dimensionless;

Dup_{HRU}^m is the antecedent volume per unit area of water in capillary reservoir that is available for evaporation during time step m, in inches;

$Dupmx_{HRU}$ is the maximum quantity of water in the capillary reservoir—parameter soil_rechr_max, in inches;

$Smidx_{HRU}^m$ is the soil moisture index of the capillary reservoir for time step m, in inches;

D_{CPR}^{m-1} is the volume per unit area of water in the capillary reservoir at the last iteration of time step m-1, in inches;

Smc_{HRU} is a coefficient used to calculate decimal fraction of pervious surfaces —parameter smidx_coef, dimensionless; and

$Smex_{HRU}$ is an exponent used to calculate the decimal fraction of pervious surfaces —parameter smidx_exp, in per inch.

The runoff from the pervious part of an HRU is calculated as:

$$ROh_{perv}^m = Fperv_{HRU}^m \left(ROhup_{HRU}^m + Pnet_{HRU}^m \right) , \qquad (35)$$

where

ROh_{perv}^m is the runoff per unit area from the pervious part of the HRU for time step m, in inches.

Hortonian runoff from an HRU is routed to one or more downslope HRUs and(or) to one or more stream segments with the cascading-flow procedure (see section "Cascading-Flow Procedure") and connections defined by parameters input to the PRMS Cascade Module (cascade_prms, table 8). The fraction of runoff from an upslope HRU that flows to a downslope HRU or stream segment is calculated as:

$$ROh_{HRU,dwn}^m = \left(ROh_{perv,up}^m + ROh_{impr,up}^m \right)$$
$$Fcontrib_{HRU,up} \frac{A_{HRU,up}}{A_{HRU,dwn}} , \qquad (36)$$

where

$ROh_{HRU,dwn}^m$ is volume per unit area of Hortonian runoff to a downslope HRU or stream segment during time step m, in inches;

$ROh_{perv,up}^m$ is the runoff per unit area from the pervious part of the upslope HRU for time step m, in inches;

$ROh_{impr,up}^m$ is the runoff per unit area from the impervious part of the upslope HRU for time step m, in inches;

$Fcontrib_{HRU,up}$ is the decimal fraction of area in the upslope HRU that contributes Hortonian runoff to the downslope HRU—parameter hru_pct_up, dimensionless;

$A_{HRU,up}$ is area of the downslope HRU— parameter hru_area, in acres; and

$A_{HRU,dwn}$ is area of the downslope HRU— parameter hru_area, in acres.

Infiltration occurs on the pervious areas of each HRU and includes Hortonian runoff from upslope HRUs, snowmelt, and rain throughfall. Hortonian runoff from the HRU is subtracted from the available water for infiltration. Infiltration is calculated as:

$$C4_{HRU}^m = ROhup_{perv}^m + \frac{Vsm_{HRU}^m}{A_{HRU}} ,$$

$$qsi_{perv}^m = \left(C4_{HRU}^m + Pnet_{HRU}^m - ROh_{perv}^m \right) A_{perv} \qquad (37)$$
$$\text{when } C4_{HRU}^m < qsnmx_{HRU}$$

$$qsi_{perv}^m = \left(qsnmx_{HRU} + Pnet_{HRU}^m \right)$$
$$\text{when } C4_{HRU}^m \geq qsnmx_{HRU} ,$$

where

qsi_{perv}^m is the soil infiltration over the pervious part of the HRU for time step m, in acre-inch;

$qsnmx_{HRU}$ is the daily maximum snowmelt infiltration for the HRU—parameter snowinfil_max, in inches; and

A_{perv} is the pervious area of the HRU, in acres.

Soil Zone

The soil zone is an important link in the integration of PRMS with MODFLOW-2005 and its function was changed considerably from the original PRMS module for use in GSFLOW. The new Soil-Zone Module (soilzone_gsflow, table 1) is used to calculate storage and all inflows and outflows in the soil zone for each HRU. The soil zone is conceptualized as three reservoirs: the capillary, gravity, and preferential-flow reservoirs (fig. 19). These reservoirs occupy the same physical space and represent different soil-water processes at different soil-water content thresholds. The capillary reservoir is limited by the field-capacity threshold. The soil-water content between field capacity and soil saturation defines the gravity reservoir. The preferential-flow reservoir is that part of the gravity reservoir from which fast interflow occurs and is defined by the preferential-flow threshold.

The only outflows from the capillary reservoir are to evapotranspiration and gravity reservoirs. Outflow from the preferential-flow reservoir can be fast interflow and saturation excess surface runoff, hereafter referred to as Dunnian runoff. Outflow from the gravity reservoir can be slow interflow, flow to the preferential-flow reservoir, replenishment of the capillary reservoir, gravity drainage, and Dunnian runoff. Flow in the soil zone is dependent on the simultaneous solution of flow in the soil zone and ground-water heads. Storage and all inflows and outflows in the soil zone are calculated according to the sequence of steps illustrated in figure 19 and listed in table 9.

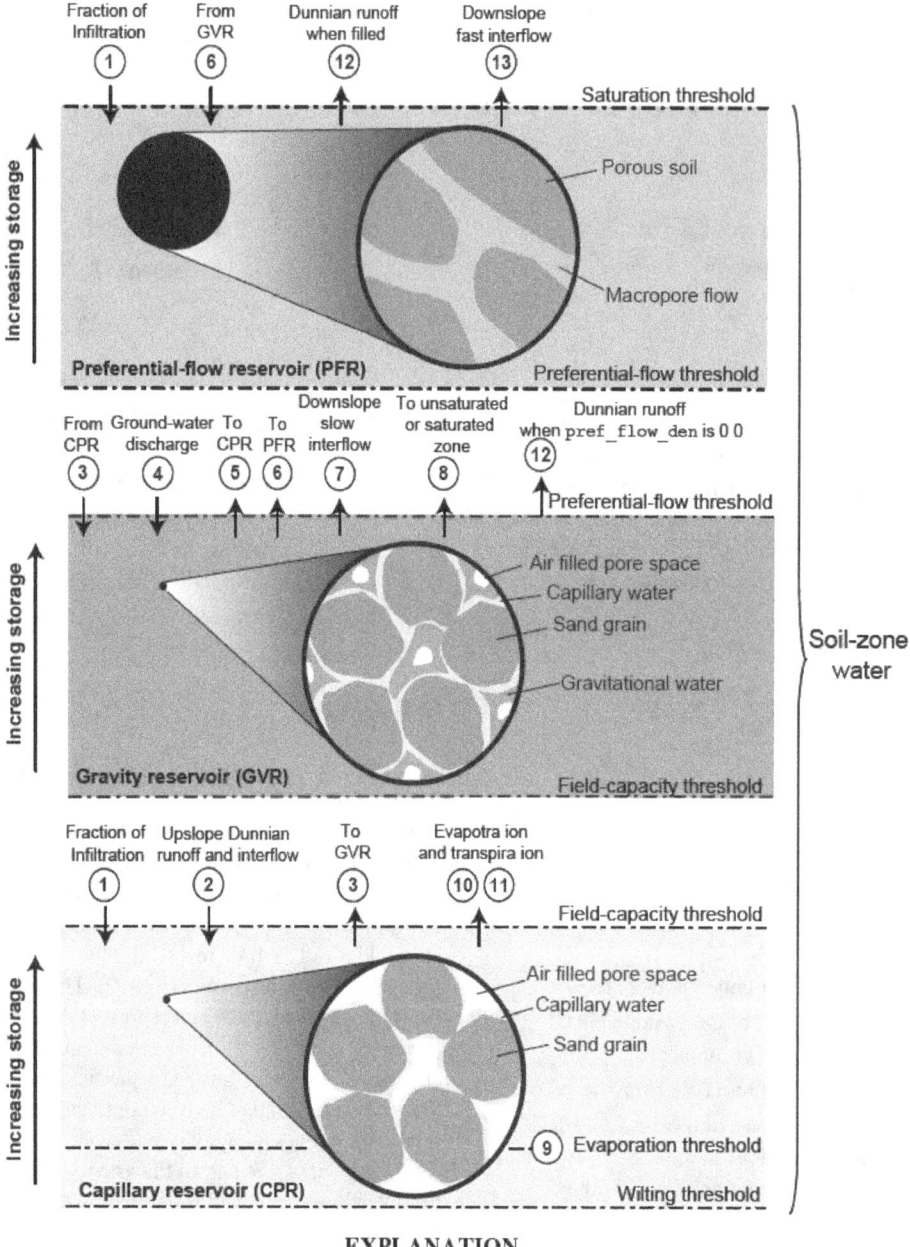

EXPLANATION

(8)   Computational sequence listed in table 9

Figure 19.   Computational sequence used in computing flow to and from a soil zone in a hydrologic response unit. Water in the soil zone consists of three reservoirs—capillary, gravity, and preferential-flow—that represent pore-space volumes for a given volume of soil.

Table 9. Sequence of steps used in the computation of flow into and out of the soil zone used in GSFLOW.

[Input parameters are specified for each hydrologic response unit]

Sequence No.	Description of flow into and out of soil zone
1	Partition infiltration of precipitation, snowmelt, and Hortonian runoff (Horton, 1933) between capillary and preferential-flow reservoirs on basis of the input parameter `pref_flow_den`.
2	Add upslope Dunnian runoff (Dunne and Black, 1970) and interflow to capillary reservoir.
3	Add excess water above field capacity in capillary reservoirs to gravity reservoirs on basis of input parameter `soil_moist_max`.
4	Add ground-water discharge from MODFLOW-2005 to gravity reservoir.
5	Replenish capillary reservoirs from gravity reservoirs if water is available and water in capillary reservoirs is below field-capacity threshold. A fraction of the excess water can be given directly to the associated ground-water reservoir for PRMS-only simulations when the input parameter `soil2gw_max` is greater than zero.
6	Add a fraction of water in gravity reservoirs to preferential-flow reservoirs on basis of input parameters `pref_flow_den` and `sat_threshold`. Parameter `sat_threshold` is the volume of water per unit area between field capacity and saturation thresholds. The value can be greater than the available porosity above field-capacity to account for surface-depression storage on land surface.
7	Compute slow interflow from gravity reservoirs on basis of input parameters `slowcoef_lin` and `slowcoef_sq` and the volume of water per unit area stored in the reservoir.
8	Compute gravity drainage from gravity reservoirs to unsaturated zone and/or saturated zone in MODFLOW-2005 for integrated simulations or to ground-water reservoirs for PRMS-only simulations on basis of input parameters `ssr2gw_rate`, `ssr2gw_exp`, and `ssrmax_coef` and the volume of water per unit area stored in the reservoir.
9	Partition capillary reservoirs into two zones on basis of input parameter `soil_rechr_max`: upper zone stores water available for evaporation and transpiration and lower zone stores water available for transpiration only.
10	Compute transpiration from lower zone of capillary reservoir.
11	Compute evaporation and transpiration from upper zone of capillary reservoirs on basis of input parameters `soil_moist_max`, `soil_rechr_max`, and `soil_type`.
12	Compute Dunnian runoff (Dunne and Black, 1970) from gravity reservoirs when parameter `pref_flow_den` is 0.0, that is, when the preferential-threshold equals the saturation threshold, or from preferential-flow reservoirs when the volume per unit area in the reservoirs exceeds the saturation threshold defined by input parameter `sat_threshold`.
13	Compute fast interflow from preferential-flow reservoirs on basis of input parameters `fastcoef_lin` and `fastcoef_sq` and the volume of water per unit area stored in the reservoir.

Infiltration to Preferential-Flow and Capillary Reservoirs

A fraction of the infiltration ($Dhru_{PFR}^{m}$) is apportioned to the preferential-flow reservoir to account for fast interflow through large openings in the soil zone near land surface, whereas the remainder of the infiltration ($Drem_{CPR}^{m}$) is added to the capillary reservoir:

$$Dhru_{PFR}^{m} = \frac{Fden_{HRU}\,qsi_{perv}^{m}}{A_{perv}}, \text{ and} \qquad (38a)$$

$$Drem_{CPR}^{m} = (1 - Fden_{HRU})\frac{qsi_{perv}^{m}}{A_{perv}}, \qquad (38b)$$

where

$Dhru_{PFR}^{m}$ is the volume of water per unit area added to all connected preferential-flow reservoirs for time step m, in inches;

$Drem_{CPR}^{m}$ is the volume of water per unit area of the pervious part of the HRU that infiltrates into the capillary reservoir at time step m, in inches; and

$Fden_{HRU}$ is the decimal fraction of the soil zone available for preferential flow—parameter `pref_flow_den`, dimensionless.

All inflows to the capillary reservoir are added to storage in the capillary reservoir prior to calculating outflows (table 9; sequence Nos. 1 and 2). Volume per unit area in a capillary reservoir is computed as:

$$D_{CPR}^{m,n} = \frac{Drem_{CPR}^{m} A_{perv}}{A_{HRU}} + ROdup_{HRU}^{m,n} \\ + Dslup_{HRU}^{m,n} + Dfup_{HRU}^{m,n} + D_{CPR}^{m-1},$$

(39)

where

$D_{CPR}^{m,n}$ is the volume of water in the capillary reservoir at time step m and iteration n, in inches;

D_{CPR}^{m-1} is the volume of water in the capillary reservoir at time step m-1, in inches;

$ROdup_{HRU}^{m,n}$ is the volume of Dunnian runoff per unit area into the capillary reservoir from all contributing HRUs at time step m and iteration n, in inches;

$Dslup_{HRU}^{m,n}$ is the volume of slow interflow per unit area into the capillary reservoir from all contributing HRUs at time step m and iteration n, in inches; and

$Dfup_{HRU}^{m,n}$ is the volume of fast interflow per unit area into the capillary reservoir from all contributing HRUs at time step m and iteration n, in inches.

Inflow to Gravity Reservoirs

Water in excess of the field-capacity threshold in a capillary reservoir is added to connected gravity reservoirs (table 9; sequence No. 3). Ground-water discharge from the connected finite-difference cell also is added to the gravity reservoir (table 9, sequence No. 4). The volume per unit area in the gravity reservoir is calculated as:

$$D_{GVR}^{m,n} = Dexcess_{CPR}^{m,n} + \left(\frac{Q_{gw}^{m,n} C_{mf2prms} \Delta t}{A_{fdc}} \right) + D_{GVR}^{m-1},$$

(40)

where

$D_{GVR}^{m,n}$ is the volume of water per unit area in the gravity reservoir at time step m and iteration n, in inches;

$Dexcess_{CPR}^{m,n}$ is the volume of water per unit area of excess water in the capillary reservoir for time step m and iteration n, in inches;

$Q_{gw}^{m,n}$ is the volumetric-flow rate of water that discharges from ground water to the soil zone at time step m, iteration n, in cubic length per time;

$C_{mf2prms}$ is the conversion from MODFLOW-2005 length per time to PRMS inches per day (table 7);

Δt is the GSFLOW time step, in one day;

A_{fdc} is the top area of the finite-difference cell, in length squared; and

D_{GVR}^{m-1} is the volume of water per unit area in the gravity reservoir at time step m-1, in inches.

Ground-water discharge from the saturated zone to connected gravity reservoirs is computed when $celtop - 0.5D_{usz} < h_{fdc}^{m,n-1}$ as:

$$Q_{gw}^{m,n} = CND_{sz}\left(h_{fdc}^{m,n} - \left(celtop - 0.5D_{usz}\right)\right), \qquad (41)$$

where

$celtop$ is the top altitude of the finite-difference cell, in length;

D_{usz} is the depth of undulations at soil-zone base, in length;

$h_{fdc}^{m,n-1}$ is the ground-water head in the finite-difference cell for time step m, iteration n-1, in length;

$h_{fdc}^{m,n}$ is the ground-water head in the finite-difference cell for time step m and iteration n, in length;

CND_{sz} is the conductance across the soil-zone base equal to

$$\frac{K_v A_{fdc}}{0.5celthkD_{usz}}\left(h_{fdc}^{m,n-1} - celtop + 0.5D_{usz}\right),$$

in length squared per time;

K_v is the vertical hydraulic conductivity of the finite-difference cell, in length per time; and

$celthk$ is the thickness of the finite-difference cell, in length.

Ground-water discharge across the soil-zone base is dependent on the ground-water head during iteration n of time step m, as described in equation 41. However, the conductance (CND_{sz}) also is dependent on ground-water head, and is linearized by using the ground-water head from iteration n-1.

Total volume per unit area from the finite-difference cell to the connected gravity reservoir is computed as:

$$D_{GVR}^{m,n**} = D_{GVR}^{m,n*} + D_{mf2GVR}^{m,n}, \qquad (42)$$

where

$D_{GVR}^{m,n**}$ is the revised volume per unit area in a gravity reservoir after gravity drainage and before Dunnian runoff for time step m and iteration n, in inches;

$D_{mf2GVR}^{m,n}$ is the volume per unit area from the finite-difference cell to a connected gravity reservoir for time step m and iteration n, in inches; and

$D_{GVR}^{m,n*}$ is the volume per unit area in a gravity reservoir after slow interflow and before gravity drainage for time step m and iteration n, in inches.

The difference between potential and net gravity drainage is returned to the soil zone. The returned water is redistributed back to each gravity reservoir as:

$$D_{GVR}^{m,n**} = D_{GVR}^{m,n*} + D_{rej}^{m,n}, \qquad (43)$$

where

$D_{rej}^{m,n}$ is gravity drainage volume per unit area that is rejected by the finite-difference cell, in inches.

Table 10.   Variables used to transfer water from reservoirs in PRMS to finite-difference cells and stream reaches in MODFLOW-2005 used in GSFLOW.

[HRU, hydrologic response unit; ngwcell, number of finite-difference cells; nhru, number of HRUs; nhrucell, product of HRUs and finite-difference cells; and nreach, number of stream reaches]

Variable name	Array dimension	Description
Transfer of gravity drainage from gravity reservoirs to finite-difference cells		
basin_szreject	1	Watershed area-weighted average gravity drainage not added to connected finite-difference cells.
cell_drain_rate	ngwcell	Gravity drainage from gravity reservoirs added to connected finite-difference cells.
gw_rejected	nhru	HRU area-weighted gravity drainage from gravity reservoirs not added to connected finite-difference cells.
gw_rejected_grav	nhrucell	Gravity drainage from gravity reservoirs not added to connected finite-difference cells.
Transfer of surface runoff and interflows from gravity and preferential-flow reservoirs to stream reaches		
basin_reach_latflow	1	Watershed area-weighted average lateral inflow to stream reaches.
reach_latflow	nreach	Lateral inflow to each stream reach.
Unused potential evapotranspiration transferred to finite-difference cells		
unused_potet	nhru	Unsatisfied potential evapotranspiration in each HRU available as evapotranspiration from unsaturated and saturated zone of finite-difference cells.
Transfer of ground-water discharge from finite-difference cells to gravity reservoirs		
basin_gw2sm	1	Watershed area-weighted average ground-water discharge from all connected finite-difference cells.
gw2sm	nhru	HRU area-weighted average ground-water discharge from connected finite-difference cells.
gw2sm_grav	nhrucell	Ground-water discharge from connected finite-difference cell to gravity reservoir.
actet_gw	nhru	HRU area-weighted evapotranspiration from finite-difference cells.
actet_tot_gwsz	nhru	HRU area-weighted evapotranspiration from finite-difference cells and capillary reservoirs in the soil zone.

GSFLOW variables that are associated with the exchange of water between gravity reservoirs and connected finite-difference cells are listed in table 10.

Outflow from Gravity Reservoirs

There are five possible processes by which outflow from gravity reservoirs can occur: flow to a capillary reservoir (sequence No. 5, table 9), flow to a preferential-flow reservoir (sequence No. 6, table 9), flow to downslope HRUs and(or) stream segments (sequence No. 7, table 9), gravity drainage to a finite-difference cell (sequence No. 8, table 9), and Dunnian runoff when pref_flow_den is zero (sequence No. 12, table 9). Each of these is described below.

Flow from Gravity Reservoirs to Capillary Reservoirs

Gravity reservoirs replenish the capillary reservoir up to field capacity whenever storage in the capillary reservoir is below the field-capacity threshold (fig. 19; and table 9, sequence No. 5); however, flow from a gravity reservoir

to a capillary reservoir is limited by storage in the gravity reservoir. Flow from gravity reservoirs to a capillary reservoir is determined by first computing the capillary reservoir deficit, which is the volume of water needed to replenish the capillary reservoir to field capacity. The deficit in the capillary reservoir is calculated according to:

$$Ddeficit_{CPR}^{m,n} = Dfct_{HRU} - D_{CPR}^{m,n} , \qquad (44)$$

where

$Ddeficit_{CPR}^{m,n}$ is the volume of water per unit HRU area required to replenish the capillary reservoir to the field-capacity threshold for time step m and iteration n, in inches; and

$Dfct_{HRU}$ is the maximum volume of water per unit area in the capillary reservoir-parameter soil_moist_max, in inches.

If a capillary reservoir has a deficit, water is removed equally from each gravity reservoir associated with the capillary reservoir to replenish the capillary reservoir to the field-capacity threshold. If the volume of water added to a capillary reservoir is limited by storage in the gravity reservoir then the deficit in the capillary reservoir will not be replenished to field capacity. Water is removed from each associated gravity reservoir according to:

$$Drpl_{GVR}^{m,n} = \frac{Ddeficit_{CPR}^{m,n} A_{GVR}}{A_{HRU}}$$

$$\text{when} \quad \frac{Ddeficit_{CPR}^{m,n} A_{GVR}}{A_{HRU}} < D_{GVR}^{m,n-1}$$

$$Drpl_{GVR}^{m,n} = D_{GVR}^{m,n-1}$$

$$\text{when} \quad \frac{Ddeficit_{CPR}^{m,n} A_{GVR}}{A_{HRU}} \geq D_{GVR}^{m,n-1},$$

(45)

where

$\quad D_{GVR}^{m,n-1}$ is the volume per unit area of water in the gravity reservoir during time step m and iteration n-1, in inches; and

$\quad Drpl_{GVR}^{m,n}$ is the volume of water per unit area that is removed from the gravity reservoir to replenish the capillary reservoir during time step m and iteration n, in inches.

Water is added to the capillary reservoir according to:

$$D_{CPR}^{m,n*} = D_{CPR}^{m,n} + \sum_{L=1}^{LL}\left[Drpl_{GVR,L}^{m,n} \frac{A_{GVR,L}}{A_{HRU}} \right]$$

$$\text{when} \quad D_{CPR}^{m,n} < Dfct_{HRU},$$

(46)

where

$\quad D_{CPR}^{m,n*}$ is the revised volume per unit area of water in the capillary reservoir for time step m, iteration n, in inches;

$\quad L$ is the counter for the gravity reservoir number, dimensionless;

$\quad LL$ is the total number of gravity reservoirs, dimensionless;

$\quad Drpl_{GVR,L}^{m,n}$ is the volume of water per unit area that is removed from gravity reservoir L to replenish the capillary reservoir during time step m and iteration n, in inches;

$\quad A_{GVR,L}$ is the area of gravity reservoir L, in acres.

A deficit in the capillary reservoir can occur following periods without precipitation and can be replenished from gravity reservoirs that have inflow from ground-water discharge. If the HRU area is much greater than the top area of the finite-difference cell, focused ground-water discharge can result in underestimates of soil-zone saturation and overestimates of areas affected by ground-water discharge. This problem is illustrated by considering flow of water from a finite-difference cell to a gravity reservoir that represents spring flow. This condition may result in a wetland in a particular watershed. However, in the model, the water will be averaged over the large volume of the capillary reservoir if there is a deficit in the reservoir, resulting in a low saturation in the capillary reservoir that is not indicative of a wetland. This problem can be avoided by delineation of HRUs that correspond to wetlands and riparian areas.

Flow from Gravity Reservoirs to Preferential-Flow Reservoirs

Water in the gravity reservoirs is added to preferential-flow reservoirs whenever the preferential-flow threshold ($Dpft_{HRU}$) is exceeded (fig. 19 and table 9, sequence number 6). The threshold is determined for each HRU as:

$$Dpft_{HRU} = (Dsat_{HRU} - Dfct_{HRU})(1 - Fden_{HRU}) , \qquad (47)$$

where

$Dpft_{HRU}$ is the preferential-flow threshold as volume per unit area, in inches; and

$Dsat_{HRU}$ is the maximum volume of water per unit area in the soil zone-parameter `sat_threshold`, in inches.

Any water in the gravity reservoir above the preferential-flow threshold ($Dpft_{HRU}$) is subtracted from the gravity reservoir and added to the preferential-flow reservoir. The volume of water per unit area in the preferential-flow reservoir in inches ($D_{PFR}^{m,n}$) is calculated as:

$$
\begin{aligned}
D_{PFR}^{m,n} &= D_{PFR}^{m-1} + D_{GVR}^{m,n} - Dpft_{HRU} & D_{GVR}^{m,n} - Dpft_{HRU} > 0 \\
D_{PFR}^{m,n} &= D_{PFR}^{m-1} & D_{GVR}^{m,n} - Dpft_{HRU} \le 0
\end{aligned}
\qquad (48)
$$

where

$D_{PFR}^{m,n}$ is the volume per unit area in the preferential-flow reservoir for time step m, iteration n, in inches; and

D_{PFR}^{m-1} is the volume per unit area in the preferential-flow reservoir at the last iteration of time step m-1, in inches.

Flow from Gravity Reservoirs to Slow Interflow

Slow interflow from the gravity reservoirs represents the perching of water in the soil zone above the water table that can occur as a result of mineralization near the bottom of the soil zone or when soil develops over fine-grained material. Slow interflow can occur when the water content in the soil zone exceeds the field-capacity threshold (fig. 19 and table 9, sequence number 7). The slow interflow equation is the same as the fast interflow equation from preferential-flow reservoirs.

The interflow (slow and fast) is developed from continuity and an empirical equation written as (Leavesley and others, 1983):

$$\frac{dD_{GVR}^{m,n}}{dt} = q_{GVR,in}^{m,n} - q_{GVR,sif}^{m,n} , \text{ and} \qquad (49)$$

$$q_{GVR,sif}^{m,n} = slwcoef_{lin} D_{GVR}^{m,n} + slwcoef_{sq} (D_{GVR}^{m,n})^2 , \qquad (50)$$

where

$q_{GVR,in}^{m,n}$ is the volumetric inflow rate per unit area to the gravity reservoir, in inches per day;

$q_{GVR,sif}^{m,n}$ is the slow interflow rate per unit area from the gravity reservoir, in inches per day;

$slwcoef_{lin}$ is the linear flow routing coefficient for slow interflow—parameter `slowcoef_lin`, in per day; and

$slwcoef_{sq}$ is the non-linear preferential-flow reservoir routing coefficient for slow interflow—parameter `slowcoef_sq`, in per inch-day.

The right-hand side of equation 50 is substituted for $q_{GVR,sif}^{m,n}$ in equation 49 and the result is integrated indefinitely:

$$\int dt = \int \frac{dD_{GVR}^{m,n}}{q_{GVR,in}^{m,n} - slwcoef_{lin} D_{GVR}^{m,n} - slwcoef_{sq} (D_{GVR}^{m,n})^2} . \qquad (51)$$

The solution to equation 51 takes the form:

$$t = \frac{-2\tan^{-1}\left[\dfrac{slwcoef_{lin} + 2slwcoef_{sq} D_{GVR}^{m,n}}{cx} \right]}{cx} + Ci , \qquad (52)$$

where

$$cx = \sqrt{(slwcoef_{lin})^2 + 4slwcoef_{sq} q_{GVR,in}^{m,n}} ; \text{ and}$$

Ci is a constant of integration, in days.

The arctangent function (tan⁻¹) in equation 52 can be rewritten in terms of a logarithm using the identity (Selby, 1970; p. 166):

$$\tan^{-1}[x] = \ln\left(\frac{1+x}{1-x}\right), \tag{53}$$

where

$$x = \frac{slwcoef_{lin} + 2slwcoef_{sq}D_{GVR}^{m,n}}{cx}$$

The constant of integration in equation 52 can be determined by setting time to zero and the storage ($D_{GVR}^{m,n}$) equal to the initial storage (D_{GVR}^{m-1}) and replacing the arctan with the identity in equation 53. Equation 52 then becomes:

$$t = \frac{-\ln\left[\dfrac{1+\left(\dfrac{slwcoef_{lin} + 2slwcoef_{sq}D_{GVR}^{m,n}}{cx}\right)}{1-\left(\dfrac{slwcoef_{lin} + 2slwcoef_{sq}D_{GVR}^{m,n}}{cx}\right)}\right]}{cx}$$

$$+ \frac{\ln\left[\dfrac{1+\left(\dfrac{slwcoef_{lin} + 2slwcoef_{sq}D_{GVR}^{m-1}}{cx}\right)}{1-\left(\dfrac{slwcoef_{lin} + 2slwcoef_{sq}D_{GVR}^{m-1}}{cx}\right)}\right]}{cx} . \tag{54}$$

Equation 54 is rearranged in terms of $D_{GFR}^{m,n}$ and substituted into equation 49. The time derivative in equation 49 is approximated as:

$$\frac{dD_{GVR}^{m,n}}{dt} \cong \frac{D_{GVR}^{m,n} - D_{GVR}^{m-1}}{\Delta t} , \tag{55}$$

and the result is rearranged in terms of slow interflow from the gravity reservoir for iteration n, time step m as (Leavesley and others, 1983, p. 31-33):

$$q_{GVR,sif}^{m,n} = q_{GVR,in}^{m,n}\Delta t + cs\frac{\left[1+\dfrac{slwcoef_{lin}}{cx}cs\right]\left(1-e^{-cx\Delta t}\right)}{1+\dfrac{slwcoef_{sq}}{cx}cs\left(1-e^{-cx\Delta t}\right)} , \tag{56}$$

where

$$cs = \frac{D_{GVR}^{m-1}}{A_{GVR}} - \left[\frac{cx - slwcoef_{lin}}{2slwcoef_{sq}}\right].$$

Slow interflow ($Dsif_{HRU}^{m,n}$) is summed for all gravity reservoirs in the HRU at the end of the iteration as:

$$Dsif_{HRU}^{m,n} = \sum_{i=1}^{k} q_{i,sif}^{m,n}\Delta t , \tag{57}$$

where

$Dsif_{HRU}^{m,n}$ is the volume per unit area of slow interflow from the gravity reservoirs of the HRU at time step m, iteration n, in inches;

k is the total number of gravity reservoirs in the HRU, dimensionless; and

$q_{i,sif}^{m,n}$ is the slow interflow from gravity reservoir i for time step m, iteration n, in inches per day.

Slow interflow is partitioned to downslope HRUs and(or) stream segments during the iteration loop according to the cascading-flow procedure and connections defined by parameters input to the Cascade Module (cascade_prms, table 8). Slow interflow to a downslope HRU is added to the capillary reservoir (equation 46) and(or) the stream segment at the beginning of the next iteration. Slow interflow decreases the volume of storage in a gravity reservoir prior to computing gravity drainage, such that:

$$D_{GVR}^{m,n*} = D_{GVR}^{m,n} A_{HRU} - q_{GVR,sif}^{m,n} \Delta t. \tag{58}$$

The linear coefficient of slow interflow can be estimated using the recession constant determined from hydrograph separation (Linsley and others, 1975, p. 227).

Flow from Gravity Reservoirs to Gravity Drainage

Potential gravity drainage from the soil zone is computed as a function of storage in the gravity reservoirs after slow interflow has been removed (fig. 19; and table 9, sequence number 8). Net gravity drainage from each gravity reservoir is dependent on the ground-water heads and the vertical hydraulic conductivity of the connected finite-difference cell. Potential gravity drainage per unit area is computed as:

$$q_{gd,pot}^{m,n} = coeflin_{HRU} \left[\frac{D_{GVR}^{m,n}}{Dmx_{HRU}} \right]^{coefex_{HRU}}, \tag{59}$$

where

$q_{gd,pot}^{m,n}$ is the potential gravity drainage per unit area for time step m and iteration n, in inches per day;

$coeflin_{HRU}$ is the linear coefficient in the equation used to compute gravity drainage from the gravity reservoir—parameter ssr2gw_rate, in inches per day;

$coefex_{HRU}$ is the exponent in the equation used to compute gravity drainage from the gravity reservoir—parameter ssr2gw_exp, dimensionless; and

Dmx_{HRU} is the maximum amount of gravity drainage from the gravity reservoir—parameter ssrmax_coef, in inches.

Potential gravity drainage is averaged for iterations n and n-1 after the second iteration, and is summed for all gravity reservoirs contributing to a single finite-difference cell according to:

$$q_{gdc,pot}^{m,n} = \frac{\left(\sum_{L=1}^{LL} \left[w q_{gd,L}^{m,n} + (1-w) q_{gd,L}^{m,n-1} \right] A_{GVR,L} \right)}{A_{fdca}} C_{prms2mf}, \tag{60}$$

where

$q_{gdc,pot}^{m,n}$ is the potential gravity drainage for time step m and iteration n from all gravity reservoirs connected to a finite-difference cell, in length per time;

$q_{gd,L}^{m,n}$ is the potential gravity drainage per unit area of gravity reservoir L for time step m and iteration n, in inches per day;

$q_{gd,L}^{m,n-1}$ is the potential gravity drainage per unit area of gravity reservoir L for time step m and iteration n-1, in inches per day;

w is a weighting factor used to average gravity drainage between iterations, set to 0.5; and

$C_{prms2mf}$ is the conversion from PRMS inches per day to MODFLOW-2005 length per time (table 7).

Net gravity drainage ($q_{gdc,pot}^{m,n}$) to the underlying finite-difference cell is dependent on conditions in the unsaturated and saturated zones. First, net gravity drainage is set equal to the potential gravity drainage calculated above, or in the case where this value exceeds the vertical hydraulic conductivity of the unsaturated zone (K_s), it is set to the value of K_s. Because changes in gravity drainage to the unsaturated zone are represented by waves (see discussion in section "Unsaturated-Zone Flow"), small changes in gravity drainage can result in numerous waves in the unsaturated zone causing an unnecessary computational burden. This problem is avoided by grouping net gravity drainage into a set of individual values. Individual values are determined by dividing a span of values that range between 1×10^{-4} and 86,400 m/d into 51 log-normally distributed numbers, which represent most or all possible values of vertical hydraulic conductivity of natural porous materials (Freeze and Cherry, 1979).

The volumetric water content at the top of the unsaturated zone is computed from net gravity drainage and the Brooks and Corey (1966) relation as:

$$\theta^{m,n} = \left(\frac{q_{gdc,net}^{m,n}}{K_s} \right)^{1/\varepsilon} (S_y) + \theta_r \quad 0 < q_{gdc,net}^{m,n} \leq K_s \tag{61a}$$

$$\theta^{m,n} = \theta_s \qquad\qquad K_s < q_{gdc,net}^{m,n} , \tag{61b}$$

where

$\quad\theta^{m,n}$ is the water content at the top of the unsaturated zone for time step m, and iteration n, dimensionless;

$\quad q_{gdc,net}^{m,n}$ is the net gravity drainage to the connected finite-difference cell for time step m, and iteration n, in length per time;

$\quad K_s$ is the vertical saturated hydraulic conductivity of the unsaturated zone, in length per time;

$\quad\varepsilon$ is the Brooks-Corey exponent, dimensionless;

$\quad S_y$ is the specific yield that is used to approximate $\theta_s - \theta_r$, dimensionless;

$\quad\theta_r$ is the residual water content of the unsaturated zone, in volume of water per volume of rock; and

$\quad\theta_s$ is the saturated water content of the unsaturated zone, in volume of water per volume of rock.

Specific yield is used to approximate $\theta_s - \theta_r$ to maintain continuity across the boundary between the unsaturated and saturated zones (Niswonger and others, 2006a).

Net gravity drainage is applied to the saturated zone instead of the unsaturated zone if the ground-water head is above the soil-zone base minus one-half the undulation depth (fig. 20). Net gravity drainage to the saturated zone decreases as a function of ground-water head and is calculated according to:

$$q_{gdc,net}^{m,n*} = 0 \qquad\qquad\qquad h_{fdc}^{m,n} > celtop + 0.5 D_{usz}$$

$$q_{gdc,net}^{m,n*} = \frac{q_{gdc,net}^{m,n}}{D_{usz}} \left[celtop + 0.5 D_{usz} - h_{fdc}^{m,n} \right] \quad celtop - 0.5 D_{usz} \leq h_{fdc}^{m,n} \leq celtop + 0.5 D_{usz} \tag{62}$$

$$q_{gdc,net}^{m,n*} = q_{gdc,net}^{m,n} \qquad\qquad\quad celtop - 0.5 D_{usz} > h_{fdc}^{m,n} ,$$

where

$\quad q_{gdc,net}^{m,n*}$ is the net gravity drainage to the saturated zone for time step m and iteration n, in length per time.

The undulation depth defines where drainage from the soil zone becomes dependent on the ground-water head. The default value of D_{usz} is 1.0 and can be changed to other values on the basis of input specifications in the Unsaturated-Zone Flow Package.

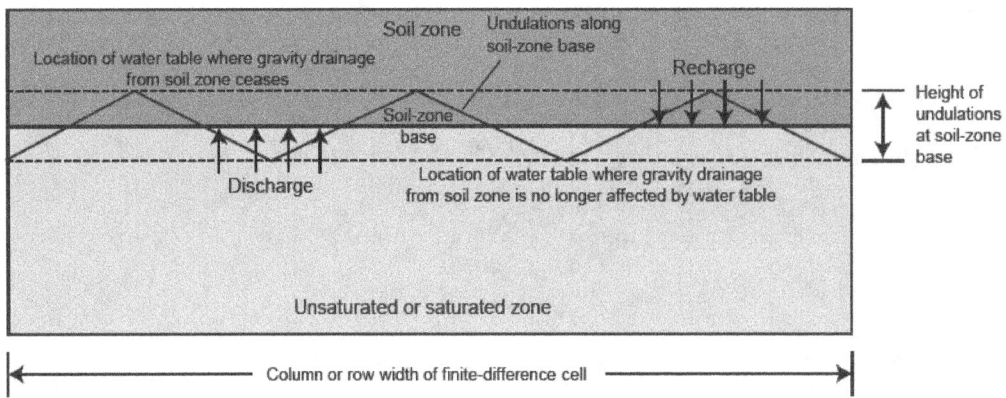

Figure 20. Effect of undulations at the soil-zone base on gravity drainage in relation to the water table in a finite-difference cell.

Evapotranspiration

Evaporation from bare soil and transpiration from plants is computed from the soil zone through the capillary reservoir (**table 9**, sequence Nos. 9-11). Evaporation and transpiration are subtracted from the reservoir as long as (1) there is water in storage, and (2) the rate of potential evapotranspiration is greater than zero after evaporation from the plant canopy, evaporation from impervious surfaces, and snow sublimation has occurred. If sufficient water is available to satisfy all remaining potential evapotranspiration after surface losses, then that volume is removed from the capillary reservoir. Water is removed from the capillary reservoir during the iteration loop because ground-water discharge into the gravity zone could affect the volume of water stored in the capillary reservoir. For the purposes of computing evapotranspiration, the soil water in the capillary reservoir is partitioned into two parts. The first, termed "recharge soil water," is available for evaporation and transpiration. The second, termed "lower zone soil water," is available for transpiration only, and is active only during the seasonal transpiration period. Evapotranspiration and transpiration from both partitions of the capillary reservoir are computed the same way:

$$DR_{CPR}^{m,n} = \frac{D_{CPR}^{m-1}}{Dfct_{HRU}} \ , \tag{63a}$$

$$EAPR_{CPR}^{m,n} = \begin{cases} 1.0 \\ DR_{CPR}^{m,n} \quad \text{and}, \\ \tfrac{1}{2} DR_{CPR}^{m,n} \end{cases} \tag{63b, c, d}$$

$$ET_{CPR}^{m,n} = EAPR_{CPR}^{m,n} PET_{HRU}^{m} \ , \tag{63e}$$

where

$DR_{CPR}^{m,n}$ is the soil-water content ratio in the capillary reservoir for time step m and iteration n, dimensionless;

$EAPR_{CPR}^{m,n}$ is the actual to potential evapotranspiration ratio in the capillary reservoir for time step m and iteration n, dimensionless; and

$ET_{CPR}^{m,n}$ is the actual evapotranspiration removed from the capillary reservoir for time step m and iteration n, in inches per day.

The calculation method of $EAPR_{CPR}^{m,n}$ is dependent on the value of D_{CPR}^{m-1} and soil type (Zahner, 1967; Leavesley and others, 1983, p. 22-23). If the predominate soil type of the capillary reservoir is sand and D_{CPR}^{m-1} is greater than 0.25, equation 63b is used; otherwise equation 63d is used. If the predominate soil type is loam and D_{CPR}^{m-1} is greater than 0.5, equation 63b is used; otherwise equation 63c is used. If the predominate soil type is clay and D_{CPR}^{m-1} is greater than 0.67, equation 63b is used; if D_{CPR}^{m-1} is greater than 0.33, but less than 0.67, equation 63c is used; otherwise equation 63d is used.

Evapotranspiration from the capillary reservoir is subtracted from the remaining potential evapotranspiration for each HRU. The unused potential evapotranspiration is made available for the unsaturated zone and the ground-water system in MODFLOW-2005 when the rooting depth of plants is beneath the soil zone.

Dunnian Runoff

Dunnian runoff is simulated from the soil zone in an HRU when storage as a volume per unit area in the preferential-flow (or gravity reservoirs when `pref_flow_den` is zero) exceeds the depth defined by the saturation minus field-capacity thresholds (table 9, sequence No. 12) as:

$$D_{SZfc}^{m,n} = D_{GVR,tot}^{m,n^{**}} + D_{PFR}^{m,n-1} \, , \qquad (64)$$

where

$D_{SZfc}^{m,n}$ is the volume of water per unit area of the soil zone above field capacity for time step m and iteration n, in inches;

$D_{GVR,tot}^{m,n^{**}}$ is the volume of water in all gravity reservoirs after interflow and gravity drainage in an HRU, divided by the area of the HRU for time step m and iteration n, in inches; and

$D_{PFR}^{m,n-1}$ is the volume per unit area of water in the preferential-flow reservoir before fast interflow for time step m and iteration n-1, in inches.

Dunnian runoff from each HRU is computed as:

$$ROd_{HRU}^{m,n} = D_{SZfc}^{m,n} - \left(Dsat_{HRU} - Dfct_{HRU} \right)$$
$$\text{when } D_{SZfc}^{m,n} \geq Dsat_{HRU} - Dfct_{HRU} \, , \qquad (65)$$

where

$ROd_{HRU}^{m,n}$ is the volume of Dunnian runoff per unit area from an HRU for time step m and iteration n, in inches.

Dunnian runoff from an HRU comes from either the preferential-flow reservoir when `pref_flow_den` is greater than zero or from the gravity reservoirs otherwise. Dunnian runoff from gravity reservoirs in an HRU is computed as:

$$ROd_{HRU}^{m,n} = \sum_{L=1}^{LL} ROd_{L}^{m,n} \, , \qquad (66)$$

where

$ROd_{L}^{m,n}$ is the Dunnian runoff from gravity reservoir L for time step m, iteration n, in inches.

Dunnian runoff from an HRU is routed to the capillary reservoir of downslope HRUs (equation 39) and(or) to stream segments with the cascading-flow procedure and connections defined by parameters input to the PRMS Cascade Module (cascade_prms, table 8). Dunnian runoff typically occurs from HRUs that define riparian areas next to streams or lakes or depressions in the topography.

Fast Interflow from Preferential-Flow Reservoirs

Fast interflow is simulated whenever water is stored in the preferential-flow reservoir (table 9, sequence No. 13). It is computed every iteration of a time step using the same empirical equation as slow interflow from the gravity reservoirs (equation 56). The volumetric-flow rate per unit area from the preferential-flow reservoir ($q_{PFR,fif}^{m,n}$) is computed using the fast interflow coefficients (`fastcoef_lin` and `fastcoef_sq`) in equation 56 as:

$$q_{PFR,fif}^{m,n} = q_{PFR,in}^{m,n} \Delta t + cs \frac{\left[1 + \dfrac{fstcoef_{lin}}{cx} cs \right]\left(1 - e^{-cx\Delta t} \right)}{1 + \dfrac{fstcoef_{sq}}{cx} cs \left(1 - e^{-cx\Delta t} \right)} \, , \qquad (67a)$$

where

$fstcoef_{lin}$ is the linear flow routing coefficient for fast interflow—parameter `fast_coef_lin`, in per day; and

$fstcoef_{sq}$ is the non-linear flow routing coefficient for fast interflow—parameter `fast_coef_sq`, in per inch-day; and

$$cs = \frac{D_{PFR}^{m-1}}{A_{HRU}} - \left[\frac{cx - fstcoef_{lin}}{2 \, fstcoef_{sq}} \right].$$

Flow from the preferential flow reservoir decreases the volume of storage according to:

$$D_{PFR}^{m} = D_{PFR}^{m-1} - q_{PFR,fif}^{m,n} \Delta t \, , \qquad (67b)$$

where

D_{PFR}^{m} is the volume of water in the preferential-flow reservoir per unit area for time step m, in inches.

Fast interflow from an HRU is routed to the capillary reservoir of downslope HRUs (equation 39) and(or) to stream segments with the cascading-flow procedure and connections defined by parameters input to the Cascade Module (cascade_prms, table 8).

Unsaturated-Zone Flow

Unsaturated-zone flow is simulated in GSFLOW using a kinematic-wave approximation to Richards' equation that assumes diffusive gradients (capillary pressure gradients) are negligible (Colbeck, 1972; Smith, 1983). This allows Richards' equation to be solved using the method of characteristics, which was originally done by Smith (1983) and Charbeneau (1984). A brief description of the approach as implemented in GSFLOW is provided here; the reader is referred to Niswonger and others (2006a) for a more complete narrative.

The approach of simulating unsaturated flow in GSFLOW differs from previous approaches that coupled a one-dimensional finite-difference form of Richards' equation to two- or three-dimensional ground-water flow equations (Pikul and others, 1974; Refsgaard and Storm, 1995) because it does not have a fixed-grid structure. The method adds flexibility for simulating an unsaturated zone that can change in thickness through space and time. The package simulates ground-water discharge directly to the soil zone or land surface as well as evapotranspiration from the unsaturated and saturated zones.

The kinematic-wave approximation to Richards' equation can be written to include evapotranspiration losses as:

$$\frac{\partial \theta}{\partial t} + \frac{\partial K(\theta)}{\partial z} + i = 0 \, , \tag{68}$$

where

$\quad \theta$ is the volumetric water content, in volume of water per volume of rock;

$\quad z$ is the altitude in the vertical direction, in lengt

$\quad K(\theta)$ is the unsaturated vertical hydraulic conductivity as a function of water content and is equal to the vertical flux, in length per time;

$\quad i$ is the evapotranspiration rate beneath the soil-zone base per unit depth, in length per time per length; and

$\quad t$ is time.

Application of the method of characteristics to equation 68 results in the following set of coupled ordinary differential equations:

$$\frac{dz}{dt} = \frac{\partial K(\theta)}{\partial \theta} = v(\theta) \, , \tag{69a}$$

$$\frac{d\theta}{dz} = \frac{-i}{v(\theta)} \, , \tag{69b}$$

$$\frac{d\theta}{dt} = -i \, , \tag{69c}$$

where

$\quad v(\theta)$ is the characteristic velocity restricted to the downward $\left(\text{positive } z \right)$ direction, in length per time.

Equation 69a provides the velocity of waves that represent wetting and drying in the unsaturated zone. Equation 69b provides the change in water content caused by evapotranspiration during wetting. Equation 69c provides the change in water content caused by evapotranspiration during drying. Equations 69a, b, and c are separable and can be integrated to find algebraic equations representing the characteristics of wetting and drying fronts in the unsaturated zone.

An analytic solution for the wetting-front velocity, $v(\theta)$, can be derived by considering the effects of diffusion and substituting an equivalent sharp wetting front of equal mass (Smith 1983; Charbeneau, 1984). For simplicity, evapotranspiration is left out of the derivation because it does not show up in the final equation for wave velocity. The solution to equation 68 for a wetting front that considers hydraulic diffusion can be found by integrating over a control volume containing a single wetting front according to (Phillip, 1957; Charbeneau, 1984):

$$\frac{d}{dt}\int_{z_1}^{z_2} \theta dz + \left(K(\theta) - D(\theta)\frac{\partial \theta}{\partial z} \right)\Big|_{z_1}^{z_2} = 0 \, , \tag{70}$$

where

$\quad D(\theta)$ is the hydraulic diffusion coefficient, in length squared per time; and

$\quad z_1$ and z_2 are points above and below the wetting front at distances far enough that $\frac{\partial \theta}{\partial z} \approx 0$, respectively, in length.

The term $D(\theta)\frac{\partial \theta}{\partial z}$, can be neglected because $\frac{\partial \theta}{\partial z} \approx 0$ at z_1 and z_2 such that:

$$\frac{d}{dt}\int_{z_1}^{z_2} \theta dz + K(\theta_{z_2}) - K(\theta_{z_1}) = 0 \, , \tag{71}$$

where

$\quad K(\theta_{z_1})$ and $K(\theta_{z_2})$ are the values of $K(\theta)$ at z_1 and z_2, respectively.

Integrating over a profile containing a sharp front with equivalent mass gives:

$$\int_{z_1}^{z_2} \theta dz = \theta_{z_1}(z_f - z_1) + \theta_{z_2}(z_2 - z_f) , \qquad (72)$$

where

θ_{z_1} is the volumetric water content at depth z_1, in volume of water per volume of rock;

θ_{z_2} is the volumetric water content at depth z_2, in volume of water per volume of rock; and

z_f is the depth of the sharp front.

Combining equations 71 and 72 gives:

$$\frac{dz_f}{dt} = u_z(\theta_{z_1}, \theta_{z_2}) = \frac{K(\theta_{z_1}) - K(\theta_{z_2})}{\theta_{z_1} - \theta_{z_2}} , \qquad (73)$$

where

u_z is the velocity of a sharp wetting front, in length per time.

Equations 69c and 73 must be solved simultaneously because z_f and θ_{z_1} are both unknown when ET losses are considered.

The Brooks-Corey unsaturated hydraulic conductivity function is used to represent $K(\theta)$ and can be expressed as (Brooks and Corey, 1966):

$$K(\theta) = K_z \left[\frac{\theta - \theta_r}{\theta_z - \theta_r} \right]^c , \qquad (74)$$

where

K_z, θ_r, and θ are defined in equation 61.

Wetting and Drying Fronts in Unsaturated Zone

An increase in the infiltration rate from the soil zone into the unsaturated zone will cause a wetting front to form, which is represented by a lead wave. A decrease in the infiltration rate will cause a drying front to occur, which is represented by a trailing wave. Thus, waves are used to represent both wetting and drying fronts. Attenuation of a lead wave occurs as a trailing wave of higher velocity overtakes it. When a trailing wave overtakes a lead wave, the water content of the lead wave becomes equal to the water content of the trailing wave. Consequently, this reduces the velocity and water content of the lead wave. Conversely, when a lead wave overtakes a trailing wave or another lead wave of lower velocity, the overtaken wave is removed, and

the water content and flux of the uppermost lead wave are maintained, resulting in rewetting. A depiction of how wetting and drying fronts are approximated by kinematic waves is shown in **figure 21**.

A lead wave's velocity and water content will decay during a subsequent period of less infiltration as trailing waves overcome the lead wave. Analytical equations that represent the velocity of a trailing wave can be derived. However, these equations become complicated when trying to represent a trailing wave intercepting a lead wave. A simpler approach is to divide the trailing wave into steps or increments and calculate velocities ($v(\theta)$) of each increment on the basis of a finite-difference approximation (Smith, 1983):

$$v(\theta) = \frac{K(\theta) - K(\theta - \Delta\theta)}{\Delta\theta} . \qquad (75)$$

where $\Delta\theta$ is the change in water content between two adjacent locations along a trailing wave, in volume of water per volume of rock. An analytic solution of equation 69c, as shown in the next section, is solved simultaneously with equation 65 during evapotranspiration.

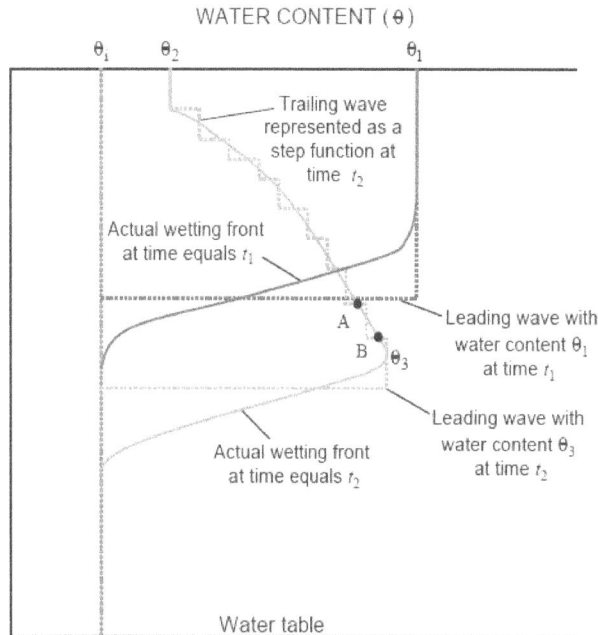

Figure 21. A wetting front moving through a uniform column of unsaturated material affected by a decrease in flux (volumetric flow per unit area), and results from a kinematic-wave solution of the wetting front represented by leading and trailing waves (from Niswonger and others, 2006a, fig. 2 and modified from Smith and Hebbert, 1983).

Evapotranspiration Beneath the Soil Zone

Evapotranspiration that occurs at or below land surface can be simulated with the PRMS Soil Zone Module (soilzone_gsflow, table 1) when the bottom of the soil zone is defined as the root depth. However, where soils are thin and the root depth extends beneath the soil-zone base, evapotranspiration that is not satisfied by storage in the soil zone is removed from the underlying unsaturated and saturated zones. Evapotranspiration is removed from the unsaturated zone by integrating equations 69b and 69c and evapotranspiration is removed from the saturated zone (ground water) using the same method implemented in the MODFLOW-2005 Evapotranspiration Package (Harbaugh, 2005).

Integrating equations 69b and 69c results in:

$$z_2 = \frac{[K(\theta_{z_2}) - K(\theta_{z_1})] - iz_1}{-i}, \text{ and} \quad (76a)$$

$$\theta_{\tau+\Delta t} = \theta_\tau - i\Delta t \quad (76b)$$

where

z_1 and z_2 are depths of two arbitrary points along a wetting front, where z_2 is deeper than z_1, in length;

τ and $\tau + \Delta t$ are the times at the beginning and end of the time step, respectively;

θ_τ is the water content of trailing waves above the evapotranspiration extinction depth after time t, in volume of water per volume of rock; and

$\theta_{\tau+\Delta t}$ is the water content of trailing waves above the evapotranspiration extinction depth after time $t + \Delta t$, volume of water per volume of rock.

Evapotranspiration beneath the base of the soil zone only occurs when there is an evapotranspiration deficit in the soil zone and when evapotranspiration is active in the Unsaturated-Zone Flow Package. The deficit is used to calculate transpiration from the unsaturated and saturated zones according to equations 76a and 76b. Equation 76a is used to calculate the depth of points along the leading-wave profile between θ_{z_1} and θ_{z_2}, where $\theta_{z_1} = \theta^{m,n}$ is the water content at the top of the unsaturated zone corresponding to the infiltration rate $q_{gdc,net}^{m,n}$. θ_{z_2} is unknown and must be solved for by coupling equations 73 and 76a.

Recharge to Saturated Zone

Recharge to the saturated zone is computed at the end of a time step on the basis of the volumetric flow in the unsaturated zone across the water table plus any water that may be in storage in the unsaturated zone when the water table rises. Thus, large fluctuations in the water table can occur during a time step, but the volumetric flow and storage of the unsaturated zone is kept in balance with recharge to the saturated zone even during these large fluctuations. This includes the disappearance and appearance of an unsaturated zone. A complete description of the coupling between the unsaturated and saturated zones is presented by Niswonger and others (2006a, p. 8-10).

Recharge is dependent on the location of the water table and how its position varies with time. This is because ground-water flow in the saturated zone does not account for storage in the unsaturated zone. Rather, a specific yield is used to estimate changes in ground-water storage. Thus, when the water table rises, storage in the unsaturated zone in the interval of the rise is added as recharge to the saturated zone. When the water table declines, the thickness of the unsaturated zone is increased and a wetting front must advance through the interval of the decline before there is recharge.

Streams

Streams are simulated in GSFLOW by the Streamflow-Routing Package in MODFLOW-2005. Accumulated surface runoff and interflow generated in the HRUs, and base flow generated by ground-water discharge to the stream reaches are routed by the methods described in the following section. Stream depths, intermittent streamflow conditions, and unsaturated flow conditions beneath streams also can be computed. The reader is referred to Prudic and others (2004) for a more complete narrative.

Streamflow Routing

Streamflow is computed each iteration by stream reach starting from the first reach of the first stream (farthest upstream) segment and sequentially continuing to the last reach of the last stream (farthest downstream) segment (fig. 8). Inflows to a stream reach include any user-specified inflows to the first reach of a stream segment, inflows from an upstream reach, precipitation directly on the channel, surface runoff and interflows from contiguous HRUs, and ground-water discharge across the streambed (fig. 22A). Outflows include any specified diversions into pipelines at the beginning of the first reach in a stream segment or into unlined canals from the last reach of a stream segment, evaporation directly from the channel, downward leakage across the streambed, and flow from the end of the reach (fig. 22*B*). Outflows from the last reach of a stream segment are saved and added as inflow to the connecting downstream segment or, if the segment is not connected to any downstream reaches, the outflow from its last reach exits the modeled region. Outflows from multiple upstream segments or tributary segments can be added as inflow to a downstream reach.

User-specified flows can be added to or subtracted from the beginning of the first reach of a segment. The volumetric flow rates are specified by stress periods in the data input of the Streamflow-Routing Package (appendix 1). Thus, a stream that enters the modeled region can have a specified flow at the beginning of the first reach (fig. 22*A*). If the first reach of a stream segment has no streamflow entering the reach, any specified rate less than zero is ignored. Lateral inflow from contiguous HRUs is computed to a stream reach by summing all surface runoff and interflows from each HRU as:

$$Q_{lateral}^{m,n} = \sum_{J=1}^{JJ} \frac{C_{prms\,2mf}'}{\Delta t} F_{J,sr} A_J \left(ROh_J^{m,n} + ROd_J^{m,n} + Dsif_J^{m,n} + Dfif_J^{m,n} \right), \tag{77}$$

where

$Q_{lateral}^{m,n}$ is the volumetric flow rate into a stream reach from all connected HRUs for time
step m and iteration n, in cubic length per time;

$C_{prms\,2mf}'$ is the conversion factor from PRMS acre-inch per day to MODFLOW-2005 cubic
length per time (table 7);

$F_{J,sr}$ is the decimal fraction of the total area of HRU J that contributes runoff and interflow
to a particular stream reach defined by GSFLOW parameter `segment_pct_area`
in table 4, dimensionless;

A_J is the area of HRU J, in acres;

$ROh_J^{m,n}$ is the Hortonian runoff per unit area from HRU J for time step m and iteration n,
in inches;

$ROd_J^{m,n}$ is the Dunnian runoff per unit area from HRU J for time step m and iteration n,
in inches;

$Dfif_J^{m,n}$ is the fast interflow from preferential flow reservoirs per unit area of HRU J for time
step m and iteration n, in inches;

$Dsif_J^{m,n}$ is the slow interflow from gravity reservoirs per unit area of HRU J for time step m and
iteration n, in inches;

J is the counter for the HRU number, dimensionless; and

JJ is the total number of HRUs that contribute surface runoff and interflow to a particular
stream reach.

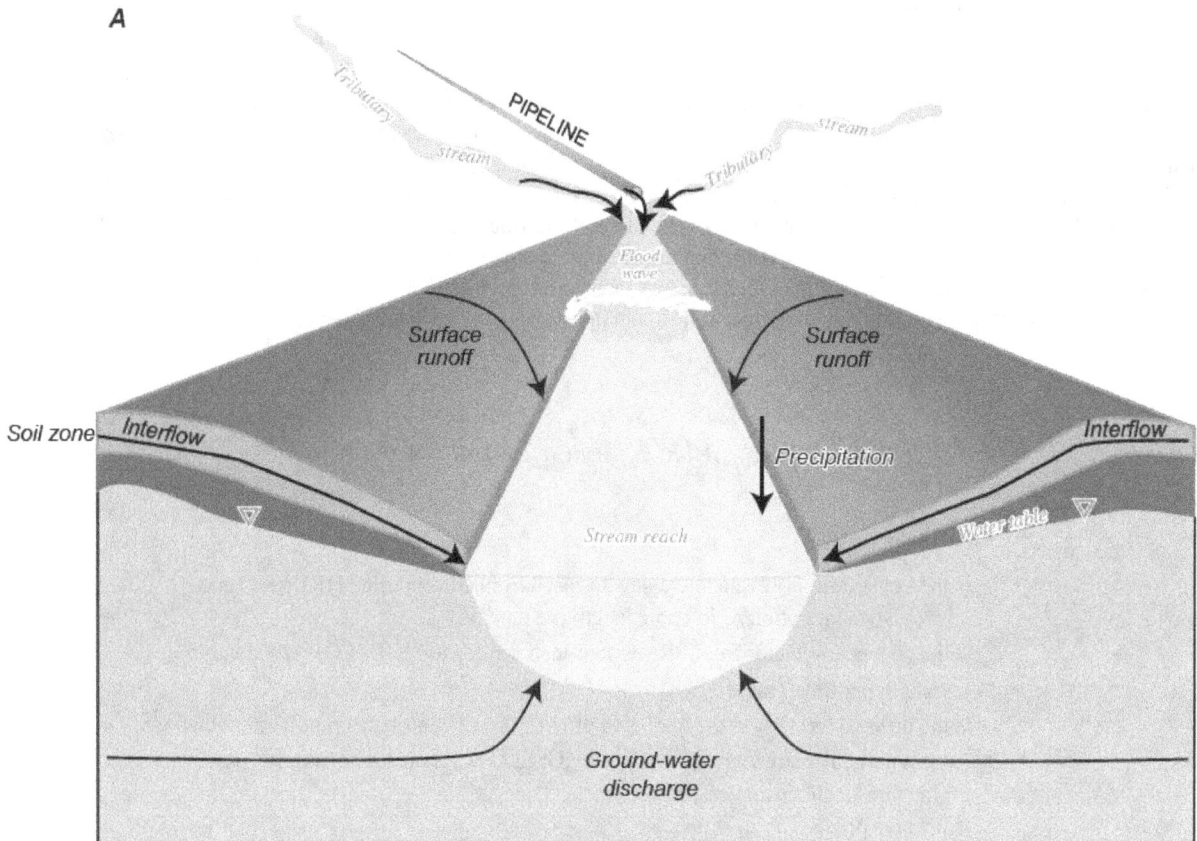

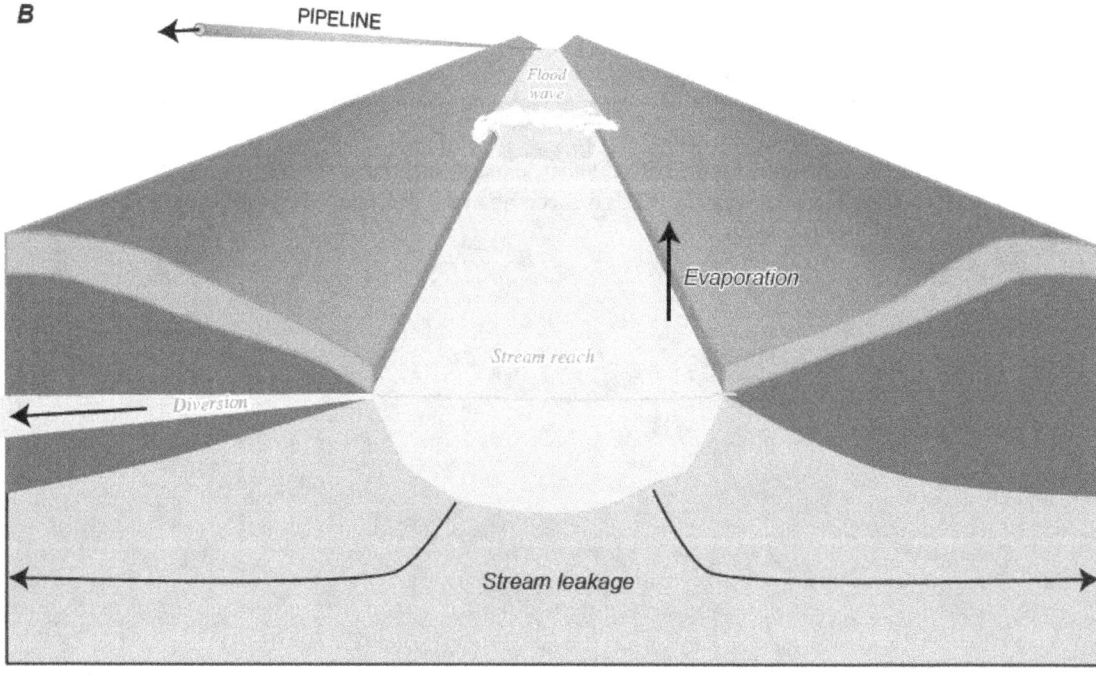

Figure 22.   (*A*) Inflows to and (*B*) outflows from a stream reach.

Ground-water discharge across the streambed is computed when the underlying ground-water head in the finite-difference cell connected to the stream is greater than the stream head at the midpoint of the reach, whereas downward leakage is computed when the ground-water head is less than the stream head. Methods for determination of the stream head are discussed in section "Computation of Stream Depth." The volumetric rate of flow across the streambed is computed as (Prudic and others, 2004, p. 4):

$$Q_{srleak}^{m,n} = K_{strbed}\, wetper_{str}\, length_{str} \left(\frac{h_{str}^{m,n} - h_{fdc}^{m,n}}{thick_{strbed}} \right), \qquad (78)$$

where

$Q_{srleak}^{m,n}$ is the volumetric-flow rate across the streambed for time step m and iteration n and is downward leakage when positive and ground-water discharge to the stream when negative, in cubic length per time;

K_{strbed} is the hydraulic conductivity of the streambed, in length per time;

$wetper_{str}$ is the wetted perimeter of the streambed, in length;

$length_{str}$ is the length of the stream reach, in length;

$h_{str}^{m,n}$ is the stream head at the midpoint of the stream reach for time step m and iteration n, in length; and

$thick_{strbed}$ is the streambed thickness in the stream reach, in length.

Transient leakage across the streambed can change during a time step and is dependent on both the stream and ground-water heads. When the ground-water head in the finite-difference cell is less than the bottom of the streambed (that is, the top of streambed minus the thickness of the streambed, commonly called "percolating" conditions), downward leakage across the streambed is no longer dependent on ground-water head; in this case, the ground-water head in equation 78 is replaced by the altitude of the bottom of the streambed.

Streamflow can be diverted from the last reach of a segment and routed downstream to other segments (Prudic and others, 2004, p. 5 and 9). The diverted flow is subtracted from the end of the last reach of a segment and any remaining flow in that segment is added as inflow to the beginning of the first reach of the next downstream segment. Four different options are available for diverting flow into another stream segment (variable IPRIOR in the input instructions for the Streamflow-Routing Package in appendix 1). The first option allows for the diversion of all available flow up to the specified rate; the second option allows for the diversion of the specified rate only when flow is greater than the specified rate; the third

option allows for the diversion of a specified fraction of flow; and the fourth option allows for diversion of all flow in excess of the specified rate. Flow from the channel at the end of a stream reach that may become inflow to a downstream reach is the difference in flow that enters and leaves the stream reach during a time step.

Two options are available to route streamflow through a network of channels. The first is based on steady, uniform flow for computing flow through rectangular or non-prismatic channels (Prudic and others, 2004, p. 6). Outflow from the end of a stream reach is set equal to the sum of all inflows to the reach minus any upstream outflows. Precipitation, evaporation, and surface runoff were originally defined by data input to the Streamflow Routing Package (Niswonger and Prudic, 2005, p. 28 and described in appendix 1). However, these processes, as well as interflow through the soil zone, are computed in GSFLOW using PRMS and, therefore, these input variables should be set to zero for an integrated simulation. However, stream inflows and outflows other than those calculated by PRMS (for example stream diversions) can be specified in the Streamflow Routing Package input file.

Outflow at the end of the reach can be expressed as:

$$Q_{srout}^{m,n} = Q_{srup}^{m,n} + Q_{srin}^{m} + Q_{lateral}^{m,n}$$
$$+ Q_{srpp}^{m,n} - Q_{srevp}^{m,n} - Q_{srleak}^{m,n} - Q_{srdvr}^{m,n}, \qquad (79)$$

where

$Q_{srout}^{m,n}$ is the volumetric flow rate from the end of a stream reach for time step m and iteration n, in cubic length per time;

$Q_{srup}^{m,n}$ is the sum of the volumetric flow rate that enters the stream reach from outflow of upstream reaches for time step m and iteration n, in cubic length per time;

Q_{srin}^{m} is the specified volumetric flow rate at the beginning of a stream reach for time step m, in cubic length per time;

$Q_{lateral}^{m,n}$ is the specified volumetric flow rate added to a stream reach as lateral inflow plus any runoff and interflow coming from adjacent HRUs, in cubic length per time;

$Q_{srpp}^{m,n}$ is the specified precipitation rate on the stream surface multiplied by the wetted plan-view area for time step m and iteration n, in cubic length per time;

$Q_{srevp}^{m,n}$ is the specified evaporation rate from the stream surface multiplied by the wetted plan-view area for time step m and iteration n, in cubic length per time; and

$Q_{srdvr}^{m,n}$ is the volumetric flow rate diverted from the end of the stream reach for time step m and iteration n, in cubic length per time.

The second option is based on a kinematic-wave equation that approximates the Saint-Venant equations for computing flow through rectangular or non-prismatic channels. The kinematic-wave equation for routing flow in streams can be written (Lighthill and Whitham, 1955):

$$\frac{\partial Q^{msfr}_{dwn}}{\partial x_{channel}} + \frac{\partial A^{msfr}_{dwn}}{\partial t} = q_{str} \; , \tag{80}$$

where

Q^{msfr}_{dwn} is volumetric-flow rate at the downstream end of a stream reach for the current streamflow-routing time step *msfr*, in length squared;

A^{msfr}_{dwn} is the cross-sectional area at the downstream end of a stream reach for the current streamflow-routing time step *msfr*, in length squared;

q_{str} is the sum of all lateral inflows and outflows per unit length of stream listed on the right side of equation 79, in length squared per time;

$x_{channel}$ is the distance along the stream channel, in length; and

t is time.

Momentum in the kinematic-wave approximation assumes that gravitational forces are balanced by frictional forces such that:

$$Slope_0 = Slope_f \; , \tag{81}$$

where

$Slope_0$ is the slope of the channel in the longitudinal profile, dimensionless; and

$Slope_f$ is the friction slope of the channel in the longitudinal profile, dimensionless.

Several forms of the momentum equation are combined with equation 80 to route flow to downstream channels. The form of the momentum equation differs in how the channel cross-sectional dimensions and channel slope are represented in Manning's equation. Another option represents momentum using tables of specified values from streamflow gages, or the power-law equations described by Prudic and others (2004, p. 6-9). Momentum caused by the water-surface slope, velocity head, and acceleration is neglected.

The kinematic-wave approximation neglects the dynamic components of flow that are represented by the derivative terms in the more complete form of the momentum equation referred to as the Saint-Venant equations. Another less-simplified approximation of the Saint-Venant equations combines the spatial derivative from the momentum equation with the continuity equation, which results in a second derivative term in the continuity equation (Lighthill and Whitham, 1955). The second derivative term in the continuity equation causes the flood wave to spread upstream slightly, and is commonly referred to as a diffusion analogy for the dynamic component of the momentum equation. This approach is the basis for the surface-water model called DAFLOW that was coupled with MODFLOW (Jobson and Harbaugh, 1999).

The kinematic-wave equation (equation 80) was solved numerically by stream reaches using an implicit four-point, finite-difference solution technique (Fread, 1993). Because the movement of a flood peak in a channel may be of interest, time steps for streamflow routing can have durations less than the time steps for GSFLOW or MODFLOW-2005. An option was added that divides the GSFLOW or MODFLOW-2005 time step in a specified number of increments. The time derivative ($\partial A_{str} / \partial t$) was approximated by a forward-difference quotient centered between the end of the previous and current time steps used for routing streamflow:

$$\frac{\partial A^{msfr}_{dwn}}{\partial t} \approx \frac{(A^{msfr}_{up} - A^{msfr-1}_{up}) + (A^{msfr}_{dwn} - A^{msfr-1}_{dwn})}{2\Delta t_{sfr}} \; , \tag{82}$$

where

A^{msfr}_{up} is cross-sectional area at the upstream end of a stream reach for the current streamflow-routing time step *msfr*, in length squared;

A^{msfr-1}_{up} is cross-sectional area at the upstream end of a stream reach for the previous streamflow-routing time step *msfr*-1, in length squared;

A^{msfr}_{dwn} is cross-sectional area at the downstream end of a stream reach for the current streamflow-routing time step *msfr*, in length squared;

A^{msfr-1}_{dwn} is cross-sectional area at the downstream end of a stream reach for the previous streamflow-routing time step *msfr*-1, in length squared; and

Δt_{sfr} is the time-step for routing streamflow, in time.

The spatial derivative in equation 80, $\partial Q_{str} / \partial x_{channel}$, is approximated by a forward-difference quotient that includes time weighting of the quotient at the previous time step:

$$\frac{Q_{dwn}^{msfr}}{\partial x_{channel}} \approx \frac{\varpi(Q_{dwn}^{msfr-1} - Q_{up}^{msfr-1}) + (\varpi - 1)(Q_{dwn}^{msfr} - Q_{up}^{msfr})}{\Delta x} , \quad (83)$$

where

ϖ is the time-weighting factor that ranges between 0.5 and 1, dimensionless;

Q_{up}^{msfr} is the volumetric flow at the upstream end of the stream reach for the current streamflow-routing time step *msfr*, in cubic length per time;

Q_{up}^{msfr-1} is the volumetric flow at the upstream end of the stream reach for the previous streamflow-routing time step *msfr*-1, in cubic length per time; and

Q_{dwn}^{msfr} is the volumetric flow at the downstream end of the stream reach for the current streamflow-routing time step *msfr*, in cubic length per time;

Q_{dwn}^{msfr-1} is the volumetric flow at the downstream end of the stream reach for the previous streamflow-routing time step *msfr*-1, in cubic length per time;

Δx is the spatial increment for routing streamflow, equal to the reach length.

Equations 82 and 83 are combined and solved iteratively using Newton's method (Burden and Faires, 1997, p. 65). Streamflow for a particular stream reach for all time increments computed with the kinematic-wave equation (including those smaller than the daily time step used by the rest of GSFLOW) can be saved to a file using the Stream Gaging Station or Gage Packages (appendix 1).

Because streamflow routing during each subdivision of a GSFLOW time step is a function of the head difference between the stream and ground water, the ground-water head at the beginning of the time step is used to compute the exchange of water across the streambed and the stream depth for each time subdivision. If the ground-water head at the beginning of the time step is at or above the bottom of the streambed, the computed stream depths for all time subdivisions are averaged and used in the ground-water flow equation to compute ground-water heads at the end of the GSFLOW time step. However, the depth at the end of the time step is used for subsequent stream calculations. If the ground-water head at the beginning of the time step is beneath the bottom of the streambed, downward leakage and recharge to the saturated zone is summed for all time subdivisions and added to the ground-water flow equation to compute ground-water heads.

Computation of Stream Depth

Stream depth varies as a function of streamflow. It is added to the top of the streambed to determine the stream head, which is used to compute ground-water discharge and downward leakage across the streambed (equation 78). Stream depth as a function of streamflow is computed at the midpoint of each reach. Streamflow at the midpoint is computed as:

$$Q_{mdpt}^{m,n} = Q_{srup}^{m,n} + Q_{srin}^{m} + 0.5(Q_{lateral}^{m,n} + Q_{srpp}^{m,n} - Q_{srevp}^{m,n} - Q_{srleak}^{m,n}) , \quad (84)$$
$$- Q_{srdvr}^{m,n}$$

where

$Q_{mdpt}^{m,n}$ is the volumetric-flow rate at the midpoint of a stream reach for time step *m* and iteration *n*, in cubic length per time.

This equation is nonlinear because flow at the midpoint of a reach is dependent on streambed leakage, which is dependent on stream depth. Thus, the equation is solved iteratively using a mixed bisection-Newton method until the calculated stream stage changes between iterations is less than a specified tolerance. This tolerance is specified in the input for the Streamflow-Routing Package (appendix 1). A warning message is written in the main MODFLOW Listing File for each stream reach whenever the solution error is greater than the specified tolerance after 100 iterations.

One of five options can be used to compute stream depth in a reach as specified in the input for the Streamflow-Routing Package (appendix 1). The option used to calculate stream depth is defined for each segment such that the depths in all reaches within a segment are calculated the same. The first option (ICALC=0) allows the user to specify a stream depth at the beginning and end of each segment; these two depths are used to interpolate the depth at the midpoint of each reach in the segment on the basis of the distance downstream from the beginning of the segment. This option for stream depth is not recommended for the integrated model because stream depth remains constant and, therefore, the volume of water in storage in the stream network does not change as a function of inflows and outflows.

The second and third options (ICALC=1 and 2, respectively) use the general form of Manning's equation to relate streamflow as a function of depth for all reaches in a stream segment (Chow and others, 1988, p. 159) as:

$$Q_{mdpt}^{m,n} = \left(\frac{C_m}{Roughness} \right) A_{str} \left(R_{hydraulic} \right)^{2/3} \left(Slope_o \right)^{1/2} , \quad (85)$$

where

C_m is a constant, which is 1.0 for cubic meters per second and 1.486 for cubic feet per second;

$Roughness$ is Manning's roughness coefficient, dimensionless; and

$R_{hydraulic}$ is the hydraulic radius of the stream, which is equal to the stream area divided by the wetted perimeter, in length.

The wetted perimeter and hydraulic radius of a stream, which are needed in Manning's equation, are often complicated functions of depth, and can only be solved analytically by making assumptions to simplify the channel geometry. An analytical equation is used for ICALC=1 by assuming a wide rectangular channel (Shen and Julien, 1993, p. 12.13). This assumption approximates the wetted perimeter as equal to the stream width and results in an approximation of the hydraulic radius that is equal to stream depth. Solving equation 85 for stream depth then yields (Prudic and others, 2004, p. 7):

$$D_{mdpt}^{m,n} = \left(\frac{Q_{mdpt}^{m,n} Roughness}{C_m w_{str} \left(Slope_o \right)^{1/2}} \right)^{3/5} , \quad (86)$$

where

w_{str} is the stream width, in length; and

$D_{mdpt}^{m,n}$ is the depth of water at the midpoint of a stream reach for time step m and iteration n, in length.

The third option (ICALC=2) uses one channel cross section for a stream segment that is divided into three parts on the basis of eight paired horizontal and vertical locations (**fig. 23**). Eight horizontal distances relative to the left edge of the cross section (viewed in downstream direction) and the corresponding eight vertical altitudes relative to the specified top of streambed (lowest point or thalweg in the channel) are used for computing stream depth, top width, and wetted perimeter. Stream depth, width, and wetted perimeter also are dependent on the slope of the streambed, and two roughness coefficients—one for the center part of the cross section and another for the two outer parts that may represent overbank flow. The ends of the cross section are assumed to have vertical walls. Because the stream area and hydraulic radius for this option result in implicit functions of depth, a mixed bisection-secant method is used to compute stream depth in relation to streamflow (Prudic and others, 2004, p. 7-8).

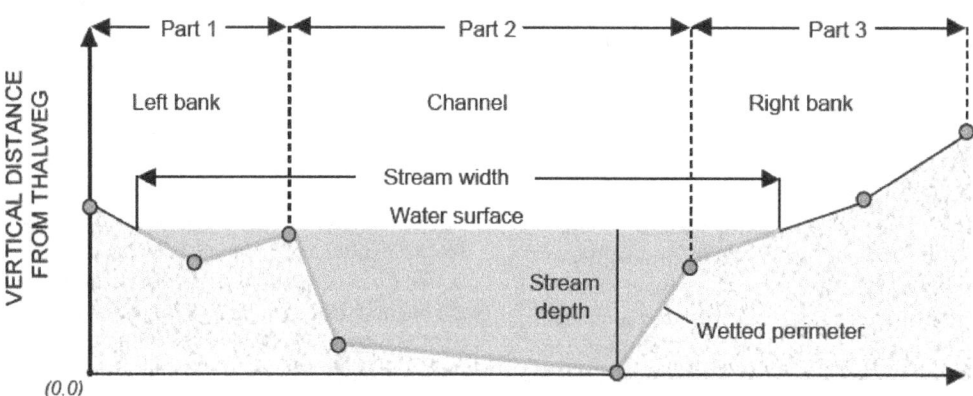

Figure 23. Eight paired horizontal and vertical locations used to compute stream depth, top width, and wetted perimeter for a stream segment (modified from Prudic and others, 2004, fig. 4).

The fourth option (ICALC=3) uses power-law equations for a stream segment that relate stream depth and width to streamflow using the following two equations (Leopold and others, 1992, p. 215):

$$D_{mdpt}^{m,n} = cdpth\left(Q_{mdpt}^{m,n}\right)^{fdpth}, \text{ and } w_{mdpt}^{m,n} = awdth\left(Q_{mdpt}^{m,n}\right)^{bwdth}, \quad (87)$$

where

$w_{mdpt}^{m,n}$ is the stream width at reach midpoint, in length;

$cdpth$ and
  $awdth$ are empirical coefficients determined from regression methods, in time per length squared; and

$fdpth$ and
  $bwdth$ are empirical exponents determined from regression methods, dimensionless.

This option assumes that the wetted perimeter is equal to the stream width. The regressions typically are determined at streamflow-gaging stations. Representative values of the numerical coefficients for selected regions of the United States are listed in Leopold and others (1992, table 7-5, p. 244).

The fifth option (ICALC=4) allows the user to enter a look-up table of stream depth, width, and corresponding streamflow for a stream segment. This option also assumes that the wetted perimeter is equal to the stream width. The values typically are determined from rating curves at streamflow-gaging stations. Stream depth and width are interpolated between two specified streamflows that bracket the computed flow at the reach midpoint. Values of depth, width, and streamflow are converted into their logarithms prior to the interpolation (Prudic and others, 2004, p. 9). A simple linear extrapolation is used to estimate stream depth and width when the computed streamflow at the midpoint of the stream reach is less than the first specified streamflow. A warning message is printed when the computed streamflow exceeds the highest specified streamflow in the table and stream depth and width are computed by extrapolation from the log values of the last two streamflows in the tabulated list.

Intermittent Streams

Stream reaches in any segment can have intermittent flow. Downward leakage from a reach during a time step is limited by the available flow in a stream reach. Water available for leakage when the ground-water head is beneath the bottom of the streambed is calculated:

$$Q_{srleak,mx}^{m,n} = Q_{srup}^{m,n} + Q_{srin}^{m} + Q_{lateral}^{m,n} + Q_{srpp}^{m,n} - Q_{srevp}^{m,n}, \quad (88)$$

where

$Q_{srleak,mx}^{m,n}$ is the total amount of water available for leakage to ground water for time step m and iteration n, in cubic length per time.

However, leakage loss across the streambed from a stream reach cannot exceed the rate of gravity drainage calculated according to:

$$Q_{srleak}^{m,n} = K_{strbed}\, wetper_{str}\, length_{str}\left(\frac{D_{mdpt}^{m,n}}{thick_{strbed}} + 1\right). \quad (89)$$

No downward leakage is computed for a reach when there are no upstream inflows or diversions and the ground-water head in the finite-difference cell is beneath the top of the streambed. However, a reach with no surface inflows may have outflow from it when ground-water discharge along the reach is greater than the evaporation rate within the reach.

Unsaturated Flow Beneath Streams

An unsaturated zone develops beneath a stream reach whenever the ground-water head in the finite-difference cell is less than the bottom of the streambed. Gravity drainage through the unsaturated zone is simulated using the kinematic-wave approximation to Richards' equation similar to that used to simulate unsaturated-zone flow beneath the soil zone. However, downward leakage across a streambed is used in place of gravity drainage for calculating infiltration into the unsaturated zone and no evapotranspiration can be simulated from an unsaturated zone beneath a reach. Maximum streambed leakage for reaches that have an underlying unsaturated zone is limited to the product of the saturated vertical hydraulic conductivity of the unsaturated zone and the streambed area (stream length multiplied by wetted perimeter). Thus, streambed leakage in reaches with an unsaturated zone is controlled either by the saturated hydraulic conductivity of the streambed, as shown in equation 89, or the saturated hydraulic conductivity of the unsaturated zone.

The ability to simulate unsaturated flow beneath streams is limited to the two options that compute stream depth on the basis of Manning's equation—the options that assume a wide rectangular channel of constant wetted perimeter (ICALC=1) or an eight-point cross section that allows the wetted perimeter to change as a function of streamflow and stream depth (ICALC=2). The unsaturated zone may be discretized into compartments beneath an eight-point cross section (**fig. 24**). Water is routed independently beneath each of the unsaturated compartments (Niswonger and Prudic, 2005, p. 6) and can be used to simulate the variability in leakage and gravity drainage through the unsaturated zone across a stream channel.

Lakes

Lakes are simulated in GSFLOW by a modified implementation of the Lake Package in MODFLOW-2005. The overall approach is described in detail by Merritt and Konikow (2000); a brief description of the overall approach, and GSFLOW-related changes to the solution routines, are described herein.

Inflow and Outflow

A lake in GSFLOW is represented as both an HRU in PRMS and a group of finite-difference cells in MODFLOW-2005. Inflows to a lake include direct precipitation; surface runoff and interflows through the soil zone from contributing areas adjacent to the lake; specified inflow, such as a pipeline (negative withdrawal; **appendix 1**); tributary streamflow; and ground-water discharge through the subsurface (**fig. 25**). Outflows from a lake include evaporation, specified withdrawal, discharge to a stream, and leakage to ground water (**fig. 25**). Precipitation, evaporation, and surface runoff were originally defined by data input to the Lake Package (Merritt and Konikow, 2000, p. 7-8 and described in **appendix 1**). However, these processes, as well as interflow through the soil zone, are computed in GSFLOW using PRMS and, therefore, these input variables for the Lake Package should be set to zero for an integrated simulation.

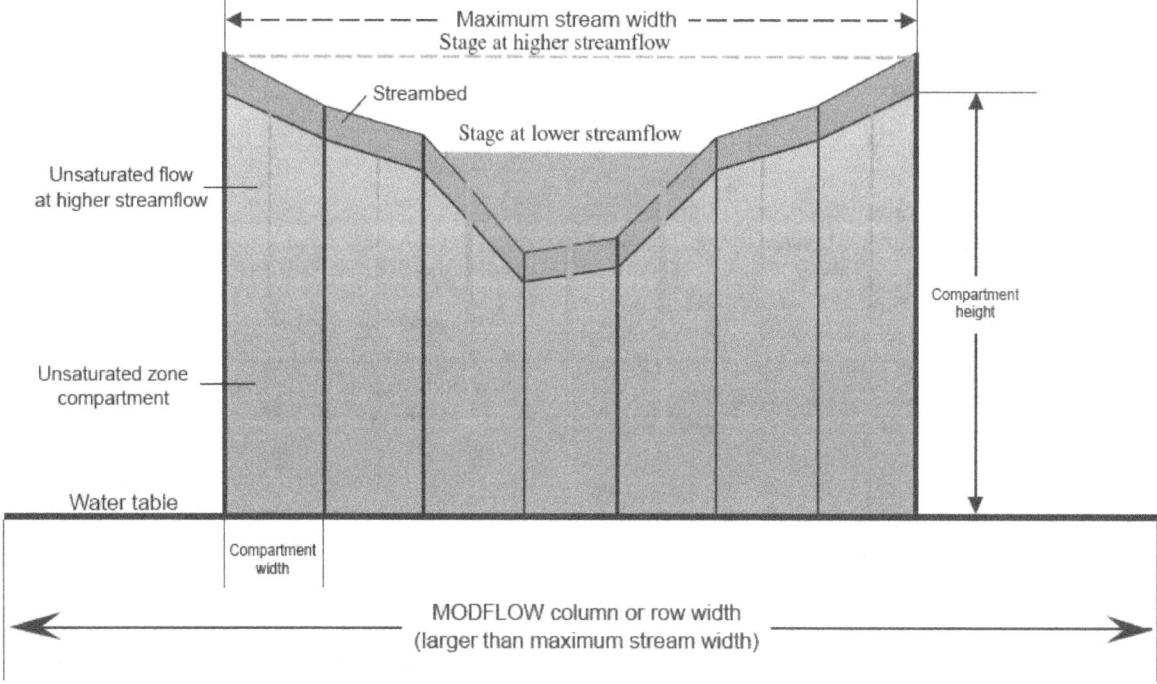

Figure 24. Discretization of unsaturated zone beneath an eight-point cross section of a stream segment within a finite-difference cell (from Niswonger and others, 2005, fig. 3).

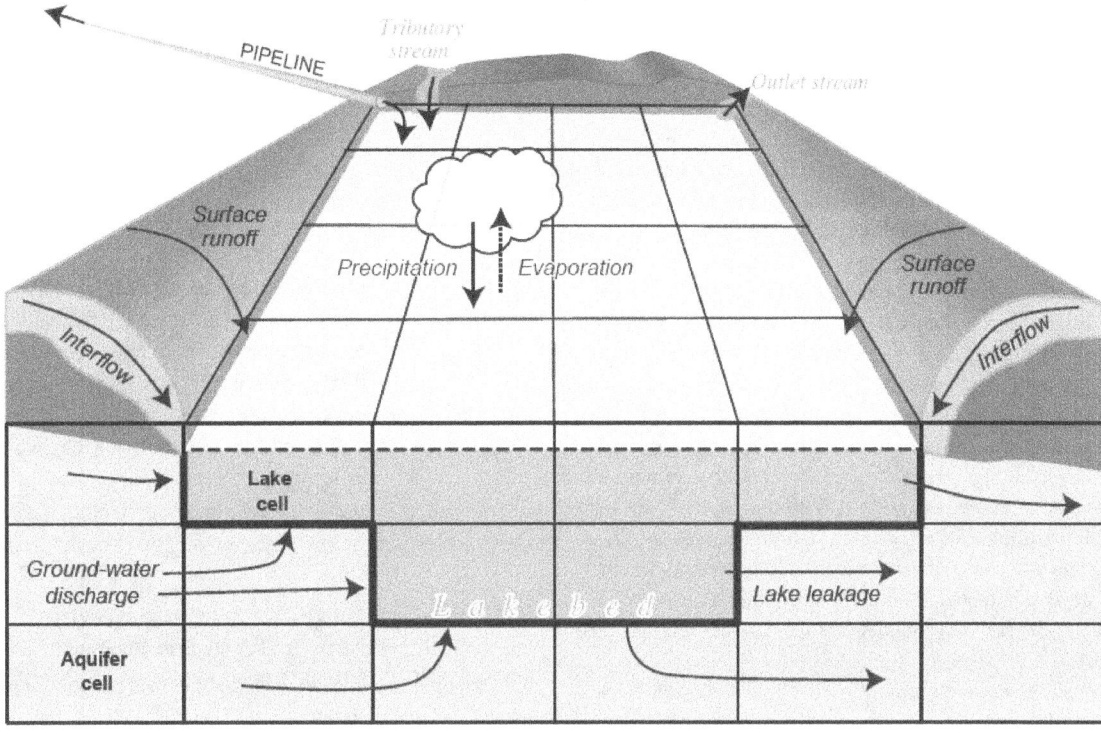

Figure 25.   Inflows to and outflows from a lake. Grid represents finite-difference cells for lakes and ground water.

Surface runoff (both Hortonian and Dunnian) and fast and slow interflow to a lake HRU are calculated for all iterations of a time step using the cascading-flow procedure as:

$$V_{lakeHRU}^{m,n} =$$

$$\sum_{J=1}^{JJ} F_{J,lakeHRU} A_J \left(ROh_J^{m,n} + ROd_J^{m,n} + Dsif_J^{m,n} + Dfif_J^{m,n} \right) , \quad (90)$$

where

$V_{lakeHRU}^{m,n}$ is the volume of surface runoff and interflow routed to a lake HRU from contributing HRUs for time step m and iteration n, in acre-inch;

$F_{J,lakeHRU}$ is the decimal fraction of the total area of HRU J that contributes runoff and interflow to a lake HRU defined by GSFLOW parameter `hru_pct_up` in table 8, dimensionless;

J is the counter for the HRU number, dimensionless; and

JJ is the total number of HRUs that contribute surface runoff and interflow to a particular lake, dimensionless.

Any snow accumulation on a lake is expressed as a liquid equivalent. Thus, the stage of all lakes is assumed to be the liquid equivalent of the solid and liquid phases. The volume of inflow from surface runoff and interflow from all contributing HRUs is added to the volume of precipitation minus evaporation on the lake HRU and converted to a volumetric flow rate with dimensions used by the Lake Package as:

$$Q_{lakeHRUtolakemf}^{m,n} = \frac{C_{prms2mf}'}{\Delta t}$$

$$\left[(P_{lakeHRU}^m - Evap_{lakeHRU}^m)A_{lakeHRU} + V_{lakeHRU}^{m,n} \right] , \quad (91)$$

where

$Q_{lakeHRUtolakemf}^{m,n}$ is the volumetric flow rate from a lake HRU to a lake represented in MODFLOW-2005 for time step m and iteration n, in cubic length per time;

$C_{prms2mf}'$ is the conversion from PRMS inch-acre to MODFLOW-2005 cubic length;

$P_{lakeHRU}^m$ is the precipitation on the lake HRU for time step m, in inches;

$Evap_{lakeHRU}^m$ is the evaporation from the lake HRU for time step m, in inches; and

$A_{lakeHRU}$ is the area of the lake HRU, in acres.

Stream inflow to a lake is computed as the sum of all outflows from the last reach of each tributary stream segment (Merritt and Konikow, 2000, p. 11; Prudic and others, 2004, p. 9). Any number of inflowing stream segments can be added to a lake. Although most lakes have only one outlet stream, stream outflow from a lake is the sum of any number of outlet stream segments. Outflow from a lake to a stream segment can be user-specified, either as a constant flow rate or by way of a lookup table, or it can vary as a function of lake stage and the outlet channel geometry. Additionally, non-stream-related withdrawals also can be specified from a lake by the user. The withdrawals are specified every stress period in the input for the Lake Package (appendix 1). A negative value for withdrawal can be used to specify the addition of water to a lake from some external source such as flow through a pipe.

The exchange of water between a lake and ground water is dependent on the lake stage and ground-water heads next to and beneath the lake. Ground water discharges to a lake when and where the ground-water head is greater than the lake stage, whereas outward and downward leakage occurs when and where the ground-water head is less than the lake stage. Ground-water discharge to a lake or leakage from a lake may occur anywhere within a lake. The distribution of these flows will depend on the lake stage and the ground-water head calculated in aquifer cells adjoining the lake. Lakes can have areas of ground-water discharge and leakage at the same time, and the areas of ground-water discharge and leakage can vary through time depending on transient variations in lake stage and ground-water heads.

A lake occupies the entire surface area of a finite-difference cell and does not overlap with an aquifer within the cell. A lake can occupy many finite-difference cells that can extend over multiple finite-difference layers (fig. 9). A lakebed is assumed to exist between the lake cells and the aquifer finite-difference cells and it can have different properties than the underlying finite-difference cells (fig. 26A). Flow between lakes and ground water is computed as a function of the lakebed thickness and hydraulic conductivity and the thickness and hydraulic conductivity of the finite-difference layer beneath the lakebed (fig. 26B; Merritt and Konikow, 2000, eqs. 2 and 4). This function is:

$$Q_{lakeleak}^{m,n} = \frac{A_{aq}^{m,n}}{\dfrac{thick_{lkbd}}{K_{lkbd}} + \dfrac{thick_{aq}}{K_{aq}}} \left[h_{lake}^{m,n} - h_{aqfdc}^{m,n} \right] , \qquad (92)$$

where

$Q_{lakeleak}^{m,n}$ is the volumetric flow rate across the lakebed to center of aquifer finite-difference cell for time step m and iteration n and is outward or downward leakage when positive and ground-water discharge when negative, in cubic length per time;

K_{lkbd} is the hydraulic conductivity of the lakebed, in length per time;

$thick_{lkbd}$ is the lakebed thickness, in length;

$thick_{aq}$ is the horizontal or vertical distance from lakebed to the center of the adjacent finite-difference cell, in length;

$A_{aq}^{m,n}$ is the area of the lakebed covering a finite-difference cell, in length squared;

K_{aq} is the horizontal or vertical hydraulic conductivity of aquifer finite-difference cell adjacent to lake cell, in length per time;

$h_{lake}^{m,n}$ is the lake stage for time step m and iteration n, in length; and

$h_{aqfdc}^{m,n}$ is the ground-water head in the finite-difference cell adjacent to the lake at the cell node (fig. 26B) for time step m and iteration n, in length.

Leakage across the lakebed can change during a time step and is dependent on the lake and ground-water heads and the cross-sectional area through which ground-water discharge and leakage is computed. The cross-sectional area of flow along a vertical face can vary as a function of the ground-water head. The altitude at the top of the finite-difference cell adjacent to the lakebed is used to calculate leakage when the ground-water head in the finite-difference cell is beneath the lakebed.

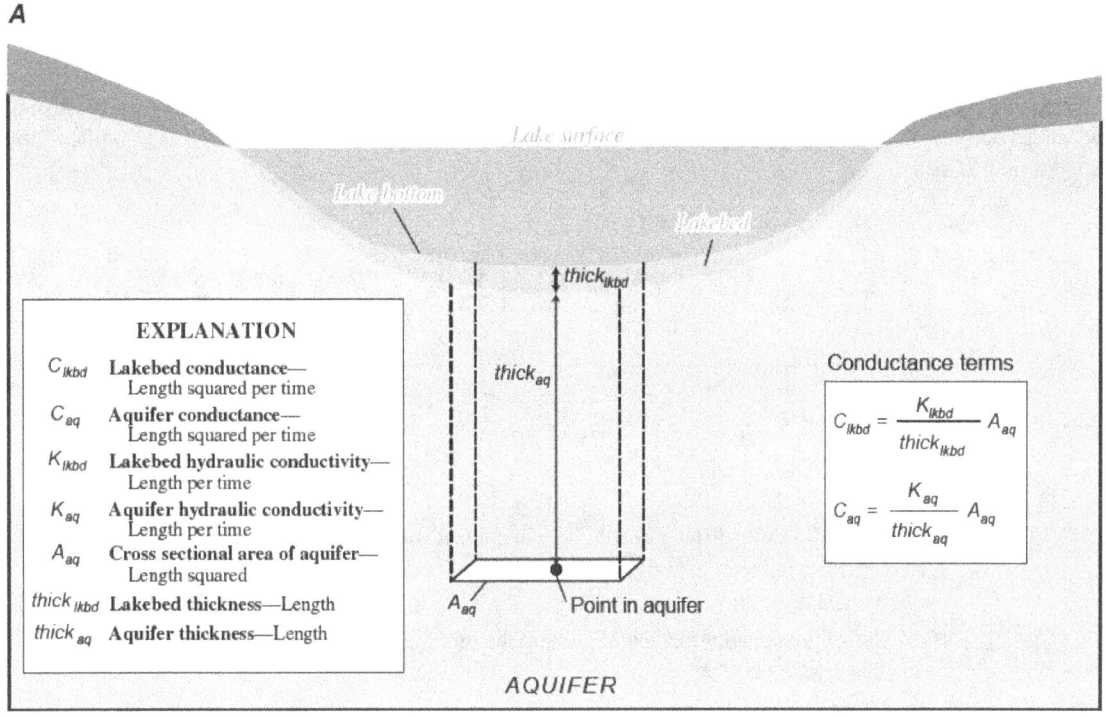

A

EXPLANATION

C_{lkbd} Lakebed conductance—
         Length squared per time

C_{aq} Aquifer conductance—
         Length squared per time

K_{lkbd} Lakebed hydraulic conductivity—
         Length per time

K_{aq} Aquifer hydraulic conductivity—
         Length per time

A_{aq} Cross sectional area of aquifer—
         Length squared

$thick_{lkbd}$ Lakebed thickness—Length

$thick_{aq}$ Aquifer thickness—Length

Conductance terms

$$C_{lkbd} = \frac{K_{lkbd}}{thick_{lkbd}} A_{aq}$$

$$C_{aq} = \frac{K_{aq}}{thick_{aq}} A_{aq}$$

B

EXPLANATION

K_h Horizontal hydraulic conductivity—Length per time

K_v Vertical hydraulic conductivity—Length per time

Figure 26.   (*A*) Conductance terms used to compute leakage from and ground-water discharge to a lake; (*B*) representation with finite-difference cells in MODFLOW-2005 (modified from Merritt and Konikow, 2000, fig. 1).

Computation of Lake Stage and Outflow to Streams

Because the Lake Package as implemented in GSFLOW differs from the original version of Merritt and Konikow (2000), an extended discussion of these differences is provided in this section. Lake stage is computed on the basis of the sum of the inflows minus outflows and a change in lake storage for each time step. The equation can be written as (modified from Merritt and Konikow, 2000, p. 9):

$$
h_{lake}^{m,n} = \frac{h_{lake}^{m-1} + \Delta t_{mf} \dfrac{Q_{lakeHRUtomf}^{m,n} + Q_{totstrin}^{m,n} - Q_{totstrout}^{m,n} - Q_{wd}^{m} + \left(\displaystyle\sum_{lc}^{LC} Cd_{lc}^{m,n} h_{afdc,lc}^{m,n-1} - (1-\varpi) h_{lake}^{m-1} \displaystyle\sum_{lc}^{LC} Cd_{lc}^{m,n} \right)}{A_{lksurf}^{m-1}}}{1 + \dfrac{\varpi \Delta t_{mf}}{A_{lksurf}^{m-1}} \displaystyle\sum_{lc=1}^{LC} Cd_{lc}^{m,n}} \,,
$$

(93)

where

LC is the number of finite-difference cell faces in contact with the submerged lakebed;

lc is a counter used to refer to individual finite-difference cell faces in contact with the submerged lakebed;

$Q_{lakeHRUtomf}^{m,n}$ is the sum of precipitation, evaporation, surface runoff, and interflow calculated in PRMS, in cubic length per time;

Δt_{mf} is the MODFLOW-2005 time step, in time;

$Q_{totstrin}^{m,n}$ is the sum of all tributary stream inflows for time step m and iteration n, in cubic length per time;

$Q_{totstrout}^{m,n}$ is the sum of lake outflows to streams for time step m and iteration n, in cubic length per time;

Q_{wd}^{m} is the specified withdrawal from a lake (negative value adds water) for time step m, in cubic length per time;

$Cd_{lc}^{m,n}$ is the lakebed conductance of cell face lc for time step m and iteration n, in length squared per time;

$h_{afdc,lc}^{m,n-1}$ is the ground-water head in finite-difference cell associated with cell face lc for time step m and iteration $n-1$, in length;

ϖ is the time-weighting factor that ranges between 0.5 and 1, dimensionless; and

A_{lksurf}^{m-1} is the surface area of the lake for the last iteration of time step $m-1$, in length squared.

The method for solving transient lake stage in the Lake Package (Merritt and Konikow, 2000, p. 9-10) was revised for GSFLOW because lake outflow originally was computed using a combination of lake stage from the previous time step and the previous iteration within the time step. This method could result in numerical oscillations of lake stage and outflow to streams that caused the simulation to not converge to its closure criterion in a finite-difference cell adjacent to a lake or beneath a stream reach downstream from the lake.

Newton's method of computing lake stage for a steady-state stress period (Merritt and Konikow, 2000, p. 12) also was revised to compute lake stage for both steady-state and transient stress periods. Previously, lake outflow to a stream and its derivative were computed once in the Streamflow-Routing Package only for steady-state stress periods (Prudic and others, 2004, p. 9-10). This approach did not allow the outflow to a stream and its derivative to change as the steady-state lake stage solution was updated. The computation of lake stage for steady-state and transient simulations was changed to solve for lake stage and outflow to streams within the Formulate Subroutine of the Lake Package (Merritt and Konikow, 2000, p. 39-40).

The revision resulted in several changes to both the Lake and Streamflow-Routing Packages. In the previous version of the Lake Package (Merritt and Konikow, 2000), the relation between lake stage and surface area was discontinuous when the lake stage fluctuated through the boundary between MODFLOW layers. This discontinuity would, in some cases, result in convergence failure for the solution of lake stage. Thus, the relation between lake stage and surface area was revised such that the lake stage would linearly transition over the boundary between MODFLOW layers. Additionally, the calculation for streamflow emanating from a lake as a function of lake stage is updated during the iterative solution of lake stage to improve convergence. A new subroutine was added to the Streamflow-Routing Package to compute a table of streamflow versus stream depth values for streams that emanate from a lake.

The relation between streamflow verses stream depth for a stream emanating from a lake can be developed using any of the options for computing stream depth in a segment: Manning's equation assuming a wide rectangular channel (ICALC=1) or an eight-point cross section (ICALC=2); power-law equations that relate stream depth and width to streamflow (ICALC=3); and lists of values (a look-up table) that relate measured streamflows to measured stream depths and widths (ICALC=4). A list is created for each outlet stream by subtracting the maximum lake stage from the top of the streambed at the beginning of the outlet stream segment, dividing that maximum depth into 200 increments, and computing flows at the beginning of the outlet segment using one of the four options for computing stream depth. The result is that lake stage and outflow to streams are solved within the Lake Package for a consistent set of ground-water heads.

Although the data input structure was not altered, the new method for solving lake stage does not use the time-weighting factor (ϖ) for steady-state stress periods, and the factor is automatically set to a value of 1.0 for all steady-state stress periods such that the new lake stage is not dependent on lake stage from the previous time step (equation 93). The original conceptualization of the time-weighting factor THETA by Merritt and Konikow (2000, p. 52) was to allow for ground-water discharge to, and leakage from, the lake to depend on a combination of lake stage from the previous and current time step.

Because Newton's method for solving lake stage and outflow to streams is used for transient stress periods, the range of the time-weighting factor is limited to values from 0.5 to 1.0. A value of 0.5 represents the stage midway between the previous time step and the end of the current time step. A value of 1.0 (fully implicit) represents the lake stage at the end of the current time step. A time-weighting factor of less than 0.5 does not perform well and a zero value is undefined when using Newton's method.

Slight errors in the transient solution of lake stage, ground-water discharge, and lake leakage may result when the time weighting factor is greater than 0.5. However, a weighting factor greater than 0.5 can be useful for dampening oscillations (Fread, 1993). A value of 0.5 represents a semi-implicit method that commonly is called Crank-Nicolson (Wang and Anderson, 1982; p. 81). Wang and Anderson (1982) present results of drawdown for a simple confined aquifer with pumping in which the Crank-Nicolson method produced the best results when compared with the Theis solution for different times. A value of 0.5 is the default value used in the solution of lake stage for transient stress periods.

Results from the revised computational method of computing lake stage were compared with the first example test simulation presented by Merritt and Konikow (2000, p. 21-23). No inflow from or outflow to streams was simulated in this example and lake stage rose through time as a result of ground-water discharge to and precipitation on the lake. The simulation was for 5,000 days and was divided into 100 time steps.

Three separate simulations were used to test differences in results between the original and revised versions of the Lake Package with time-weighting values of 0.0, 0.5, and 1.0. The solutions were consistent between the two versions of the Lake Package. A time-weighting factor of 0.5 resulted in a transition of the Lake Package stage between the initial and final solution that was midway between those computed using a time weighting factor of 0.0 and 1.0. Despite differences in the transition period, the three simulations resulted in the same final lake stage and lake inflow. A time-weighting factor of 1.0 decreased the number of iterations needed to converge on a lake stage compared to a time-weighting factor of 0.5; however, the number of iterations needed to converge on ground-water heads increased for a time-weighting factor of 1.0.

The maximum number of iterations and the closure criterion for the solution of lake stage can be specified using the Lake Package variables NSSITER and SSCNCR, respectively (appendix 1). These values must be specified for a simulation that includes a steady-state stress period and are optional for simulations that have only transient stress periods. The maximum number of iterations is set to 100 and the closure criterion is set to 0.0001 for all simulations that have only transient stress periods. An option in the Lake Package input was created in which values of NSSITER and SSCNCR can be read by specifying a negative time-weighting factor (appendix 1). A negative time-weighting factor is automatically reset to a positive value after values of NSSITER and SSCNCR are read.

The revised method for solving for lake stage in conjunction with outflow to streams may not always converge to the closure criterion because of discontinuities among lake stage, area, and outflow to streams. This is particularly so for steady-state simulations of lakes with outflow to streams. A warning message is printed that includes the last two computed lake stages of a particular lake whenever the change in lake stage between successive Newton iterations is not less than the closure criterion after reaching the maximum number of iterations.

Intermittent Lakes

The Lake Package was further revised to allow for the continued accounting of surface inflow into intermittent lakes and for the filling of an empty lake when the sum of inflows from precipitation, surface runoff, interflow, tributary streams, and ground-water discharge exceed the sum of outflows to specified withdrawal, evaporation, leakage, and outlet streams. The original version of the Lake Package (Merritt and Konikow, 2000, p. 10-11) did not maintain a water budget once a lake was empty, and the lake could only refill from ground-water discharge. An empty lake was filled to the average ground-water head in cells beneath the lowest lake area whenever the average head at the end of a time step was above the top of the lakebed. This approach for refilling the lake was discarded in favor of an approach that accounts for all inflows and outflows to an empty lake within the GSFLOW and Lake iteration loops.

Several changes were made to the lake-stage calculation to balance flow in and out of a lake. The calculations of outflow from lakes to streams were added to the Newton iteration loop in the Formulate Subroutine of the Lake Package. If a specified withdrawal from a lake exceeds the sum of computed inflow less ground-water discharge, the specified withdrawal is decreased to equal the water available in the lake. In order to simulate rewetting of a lake, ground-water discharge to an empty lake is computed assuming the lake head is equal to the top of the lakebed. If evaporation from an empty lake exceeds the sum of ground-water discharge and other inflows and outflows, evaporation is decreased to the amount of available water. If inflows exceed evaporation, leakage to ground water, and stream outflow, then the lake is allowed to rewet. The computation of lake stage and outflow to streams requires that the top of the streambed for all outlet streams must be equal to or greater than the top of the lakebed in the lowest part of the lake.

Unsaturated Flow Beneath Lakes

An unsaturated zone can develop beneath a finite-difference cell represented by a lake whenever the ground-water head in the underlying finite-difference cell is less than the bottom of the lake cell. An unsaturated zone beneath a lake can develop when the hydraulic conductivity of the lakebed is much less than that of the underlying aquifer and/or ground water is pumped from a well or wells near the lake.

The Lake Package was further revised to compute unsaturated flow beneath a lake using the Unsaturated-Zone Flow Package whenever the ground-water head in an underlying cell is beneath the bottom of the lake cell. Hydraulic properties of the unsaturated zone in cells beneath a lake are specified in the data input for the Unsaturated-Zone Flow Package in the same manner as any other cell that is used to simulate unsaturated flow from land surface to the water table (appendix 1). Unsaturated flow beneath a lake cell is computed from the bottom of the lake cell. The bottom of a lake cell can be higher than the water table in the underlying aquifer cell, in which case leakage is routed through the unsaturated zone between the lakebed and underlying water table.

Gravity drainage through the unsaturated zone beneath lakes is simulated using the same method as that used to simulate unsaturated-zone flow beneath the soil zone and streams. Unsaturated flow and water-contents beneath a particular lake can be printed using the option variable NUZGAG in the data input for the Unsaturated-Zone Flow Package (appendix 1). The maximum lake leakage through a lakebed is limited to the saturated vertical hydraulic conductivity multiplied by the top area of the finite-difference cell. Initial water contents within the unsaturated zone beneath a lake can be calculated using a steady-state simulation or specified in the input for the Unsaturated-Zone Flow Package.

Ground Water

Flow in the saturated zone is governed by a partial differential equation for three-dimensional ground-water flow of constant density (Harbaugh, 2005, p. 2-1):

$$\frac{\partial}{\partial x}\left(K_{xx}\frac{\partial h}{\partial x}\right)+\frac{\partial}{\partial y}\left(K_{yy}\frac{\partial h}{\partial y}\right)+\frac{\partial}{\partial z}\left(K_{zz}\frac{\partial h}{\partial z}\right)+W=S_s\frac{\partial h}{\partial t} \ , \ (94)$$

where

K_{xx}, K_{yy} and K_{zz} are the hydraulic conductivity tensors aligned with the x, y, and z coordinate axes, in length per time;

h is the potentiometric head, in length;

W is volumetric flow rate per unit volume representing sources and / or sinks of water, $W<0.0$ for flow out of the ground-water system, and $W>0.0$ for flow into the system, in per time; and

S_s is the specific storage of the porous rock, in per length.

The specific storage (S_s) and hydraulic conductivities (K_{xx}, K_{yy}, and K_{zz}) vary as a function of space, and the volumetric flow rate per unit volume (W) varies as a function of space and time.

Volume-Averaged Flow Equation

Equation 94 represents flow through a macroscopic region large enough to consider water and porous material as a single continuum; however, finite-difference cell volumes used in MODFLOW are usually much larger than this macroscopic region. Thus, MODFLOW solves a volume-averaged equation that is derived by integrating equation 94 over an arbitrary cell volume. The finite-difference method is applied following the derivation of the volume-averaged flow equation, which results in a system of algebraic equations. A convention on the basis of alphabetical indices is used for defining both the cell center (node) and cell faces for the derivation of the finite difference equations (fig. 27A). Both nodes and cell faces are indexed because ground-water head is calculated at a node and the flow is calculated across the cell face. Accordingly, cell faces are referenced in the finite-difference equations by adding or subtracting one-half of the node index to represent one-half the cell distance in either the positive or negative direction along the $x, y,$ or z directions (figs. 7 and 27A).

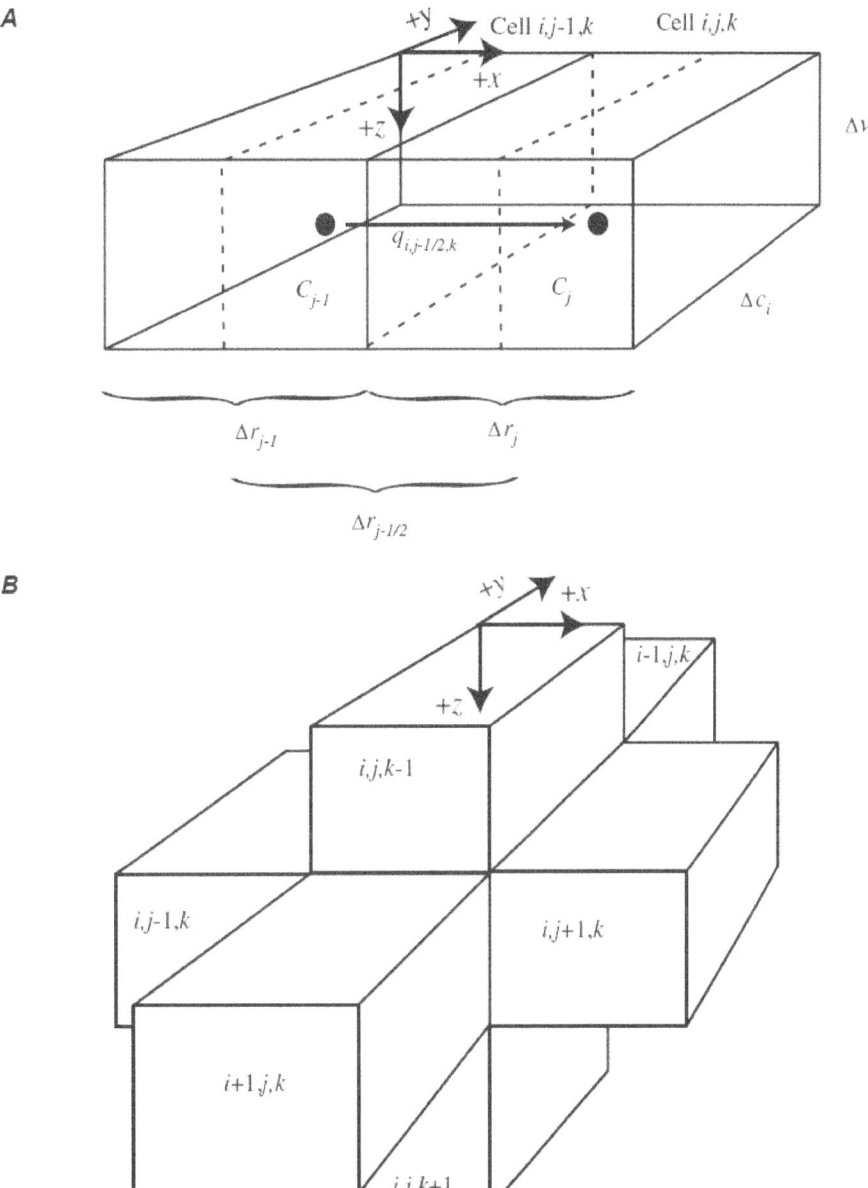

Figure 27. (*A*) Flow into finite-difference cell *i, j, k* from cell *i, j-1, k*, and (*B*) indices for six adjacent cells surrounding cell *i, j, k* (hidden) (modified from Harbaugh, 2005).

Thus, the four vertical cell faces surrounding the node i, j, k beginning with the side facing the $-x$ direction and moving clockwise are defined as $i, j-\frac{1}{2}, k$; $i-\frac{1}{2}, j, k$; $i, j+\frac{1}{2}, k$; and $i+\frac{1}{2}, j, k$. The bottom and top faces below and above node i, j, k are $i, j, k+\frac{1}{2}$, and $i, j, k-\frac{1}{2}$, respectively. However, faces will be defined by dropping the indices that remain whole, such that the cell face $i, j-\frac{1}{2}, k$ is referenced as $face_{j-1/2}$. Similarly, the distance between nodes is indexed using the same notation as for cell faces, where, for example, the distance in the x (column) direction between nodes i, j, k and $i, j-1, k$ is represented by $\Delta r_{j-1/2}$ (figs. 7 and 27*A*).

Using the notation listed in figure 27, integrating the first term of equation 94, and assuming $\frac{\partial h}{\partial x}$ is constant along the y and z directions results in:

$$\int_{\Delta c_i}\int_{\Delta r_j}\int_{\Delta v_k}\frac{\partial}{\partial x}\left(K_{xx}\frac{\partial h}{\partial x}\right)dzdydx = \int_{face_{j-1/2}}^{face_{j+1/2}}\frac{\partial}{\partial x}\left(K_{xx}\frac{\partial h}{\partial x}\right)dx\Delta r_j\Delta v_k \ , \quad (95)$$

where

Δr_j is the length along the row direction for a finite-difference cell, in length;

Δc_i is the length along the column direction for a finite-difference cell, in length;

Δv_k is the length along the layer direction for a finite-difference cell, in length;

$face_{j+1/2}$ is the cell face between cells j and $j+1$, dimensionless; and

$face_{j-1/2}$ is the cell face between cells $j-1$ and j, dimensionless.

The integral can be carried further such that:

$$\Delta c_i\Delta v_k\int_{face_{j-1/2}}^{face_{j+1/2}}\frac{\partial}{\partial x}\left(K_{xx}\frac{\partial h}{\partial x}\right)dx = \Delta c_i\Delta v_k K_{xx}\frac{\partial h}{\partial x}\bigg|_{face_{j-1/2}}^{face_{j+1/2}} \qquad (96)$$
$$= \Delta c_i\Delta v_k\left(q_{j+1/2}-q_{j-1/2}\right) ,$$

where

$q_{j-1/2}$ is the volumetric flow rate between cells $j-1$ and j, in cubic length per time; and

$q_{j+1/2}$ is the volumetric flow rate between cells j and $j+1$, in cubic length per time.

The time derivative in equation 94 can be integrated over the same volume as:

$$\int_{\Delta r_j}\int_{\Delta c_i}\int_{\Delta v_k}S_s\frac{\partial h}{\partial t}dzdydx = SS\frac{\Delta h}{\Delta t_*}\Delta v_k\Delta c_i\Delta r_j \ , \qquad (97)$$

where

SS is the volume averaged specific storage, in per length; and

$\frac{\Delta h}{\Delta t_*}$ is the change in head over a specified time interval, in length per time.

Similarly, the other terms in equation 94 can be integrated such that the volume-integrated form of the equation can be rewritten as:

$$\Delta r_j\Delta v_k\left(q_{i+1/2}-q_{i-1/2}\right)+\Delta c_i\Delta r_j\left(q_{k+1/2}-q_{k-1/2}\right)$$
$$+\Delta c_i\Delta v_k\left(q_{j+1/2}-q_{j-1/2}\right)+W = SS\frac{\Delta h}{\Delta t_*}\Delta v_k\Delta c_i\Delta r_j \ , \quad (98)$$

where

$q_{i-1/2}$ is the volumetric flow rate between cells $i-1$ and i, in cubic length per time;

$q_{i+1/2}$ is the volumetric flow rate between cells i and $i+1$, in cubic length per time;

$q_{k-1/2}$ is the volumetric flow rate between cells $k-1$ and k, in cubic length per time; and

$q_{k+1/2}$ is the volumetric flow rate between cells k and $k+1$, in cubic length per time.

Assuming the same sign convention as used by Harbaugh (2005), the volumetric flow rate into the cell containing node i, j, k can be written as the sum of flows across six cell faces as:

$$\sum Q_L = q_{i-1/2}+q_{i+1/2}+q_{j-1/2}+q_{j+1/2}+q_{k-1/2}+q_{k+1/2} \ , \quad (99)$$

where

Q_L is the volumetric flow rate into a cell (negative value is a volumetric flow out of cell), in cubic length per time.

Combining equations 98 and 99 results in equation 2-2 of Harbaugh (2005, p. 2-3):

$$\sum Q_L = SS\frac{\Delta h}{\Delta t_*}V \ , \qquad (100)$$

where

V is the volume of the cell, in cubic length.

Finite-Difference Formulation

Darcy's Law can be used to remove q from equation 98. For example, after applying the finite- difference method for flow across the side of the block facing the $-x$ direction the flow can be written as:

$$q_{j-1/2} = CR_{j-1/2}(h_{i,j-1,k} - h_{i,j,k}) , \qquad (101)$$

where

$h_{i,j-1,k}$ is head at node i, $j-1$, k, in length;

$h_{i,j,k}$ is the head at node i, j, k, in length; and

$CR_{j-1/2}$ is the conductance between nodes i, $j-1$, k and i, j, k, in length squared per time, and is equal to :

$$CR_{j-1/2} = \frac{(KR_{j-1/2}\Delta c_i \Delta v_k)}{\Delta r_{j-1/2}} , \qquad (102)$$

where

$KR_{j-1/2}$ is the hydraulic conductivity between nodes i, $j-1$, k and i, j, k, in length per time.

Equation 101 can be derived for all six faces surrounding node i, j, k (fig. 27B). These equations are presented by Harbaugh (2005).

The derivative of head with respect to time is approximated using backward in time differencing according to (Harbaugh, 2005):

$$\frac{\Delta h_{i,j,k}^m}{\Delta t_*} = \frac{h_{i,j,k}^m - h_{i,j,k}^{m-1}}{t_*^m - t_*^{m-1}} , \qquad (103)$$

where

$\dfrac{\Delta h_{i,j,k}^m}{\Delta t_*}$ is the change in head at cell i, j, k at the end of time step m over a specified time interval, in length per time;

$h_{i,j,k}^{m-1}$ is the head at node i, j, k, at end of time step $m-1$, in length;

$h_{i,j,k}^m$ is the head at node i, j, k, at end of time step m, in length;

t_*^m is the time at the end of time step m, in time; and

t_*^{m-1} is the time at the end of time step $m-1$, in time.

Substituting equations for each cell face (equation 101) and equation 103 into equation 98 results in:

$$
\begin{aligned}
&CC_{i-1/2}\left(h_{i-1,j,k}^m - h_{i,j,k}^m\right) + CC_{i+1/2}\left(h_{i+1,j,k}^m - h_{i,j,k}^m\right) \\
&+ CR_{j-1/2}\left(h_{i,j-1,k}^m - h_{i,j,k}^m\right) + CR_{j+1/2}\left(h_{i,j+1,k}^m - h_{i,j,k}^m\right) \\
&+ CV_{k-1/2}\left(h_{i,j,k-1}^m - h_{i,j,k}^m\right) + CV_{k+1/2}\left(h_{i,j,k+1}^m - h_{i,j,k}^m\right) \\
&+ W = SS_{i,j,k}\left(\frac{h_{i,j,k}^m - h_{i,j,k}^{m-1}}{t_*^m - t_*^{m-1}}\right)V ,
\end{aligned} \qquad (104)
$$

where

$CC_{i-1/2}$ is the conductance between nodes $i-1$, j, k and i, j, k, in length squared per time;

$CC_{i+1/2}$ is the conductance between nodes i, j, k and $i+1$, j, k, in length squared per time;

$CR_{j-1/2}$ is the conductance between nodes i, $j-1$, k and i, j, k, in length squared per time;

$CR_{j+1/2}$ is the conductance between nodes i, j, k and i, $j+1$, k, in length squared per time;

$CV_{k-1/2}$ the conductance between nodes i, j, $k-1$ and i, j, k, in length squared per time;

$CV_{k+1/2}$ is the conductance between nodes i, j, k and i, j, $k+1$, in length squared per time;

$h_{i-1,j,k}^m$ is the head at node $i-1$, j, k, at end of time step m, in length;

$h_{i+1,j,k}^m$ is the head at node $i+1$, j, k, at end of time step m, in length;

$h_{i,j-1,k}^m$ is the head at node i, $j-1$, k, at end of time step m, in length;

$h_{i,j+1,k}^m$ is the head at node i, $j+1$, k, at end of time step m, in length;

$h_{i,j,k-1}^m$ is the head at node i, j, $k-1$, at end of time step m, in length;

$h_{i-1,j,k+1}^m$ is the head at node i, j, $k+1$, at end of time step m, in length; and

$SS_{i,j,k}$ is the volume averaged specific storage for cell i, j, k, in per length.

Equation 104 is an algebraic equation that provides the time variability of ground-water head and flow everywhere in a three-dimensional ground-water system. Equation 104 must be solved for every variable-head cell in the model and be combined with initial and boundary conditions. This system of equations can be written in compact form as:

$$C \, \bar{h} = Ch_{bnd} + Q_W \, ,$$

(105)

where

C is a matrix of conductance values for the row, column, and layer directions (CR, CC, and CV) that are multiplied by the ground-water head in each variable-head cell and any coefficients multiplied by a variable-head cell to represent external sources and sinks, in length squared per time;

$\bar{h}$ is a vector of ground-water heads for all variable-head cells, in length;

Ch_{bnd} is a vector of conductance values multiplied by ground-water head in each constant-head cell, in cubic length per time; and

Q_W is the volumetric flow rate added to variable-head cells that represents external sources and sinks to the ground-water system, in cubic length per time.

The system of equations represented by equation 105 can be solved using several different matrix solvers that are available for MODFLOW-2005 (Harbaugh, 2005; p. 7-1).

A complication to the solution of equation 105 for unconfined ground-water conditions occurs because the horizontal conductance values shown in equation 104 are dependent on ground-water head (Harbaugh, 2005; p. 5-5). Finite-difference equations used to solve for head for unconfined conditions are solved iteratively by MODFLOW-2005 and the conductance values are approximated by the ground-water head from the previous iteration. Certain boundary conditions used to represent external flows to the ground-water system also add coefficients that are dependent on ground-water head.

Conductance

The conductance for a group of finite-difference cells will vary in the x, y, and z directions for anisotropic and heterogeneous medium. Flow is calculated across cell faces such that the conductance values used in the finite-difference equations represent averages of the conductance values for two cells (interblock conductance). The horizontal conductance of a single cell can be calculated as:

$$C_h = \frac{Tr \, Width_{fdc}}{Length_{fdc}} \, ,$$

(106)

where

C_h is the horizontal conductance for a prism, in length squared per time;

Tr is the transmissivity of the cell, in length squared per time;

$Width_{fdc}$ is the width of the prism perpendicular to the direction for which the conductance is defined, in length; and

$Length_{fdc}$ is the length of the prism, in length.

For confined conditions, transmissivity is equal to the product of the horizontal hydraulic conductivity of the cell multiplied by the saturated thickness of the cell. For unconfined conditions, transmissivity is equal to the horizontal hydraulic conductivity of the cell multiplied by the difference between the altitude of the water table in the cell and the altitude of the cell bottom.

As a result of the spatial variability of hydraulic conductivity, cell thickness, and ground-water head, there are four methods for calculating interblock conductance between cells that are supported by MODFLOW-2005. The first method for calculating interblock conductance for cells i, j-1, k and i, j, k assumes uniform transmissivity within a cell and is written as (Harbaugh, 2005; p. 5–3):

$$CR_{j-1/2} = \frac{C_{j-1} C_j}{(C_{j-1} + C_j)} \, ,$$

(107)

where

C_{j-1} is the horizontal conductance for cell $j-1$ in the row direction, in length squared per time; and

C_j is the horizontal conductance for cell j in the row direction, in length squared per time.

The other three methods for calculating horizontal conductance between finite-difference cells are described by Goode and Appel (1992) and Harbaugh (2005). These methods make different assumptions about the flow system and how transmissivity (equation 106) varies in space, including assuming that transmissivity varies linearly between nodes, the aquifer is flat and homogeneous with a water table, or the aquifer is flat with a water table in which hydraulic conductivity varies linearly between nodes.

The vertical conductance for a cell is a function of the cell thickness, horizontal area, and vertical hydraulic conductivity. The vertical conductance of a single cell is expressed as:

$$C_v = \frac{K_v \, A_{fdc}}{\Delta z} \, , \qquad (108)$$

where

C_v is the vertical conductance, in length squared per time; and

Δz is the cell thickness, or, for unconfined conditions, the thickness of the saturated region of the cell, in length.

The vertical conductance between two nodes will be the equivalent conductance of two one-half cells in a series for two cells with ground-water heads that are greater than the altitude of the cell top. The vertical conductance between cell i, j, k and $i, j, k+1$ for the condition of no semi-confining unit between the two cells is calculated according to (Harbaugh, 2005; p. 5-7):

$$CV_{k+1/2} = \frac{\Delta r_j \Delta c_i}{\left[\dfrac{1/2 thck_{i,j,k}}{VK_{i,j,k}} + \dfrac{1/2 thck_{i,j,k+1}}{VK_{i,j,k+1}} \right]} \, , \qquad (109)$$

where

$thck_{i,j,k}$ and $thck_{i,j,k+1}$ are the thicknesses of cells i, j, k and $i, j, k+1$, respectively, in length—these thicknesses are equal to cell thickness for confined conditions, and equal to the difference between altitude of the water table in the cell and altitude of the cell bottom for unconfined conditions; and

$VK_{i,j,k}$ and $VK_{i,j,k+1}$ are the vertical hydraulic conductivity of cells i, j, k and $i, j, k+1$, respectively, in length per time.

An approach for simulating the effects of semi-confining units on vertical flow through aquifers incorporates the thickness and vertical hydraulic conductivity of the semi-confining unit into the calculation of the vertical conductance. This approach commonly is referred to as the "Quasi-Three-Dimensional" approach (Bredehoeft and Pinder, 1970) and

assumes that the semi-confining unit makes no measurable contribution to the horizontal conductance or the storage capacity of two adjacent model layers. For this approach the conductance is calculated as (Harbaugh, 2005; p. 5-8):

$$CV_{k+1/2} = \frac{\Delta r_j \Delta c_i}{\left[\dfrac{1/2 thck_{i,j,k}}{VK_{i,j,k}} + \dfrac{thck_{CB}}{VKCB_{i,j,k}} + \dfrac{1/2 thck_{i,j,k+1}}{VK_{i,j,k+1}} \right]} \, , \qquad (110)$$

where

$thck_{CB}$ is the thickness of the semi-confining unit, in length; and

$VKCB_{i,j,k}$ is the vertical hydraulic conductivity of the semi-confining unit beneath cell i, j, k, in length squared per time.

Storage

Storage is formulated differently depending on the layer type designation that is specified within the Block-Centered Flow or Layer-Property Flow Packages of MODFLOW-2005. Layers in the model can be specified as "confined" or "convertible." The storage coefficient remains constant throughout a simulation for layers designated as confined, whereas the value of the storage coefficient may switch during a simulation if the ground-water head passes through the top of a layer designated as convertible. The rate of accumulation of water in a cell with a constant storage coefficient can be expressed as:

$$\frac{\Delta V}{\Delta t} = SS_{i,j,k} (\Delta v_k \Delta c_i \Delta r_j) \left[\frac{h_{i,j,k}^m - h_{i,j,k}^{m-1}}{t_*^m - t_*^{m-1}} \right] , \qquad (111)$$

where

$\dfrac{\Delta V}{\Delta t}$ is the rate of accumulation of water in the cell, in cubic length per time.

The Block-Centered Flow Package requires input of a dimensionless storage coefficient instead of specific storage, and MODFLOW-2005 converts this coefficient to units of per length by dividing by the cell thickness for confined conditions. The storage coefficient input for unconfined conditions is equal to the specific yield. The Layer-Property Flow Package also has an option for specifying a dimensionless storage coefficient instead of specific storage.

External Boundary Conditions

All boundary conditions used in MODFLOW-2005 can be generalized from two equations. The first form represents known functions that are either head dependent or head independent (Harbaugh, 2005):

$$a_{i,j,k,N} = C1_{i,j,k,N} h_{i,j,k} + C2_{i,j,k,N} , \qquad (112a)$$

where

$a_{i,j,k,N}$ flow from the *Nth* source into cell $i, j, k,$ in cubic length per time;

$C1_{i,j,k,N}$ is a constant for head-dependent boundary conditions, in length squared per time; and

$C2_{i,j,k,N}$ is a constant for head-independent boundary conditions, in cubic length per time.

Thus, for head-dependent boundary conditions (such as a stream), $C1_{i,j,k,N}$ is nonzero and $C2_{i,j,k,N}$ may be zero or nonzero. For head-independent boundary conditions (such as a well), $C1_{i,j,k,N}$ is zero and $C2_{i,j,k,N}$ is nonzero.

The Unsaturated-Zone Flow Package results in a boundary condition that is different than that described by equation 112a because recharge from the unsaturated zone is an unknown function of head. Recharge from the unsaturated zone is described by:

$$a_{i,j,k,N} = f_1^{uzf}(h_{i,j,k}^m, h_{i,j,k}^{m-1}) + f_2^{uzf}(h_{i,j,k}^{m-1}) , \qquad (112b)$$

where

f_1^{uzf} is the recharge received from the unsaturated zone that was stored in the unsaturated interval through which the water table rose during time step *m*, in length cubed per time; and

f_2^{uzf} is the recharge that flowed across the surface defined by the water table in the cell at time *m* – 1, in length cubed per time.

The functions f_1^{uzf} and f_2^{uzf} in equation 112b are considered unknown because they are dependent on the storage and vertical flow through the unsaturated zone, which are dependent on boundary conditions at the top of the unsaturated zone and ground-water head during previous time steps. However, the values of these functions are computed during each sequence of the iteration loop on the basis of the ground-water head at the previous time step, previous iteration, and the characteristics of the unsaturated zone.

Head-dependent flow as described by equation 112a causes additional terms to be added to the conductance matrix shown in equation 104, whereas equation 112b does not affect the conductance matrix.

Water Budgets

An important feature of the design of GSFLOW is the calculation of an overall water budget, as well as individual water budgets for all the storage reservoirs simulated by GSFLOW. These reservoirs include snowpack, impervious surfaces, surface-water bodies (streams and lakes), plant canopy, soil zone, unsaturated zone, and saturated zone (fig. 28).

Overall Water Budget

The overall water budget for a modeled region is calculated at the end of each daily time step using:

$$\Delta S_{total} = \Delta t (Q_{in} - Q_{out}) , \qquad (113)$$

where

ΔS_{total} is total storage change in all reservoirs from beginning to end of time step, in cubic length;

Q_{in} is total inflow to the modeled region during a time step, in cubic length per time; and

Q_{out} is total outflow from the modeled region during a time step, in cubic length per time.

Inflows include precipitation and all surface-water and ground-water flows into the modeled region. Outflows include evapotranspiration and all surface-water and ground-water flows out of the modeled region.

Ground-water heads to compute budget information are calculated at the end of a GSFLOW time step. Subsequently, flows and storages for the gravity reservoirs are calculated for integrated simulations. Budget information calculated at the end of a time step can be written to the GSFLOW Comma-Separated-Values File and GSFLOW Water-Budget File.

GSFLOW writes budget results to a user-specified output file called the GSFLOW Water-Budget File that includes a tabular summary of internal flows and storages, as well as flows into and out of the modeled region. The name of this file is specified using parameter `gsflow_output_file` in the GSFLOW Control File. An example output listing of water-budget information calculated by GSFLOW is shown in figure 29. The GSFLOW overall volumetric water budget

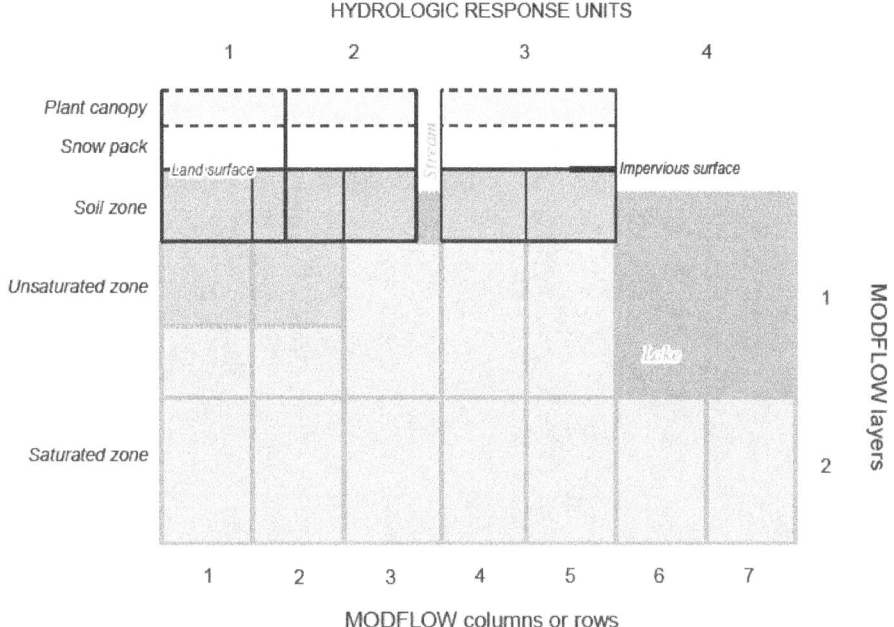

HYDROLOGIC RESPONSE UNITS

MODFLOW columns or rows

EXPLANATION

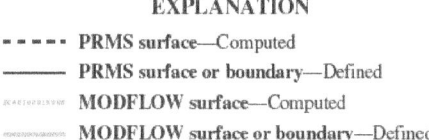

- - - - - **PRMS surface**—Computed
——————— **PRMS surface or boundary**—Defined
MODFLOW surface—Computed
MODFLOW surface or boundary—Defined

Figure 28. Storage reservoirs represented in GSFLOW include the plant canopy, snowpack, impervious surfaces, streams, and lakes at or above land surface, and the soil, unsaturated, and saturated zones in the subsurface. Adjacent reservoirs exchange water. The soil zone is composed of three reservoirs—capillary, gravity, and preferential-flow (fig. 19).

is computed by the module gsflow_sum. These budget terms are computed for every model time step (day) and also as cumulative total values for the entire period of simulation. The water-budget output is controlled by the parameter `rpt_days` specified in the GSFLOW Control File. Output only is printed to the Water-Budget File for integrated simulations. A value of 0 will turn off all reporting. The units of this report are in the MODFLOW unit of cubic length (as specified in the Discretization File) for volumes and the MODFLOW unit of cubic length per day for volumetric rates.

The input flows are: (1) precipitation, which includes rain and snowfall; (2) streamflow that flows into the watershed from outside the watershed; (3) ground-water boundary flow, which includes inflow from constant- and variable-head boundaries; and (4) flow from water injected into wells (**fig. 29**). The output flows are: (1) surface and subsurface evapotranspiration; (2) streamflow out of the watershed; (3) ground-water boundary flow, which includes outflow from constant and variable-head boundaries; and (4) flow from pumped or artesian wells. The sum of the output flows subtracted from the sum of the input flows is the total change in flow within the watershed during the time step. These flows, along with their GSFLOW budget report names are summarized in **table 11**.

```
SUMMARY VOLUMETRIC BUDGET FOR GSFLOW
DATE: 02 21 1984            CUMULATIVE TIME STEP:      1239
MODFLOW STRESS PERIOD      5    CURRENT TIME STEP:      692    ITERATIONS:      12


------------------------------------------------------------------------------


    CUMULATIVE VOLUMES           L**3   RATES FOR THIS TIME STEP          L**3/T
    ------------------                  ------------------------

                                  IN                                        IN
                                  --                                        --
         PRECIPITATION =    1.4834E+08        PRECIPITATION =          765880.63
           STREAMFLOW =                         STREAMFLOW =
        GW BOUNDARY FLOW =                   GW BOUNDARY FLOW =
                WELLS =                               WELLS =


                                  OUT                                       OUT
                                  ---                                       ---
    EVAPOTRANSPIRATION =    6.5165E+07   EVAPOTRANSPIRATION =          10703.55
           STREAMFLOW =     5.7521E+07          STREAMFLOW =           24139.96
        GW BOUNDARY FLOW =    815384.69     GW BOUNDARY FLOW =            657.31
                WELLS =                               WELLS =

    INFLOWS - OUTFLOWS =    2.4838E+07   INFLOWS - OUTFLOWS =          730379.81
    ----------                           ----------

        STORAGE CHANGE =    2.4859E+07        STORAGE CHANGE =         730400.75
    --------------                           --------------
         LAND SURFACE =    2.3072E+07          LAND SURFACE =          755174.81
            SOIL ZONE =    2481753.00             SOIL ZONE =           -1679.16
       UNSATURATED ZONE =   -2141891.25     UNSATURATED ZONE =          -8121.79
        SATURATED ZONE =    1446580.88        SATURATED ZONE =         -14973.11

OVERALL BUDGET ERROR =       -20668.00   OVERALL BUDGET ERROR =          -20.94


    PERCENT DISCREPANCY =        -0.01   PERCENT DISCREPANCY =             0.00
```

Figure 29.   Selected output from a GSFLOW Water-Budget File that shows a summary listing for December 15, 1981, from the Sagehen Creek example problem.

The change in total storage (fig. 29) is calculated as the sum of the change of four reported storage components, which are: (1) surface storage that is comprised of snowpack, intercepted precipitation, and impervious surface storage, streams and lakes; (2) soil-water storage in the root zone; (3) unsaturated-zone storage from the Unsaturated-Zone Flow Package; and (4) saturated-zone storage from MODFLOW-2005. Positive flow into storage results in a positive change in storage. These storages, along with their GSFLOW budget report names, are summarized in table 11.

Table 11. Inflow, outflow, changes in storage, and errors reported in the GSFLOW Budget Report File for GSFLOW.

GSFLOW budget report names	Description
Inflows	
Precipitation	Cumulative volume of precipitation added to modeled region and volumetric flow rate for a time step.
Streamflow	Cumulative volume of streamflow entering modeled region and volumetric flow rate for a time step.
Ground-water boundary flow	Cumulative volume of ground water entering saturated zone along external boundaries and volumetric flow rate for a time step.
Wells	Cumulative volume of water added to saturated zone by injection wells and volumetric flow rate for a time step.
Outflows	
Evapotranspiration	Cumulative volume of all evapotranspiration in modeled region and volumetric flow rate for a time step
Streamflow	Cumulative volume of streamflow leaving modeled region and volumetric flow rate for a time step.
Ground-water boundary flow	Cumulative volume of ground water leaving saturated zone along external boundaries and volumetric flow rate for a time step.
Wells	Cumulative volume of water removed from saturated zone by pumping wells and volumetric flow rate for a time step.
Inflows less outflows	Difference in cumulative volume of all inflows and outflows and difference in volumetric flow rate for a time step.
Storage changes	
Storage change	Cumulative volume of total storage change and volumetric flow rate of change for a time step.
Surface	Cumulative volume of surface storage change and volumetric flow rate of change for a time step. Surface storage includes plant canopy, snowpack, impervious surfaces, streams, and lakes.
Soil zone	Cumulative volume of soil-zone storage and volumetric flow rate of change for a time step.
Unsaturated zone	Cumulative volume of unsaturated-zone storage and volumetric flow rate of change for a time step.
Saturated zone	Cumulative volume of saturated-zone storage and volumetric flow rate of change for a time step.
Errors	
Overall budget error	Difference between inflows less outflow and total storage change.
Percent discrepancy	Difference multiplied by 100 (see equation 114).

Simulation cumulative and time step errors are computed as the difference between the total gain in flow and the total change in storage of the watershed. This difference also is reported as a percent discrepancy, according to the following equation:

$$D = 100 \frac{(in - out)\Delta t - |\Delta storage|}{((in + out)\Delta t + |\Delta storage|)/2} ,$$ (114)

where

D is the percent discrepancy, dimensionless;
in is the total inflow to the model, in cubic length per time;
out is the total outflow from the model, in cubic length per time; and
$\Delta storage$ is the total storage change, in cubic length.

The errors, along with their GSFLOW budget report names, are summarized in table 11.

Computation of Budget Errors

GSFLOW also produces output for detailed analysis of volumetric budget errors that is written to a user-specified GSFLOW Comma-Separated-Values File. The name of this file is specified using parameter csv_output_file in the GSFLOW Control File. It includes storage and flows for each major storage zone of GSFLOW for each time step. These variables and their description are listed in table 12. Although this file may be useful for exporting GSFLOW output data to other programs, it is specifically designed to be imported into a spreadsheet for detailed analysis. The units for each value in this file are in the MODFLOW unit of cubic length (as specified in the Discretization File) for volumes and the MODFLOW unit of cubic length per day for volumetric rates.

Table 12.   Names and descriptions of variables written to the GSFLOW Comma-Separated-Values File at end of each daily time step for GSFLOW.

[**Abbreviations:** L^3/T, cubic length per time; L^3, cubic length]

Variable name	Description	Units
Date	Month, day, and year designation	MONTH/DAY/YEAR
basinppt	Volumetric flow rate of precipitation on modeled region	L^3/T
basinpervet	Volumetric flow rate of evapotranspiration from pervious areas	L^3/T
basinimpervevap	Volumetric flow rate of evaporation from impervious areas	L^3/T
basinintcpevap	Volumetric flow rate of evaporation of intercepted precipitation	L^3/T
basinsnowevap	Volumetric flow rate of snowpack sublimation	L^3/T
basinstrmflow	Volumetric flow rate of streamflow leaving modeled region	L^3/T
basinsz2gw	Potential volumetric flow rate of gravity drainage from the soil zone to the unsaturated zone (before conditions of the unsaturated and saturated zones are applied)	L^3/T
basingw2sz	Volumetric flow rate of ground-water discharge from the saturated zone to the soil zone	L^3/T
gw_inout	Volumetric flow rate to saturated zone along external boundary (negative value is flow out of modeled region)	L^3/T
stream_leakage	Volumetric flow rate of stream leakage to the unsaturated and saturated zones	L^3/T
uzf_recharge	Volumetric flow rate of recharge from the unsaturated zone to the saturated zone	L^3/T
basinseepout	Volumetric flow rate of ground-water discharge from the saturated zone to the soil zone	L^3/T
sat_stor	Volume of water in the saturated zone	L^3
unsat_stor	Volume of water in the unsaturated zone	L^3
basinsoilmoist	Volume of water in capillary reservoirs of the soil zone	L^3
basingravstor	Volume of water in gravity reservoirs of the soil zone	L^3
basingwstor	Volume of water in PRMS ground-water reservoirs (PRMS-only simulation)	L^3
basinintcpstor	Volume of intercepted precipitation in plant-canopy reservoirs	L^3
basinimpervstor	Volume of water in impervious reservoir	L^3
basinpweqv	Volume of water in snowpack storage	L^3
basininterflow	Volumetric flow rate of slow interflow to streams	L^3/T
basinsroff	Volumetric flow rate of the sum of Hortonian (Horton, 1933) and Dunnian surface runoff (Dunne and Black, 1970) to streams	L^3/T
strm_stor	Volume of water in streams	L^3
lake_stor	Volume of water in lakes	L^3
obs_strmflow	Volumetric flow rate of streamflow measured at a gaging station	L^3/T
basinszreject	Volumetric flow rate of gravity drainage from the soil zone not accepted due to the conditions in the unsaturated and saturated zones	L^3/T
basinprefstor	Volume of water stores in preferential-flow reservoirs of the soil zone	L^3
uzf_et	Volumetric flow rate of evapotranspiration from the unsaturated and saturated zones	L^3/T
uzf_infil	Volumetric flow rate of gravity drainage to the unsaturated and saturated zones	L^3/T
uzf_del_stor	Change in unsaturated-zone storage	L^3
net_sz2gw	Net volumetric flow rate of gravity drainage from the soilzone to the unsaturated and saturated zones	L^3/T
sat_change_stor	Change in saturated-zone storage	L^3
streambed_loss	Volumetric flow rate of stream leakage to unsaturated and saturated zones	L^3/T
sfruz_change_store	Change in unsaturated-zone storage under streams	L^3
gwflow2strms	Volumetric flow rate of ground-water discharge to streams	L^3/T
sfruz_tot_stor	Volume of water in the unsaturated zone	L^3
lakebed_loss	Volumetric flow rate of lake leakage to the unsaturated and saturated zones	L^3/T
lake_change_stor	Change in lake storage	L^3
gwflow2lakes	Volumetric flow rate of ground-water discharge to lakes	L^3/T
basininfil	Volumetric flow rate of soil infiltration including precipitation, snowmelt, and cascading Hortonian flow	L^3/T

Table 12. Names and descriptions of variables written to the GSFLOW Comma-Separated-Values File at end of each daily time step for GSFLOW.—Continued

[**Abbreviations:** L^3/T, cubic length per time; L^3, cubic length]

Variable name	Description	Units
basindunnian	Volumetric flow rate of Dunnian runoff to streams	L^3/T
basinhortonian	Volumetric flow rate of Hortonian runoff to streams	L^3/T
basinsm2gvr	Volumetric flow rate of flow from capillary reservoirs to gravity reservoirs	L^3/T
basingvr2sm	Volumetric flow rate of replenishment of capillary reservoirs from gravity reservoirs	L^3/T
basininfil_tot	Volumetric flow rate of soil infiltration into capillary reservoirs including precipitation, snowmelt, and cascading Hortonian and Dunnian runoff and interflow minus infiltration to preferential-flow reservoirs	L^3/T
basininfil2pref	Volumetric flow rate of soil infiltration into preferential-flow reservoirs including precipitation, snowmelt, and cascading surface runoff	L^3/T
basindnflow	Volumetric flow rate of cascading Dunnian runoff and interflow to HRUs	L^3/T
basinactet	Volumetric flow rate of actual evapotranspiration from HRUs	L^3/T
basinsnowmelt	Volumetric flow rate of snowmelt	L^3/T
basinhortonianlakes	Volumetric flow rate of Hortonian runoff to lakes	L^3/T
basinlakeinsz	Volumetric flow rate of Dunnian runoff and interflow to lakes	L^3/T
basinlakeevap	Volumetric flow rate of evaporation from lakes	L^3/T
basinlakeprecip	Volumetric flow rate of precipitation on lakes	L^3/T
kkiter	Number of iterations for each time step	none

A pre-programmed Excel spreadsheet included with GSFLOW (gsflowAnalysis.xls) contains water-budget calculations and errors at the end of each daily time step for the major storage zones. Volumetric water-budgets and errors at the end of each daily time step for each of the major storage zones are computed from the continuity equation in the form:

$$\Delta s_t = s_t - s_{t-1} \, , \tag{115a}$$

$$e_t = \Delta s_t + \frac{out_t - in_t}{\Delta t} \, , \tag{115b}$$

where

Δs_t the total storage change of the time step ending at time t, in cubic length;

s_t is the storage at time t, in cubic length;

s_{t-1} is the storage at the previous time step, in cubic length;

e_t is the budget error at time t, in cubic length per time;

in_t is the total inflow at time t, in cubic length per time; and

out_t is the total outflow at time t, in cubic length per time;

Δt is one day.

The water budget and error of the surface and near surface storage zone computed by HRUs are determined when the following GSFLOW variables are substituted for terms in equations 115a-d to compute the change in flow and storage into and out of the HRUs:

$$s_t = sm_t + gz_t + ic_t + im_t + pw_t,$$
(116a)

$$s_{t-1} = sm_{t-1} + gz_{t-1} + ic_{t-1} + im_{t-1} + pw_{t-1},$$
(116b)

$$in_t = p_t + ex_t \text{, and}$$
(116c)

$$out_t = ep_t + eim_t + eic_t + esp_t + qif_t + qso_t,$$
(116d)

where

sm_t and sm_{t-1} are the volumes of water in capillary reservoirs of the soil zone (variable `basinsoilmoist`) at time t and $t-1$, respectively, in cubic length;

gz_t and gz_{t-1} are the volumes of water in gravity reservoirs of the soil zone (variable `basingravstor`) at time t and $t-1$, respectively, in cubic length;

ic_t and ic_{t-1} are the volumes of intercepted precipitation in plant canopy reservoirs (variable `basinintcpstor` at time t and $t-1$, respectively, in cubic length;

im_t and im_{t-1} are the volumes of water in impervious reservoir (variable `basinimpervstor`) at time t and $t-1$, respectively, in cubic length;

pw_t and pw_{t-1} are the volumes of water in snowpack storage (variable `basinpweqv`) at time t and $t-1$, respectively, in cubic length;

p_t is the volumetric flow rate of precipitation on modeled region (variable `basinppt`) at time t, in cubic length per time;

ex_t is the volumetric flow rate of ground-water discharge from saturated zone to the soil zone (variable `basingw2sz`) at time t, in cubic length per time;

ep_t is the volumetric flow rate of evapotranspiration from pervious areas (variable `basinpervet`) at time t, in cubic length per time;

eim_t is the volumetric flow rate of evaporation from impervious areas (variable `basinimpervevap`) at time t, in cubic length per time;

eic_t is the volumetric flow rate of evaporation of intercepted precipitation (variable `basinintcpevap`) at time t, in cubic length per time;

esp_t is the volumetric flow rate of snowpack sublimation (variable `basinsnowevap`) at time t, in cubic length per time;

qif_t is the volumetric flow rate of interflow leaving modeled region (variable `basininterflow`) at time t, in cubic length per time; and

qso_t is the volumetric flow rate of surface runoff leaving the modeled region (variable `basinsroff`) at time t, in cubic length per time.

A water budget error is computed for streams by summing the change in storage, inflows and outflows for all streams. The total change in storage in streams is estimated by subtracting the volume of water in streams (variable `strm_stor`) at the end of the current time step from that at the end of the previous time step. The total inflow to streams is computed by summing interflow (variable `basininterflow`), surface runoff (variable `basinsroff`), ground-water discharge (variable `gwflow2strms`), and any specified inflows at the end of each time step. The total outflow from streams is computed by summing stream leakage (variable `streambed_loss`) and outflow from streams that leave the watershed (variable `basinstrmflow`) at the end of each time step.

All or some of the gravity drainage from the soil zone (variable `basinsz2gw`) may be rejected (variable `basinszreject`) when the ground-water level is above an altitude of the soil-zone base less half the undulation depth. The difference between `basinsz2gw` and `basinszreject` should equal `uzf_infil`. The water budget error for the unsaturated zone is computed by summing the storage change (variable `uzf_del_stor`), inflow (variable `uzf_infil`), and outflow (variables `uzf_et` and `uzf_recharge`) for the unsaturated zone beneath the soil zone.

Assumptions and Limitations

Several assumptions and limitations are associated with each of the individual modules and packages within GSFLOW. Additionally, similar to other non-linear models, GSFLOW is limited by the possibility of non-convergences among any or all coupled dependent variables. Non-convergence of GSFLOW can be caused by specifying inappropriate input data or parameters. The number of iterations used in a GSFLOW simulation can increase when there are large changes in the number of finite-difference cells that discharge ground water into the soil zone. Other causes not discussed nor revealed through model testing also may result in an increase in the number of iterations.

Discretization of Time and Space

Because GSFLOW has a computational time step of 1 day, all flows and storages are mean daily values. In general, flows near land surface occur more abruptly than flows in the subsurface. Consequently, the daily time step will result in errors due to time averaging for near-land surface flows such as surface runoff, infiltration, and interflow, and streamflow and streambed leakage. However, simulating the sub-daily behavior of these processes is not warranted because surface runoff, infiltration, and interflow are handled conceptually in this initial version of GSFLOW. Sub-daily time steps

could be incorporated into future versions of GSFLOW, and alternate governing equations for surface runoff, interflow, and infiltration that apply at sub-daily time steps could be used, such as those used for PRMS storm mode (Leavesley and others, 1983).

Although streamflow can be routed during sub-daily time steps, lateral flow to streams is calculated on a daily time step, which results in daily-averaged hydrographs. Errors can occur in streams that have hydrographs that change abruptly over time steps much less than 1 day. This is especially true if, in actuality, a stream floodplain is inundated for a few hours but the simulated flow is unrealistically contained in the active channel due to daily averaging of the peak flow. Similarly, leakage errors can occur during sub-daily hydrographs when the flows and leakages are averaged over a daily time step.

On the basis of previous ground-water modeling studies and the test simulations presented herein, 1-day time steps are sufficiently small to simulate most ground-water flow conditions. Because the equations used to simulate unsaturated flow beneath the soil zone are analytical, other than the daily-averaged infiltration rate, their solution is unaffected by the 1-day time step used for GSFLOW. Sub-daily fluctuations in infiltration rate at land surface generally would be dampened by the soil zone. Consequently, daily changes in gravity drainage from the soil zone likely are realistic for simulating deeper unsaturated flow.

Ideally, the size of finite-difference cells are based on the spatial resolution required for providing model results that can properly address the purpose of the model and available computer resources. However, other criterion can constrain the size of finite-difference cells. For example, a large cell width relative to the width of a stream that flows over a cell can result in model errors and misrepresentation of ground-water interaction with a stream, especially if other sources or sinks are present in cells adjacent to the stream. Alternately, the model may not converge if a cell width is equal to or less than the width of a stream. Similar conditions can occur for other MODFLOW-2005 packages and the associated documentation report for each package provides more details regarding additional constraints on cell sizes.

Another constraint on the size of finite-difference cells applies when simulating unconfined aquifers. The model may not converge if the thickness of a model layer representing an unconfined aquifer is thin relative to changes in the water-table altitude. If the water table fluctuates above and beneath the bottom of a finite-difference cell then the cell will alternate between being a no-flow cell and a variable-head cell during the iteration loop and this condition can result in non-convergence of the model.

The delineation of the HRUs must be appropriate for the module algorithms. In PRMS, each HRU is assumed to be homogeneous with respect to both input parameters and computation of flows and storages, similar to a finite-difference cell.

Canopy Zone

Incomplete and dated sources of land-use and land-cover information can cause errors in GSFLOW estimates of precipitation interception and transmission of shortwave solar radiation. GSFLOW does not have a plant growth module so plant cover estimates, particularly in agricultural areas, may be problematic.

Land-Surface Precipitation and Temperature

The PRMS land-surface modules are susceptible to errors in the input time series of measured precipitation and temperature. These errors can have serious consequences on simulation of both the volume and timing of water-balance components. These errors generally are of three forms: (1) uncertainty in measurement; (2) errors in distribution of precipitation and temperature; and (3) errors in determination of the form of precipitation. Distribution errors generally will result in annual or seasonal biases, whereas precipitation-form errors generally affect individual events.

Soil Zone

The soil zone is represented as a series of reservoirs that store and transfer water among different reservoirs within the soil zone and to various surface and ground-water reservoirs (including MODFLOW finite-difference cells). Equations are used to relate the transfer rate from a reservoir in the soil zone to storage in the reservoir. The main assumption used in the soil zone is that flow into and out and changes in storage occur uniformly throughout an HRU during each daily time step. This means the parameters that define flow and storage in the soil zone remain constant in each HRU during a simulation. Thus, the delineation of HRUs is critical to a simulation in that it defines the flow into and out of a soil zone and affects gravity drainage to the unsaturated zone, ground-water discharge from the saturated zone, and runoff and interflow to streams and lakes.

Other important assumptions used to define the different reservoirs within the soil zone also may limit applicability. An important assumption is that freezing and thawing of the soil zone does not change parameters that affect flow and storage in the soil zone during a simulation. Another important assumption is that the capillary reservoir is represented by a constant value of root depth for each HRU during a simulation. Root depth is determined on the basis of the principal plant type within each HRU. Because plants grow and die, the size and types of plants in an HRU vary with time, and the root depth as well as evapotranspiration from an HRU also may vary in space and time. Finally, another assumption is that perturbations such as animal burrows, decaying roots, and leaf litter that produce preferential flow remains constant within each HRU during a simulation.

Streams

Numerous assumptions have been made to simplify the Streamflow-Routing Package. First, flow is one-dimensional in the direction of the channel. Furthermore, only the channel slope and friction loss terms in the momentum equation are represented such that both kinematic and dynamic waves are neglected for the steady-flow option, and dynamic waves are neglected in the transient-flow option. It is assumed that momentum can be represented by the Manning's equation for steady flow, and that stream depth corresponds to normal depth (Streeter and Wylie, 1985; p. 470-471). It also is assumed that the stage-discharge relation can be represented by a single-valued function. Backwater effects are not considered, and flow occurs in one direction only. Steady streamflow routing (steady-uniform flow) assumes piecewise steady (non-changing in discrete time periods), uniform (non-changing in location), and constant-density streamflow, such that during all times, volumetric inflow and outflow rates are equal and no water is added to or removed from storage in the surface channels. Unsteady routing using the kinematic-wave equation assumes flow disturbances only can move in the downstream direction with no hydrograph attenuation (diffusion of flood waves is neglected) and that energy provided by the channel slope is balanced by frictional losses along the channel bottom. Finally, flow is assumed to be turbulent.

In addition to assumptions regarding flow in stream channels, assumptions are made with regard to streambed leakage. Because this leakage is calculated on the basis of the Dupuit assumption (Anderson, 2005), head loss is neglected between the top of the streambed and the point where ground-water head is computed beneath the stream. Consequently, errors in calculated leakage can occur if the streambed conductance values do not include sufficient resistance to reflect the effects of vertical head loss between streams and aquifers. Leakage through streambeds that are not in direct contact with ground water is represented assuming a unit hydraulic gradient beneath the streambed, and leakage occurs only in the vertical direction. Finally, where streams are small relative to the finite-difference cell, substantial errors in leakage can occur if the streambed conductance is not sufficiently small.

Lakes

A lake is represented by a completely mixed volume of water within the Lake Package. Because water is not routed within lakes, water entering a lake is assumed to be instantaneously mixed with the existing water in the lake. Multiple lakes can combine into a single lake, and when this occurs the combined lakes are represented by an instantaneously mixed volume of water. An important assumption is that the lake occupies the volume of finite-difference cells within the MODFLOW grid. Thus, the position and spatial extent of the lake area is equal to the area of inundated no-flow cells used to represent the lake defined in the MODFLOW Discretization File. Consequently, cells used to represent a lake should be discretized appropriately to realistically represent the area and volume of a lake.

Similar to the Streamflow-Routing Package, leakage between lakes and ground water is calculated on the basis of the Dupuit assumption. Consequently, head loss between the top of the lakebed and the point where the ground-water head is calculated beneath the lake cell is neglected.

Ground Water

Assumptions associated with GSFLOW for simulating ground-water flow are standard assumptions corresponding to the saturated ground-water flow equation. These assumptions are that ground-water flow is laminar and inertial forces, velocity heads, temperature gradients, osmotic gradients, and chemical concentration gradients are not considered. Soils and geologic formations are assumed to be linearly and reversibly elastic and mechanically isotropic; horizontal components of deformation are assumed small compared to vertical deformations; and ground water is instantaneously released from or taken into storage.

Additional assumptions are made in the discretization of the ground-water system into a finite-difference grid. These assumptions are that the principal directions of anisotropy are assumed to be aligned with the coordinate directions of the grid, and each model layer corresponds to a distinct aquifer or permeable horizon. The finite-difference approximations of the ground-water flow equation in MODFLOW were derived assuming that the grid is rectangular; however, grids can be non-rectangular in the vertical direction (Harbaugh, 2005, p. 3-15) resulting in some errors in the computed ground-water conditions. Because hydraulic properties are uniform within individual cells, averages or integrated properties are assumed to represent the aquifer material contained in a single finite-difference cell. Flow is parallel to the coordinate axis within a finite-difference cell, nodes are at the center of cells, and discrete changes in hydraulic conductivity occur at cell boundaries. Simulations involving implicit confining units (Quasi-Three-Dimensional approach) assume the confining unit makes no measurable contribution to the horizontal conductance or the storage capacity of either model layer (Bredehoeft and Pinder, 1970). The only effect of an implicit confining unit is to restrict vertical flow between finite-difference cells.

Similar to ground-water discharge to streams, ground-water discharge to the soil zone is simulated using the Dupuit assumption and vertical head loss is neglected. Errors caused by the Dupuit assumption increase as the vertical and horizontal dimensions of the finite-difference cell increase. Recharge from the soil zone directly to ground water decreases when the water table rises into the soil zone. If the water table is at or above the soil-zone base, evapotranspiration loss from ground water occurs at the net potential ET rate. If the depth of the water table is beneath the evapotranspiration extinction depth, evapotranspiration from the water table ceases; and between these limits, evapotranspiration from the water table varies linearly with water-table altitude.

Unsaturated Zone

Flow in the unsaturated zone is simulated by the Unsaturated-Zone Flow Package that solves a vertical, one-dimensional form of Richards' equation and neglects capillary pressure gradients. An assumption is that the unsaturated zone is homogeneous in the vertical direction or else meaningful averages or integrated properties can be specified for each vertical column of unsaturated material. Because capillary pressure gradients are neglected, the capillary fringe is not simulated and evapotranspiration in the unsaturated zone cannot cause upward flow. Sediment in the unsaturated zone cannot swell or deform in any direction. Evapotranspiration is limited only by the evapotranspiration extinction depth and evapotranspiration extinction water content.

Input and Output Files

Data input for a GSFLOW simulation is specified in several files that must be prepared by the model user prior to a simulation. Outputs from a GSFLOW simulation are written to several files whose names, and in some cases whose formats, also are specified by the user as part of the simulation input data. The input and output files needed for a GSFLOW simulation are described in this section.

Input Files

Several input files are needed for a GSFLOW simulation. These files are the GSFLOW Control File, the PRMS Data File, the PRMS Parameter File, the MODFLOW Name File, and the input files for each MODFLOW-2005 package specified for use in the Name File. Because GSFLOW is based on two different models, the input format for each file depends on the genesis of the file. The format of the GSFLOW Control File is based on Modular Modeling System input structures, as are the PRMS Data File and PRMS Parameter File. Input files for each MODFLOW-2005 package are based on specifications defined for each package. Brief descriptions of each input file follow; detailed descriptions of the input data required for each input file are provided in appendix 1.

GSFLOW Control File

The GSFLOW Control File is used to set basic administrative data values related to the PRMS and GSFLOW modules for a GSFLOW simulation, such as specification of model input and output file names and simulation starting and ending dates. There are four types of parameters that are specified in the file: (1) those related to model execution; (2) those related to model input; (3) those related to model output; and (4) those related to initial conditions for the PRMS model.

The name of the GSFLOW Control File can be specified on the command line of a GSFLOW execution. For example, if the GSFLOW executable is named 'gsflow.exe' and the name of the Control File is 'sagehen.control' then the command line to execute the code would be:

gsflow.exe sagehen.control

If no control-file name is given after gsflow.exe, GSFLOW will look for a Control File named gsflow.control in the user's current directory.

PRMS Data File

The PRMS Data File is used to define available measured or simulated climate and streamflow data for use in a GSFLOW simulation. These data are specified as a time-series that defines the time extent of possible simulations. The forcing functions of the GSFLOW simulation—precipitation, maximum and minimum air temperature, and solar radiation—are specified in the file, as well as the form of precipitation (unknown, snow, or rain), pan evaporation rates, and measured streamflow. The name of the Data File is set using parameter data_file in the GSFLOW Control File. Time-series of daily precipitation and maximum and minimum air temperature are required for a GSFLOW simulation; additional data are optional.

PRMS Parameter File

The majority of input parameters used by the PRMS and GSFLOW modules are specified in the PRMS Parameter File. The file also is used to specify the dimensions of several PRMS and GSFLOW parameters, such as the number of measured precipitation, air temperature, and streamflow-gaging stations for which data are specified in the PRMS Data File and for spatial parameters, such as the number of hydrologic response units (HRUs) and stream segments. Input instructions for the PRMS Parameter File are provided in appendix 1. The name of the Parameter File is set using parameter param_file in the GSFLOW Control File.

MODFLOW-2005 Input Files

The MODFLOW-2005 packages input files consist of data values that are related to the saturated and unsaturated subsurface zones of a GSFLOW simulation (that is, beneath the soil zone), as well as the stream and lake components of a simulation, and solver and output control input files. Unlike the PRMS modules, in which most model-input data are specified in a single file (the PRMS Parameter File), MODFLOW-2005 packages require model input to be gathered from different files. For each of the MODFLOW-2005 packages that are used in a GSFLOW simulation, there are one or more corresponding input files. Table 1 summarizes the MODFLOW-2005 packages that can be used with GSFLOW; specific input instructions for each input file are provided in appendix 1.

The MODFLOW Name File is used to specify the input and output files associated with each MODFLOW-2005 package that will be used in a GSFLOW simulation. Each line in the Name File defines one file and specifies a file type (Ftype), unit number (Nunit), and name (Fname) as shown in figure 30. The parameter Ftype specifies the package or the purpose of a file. Values of Ftype correspond to a major MODFLOW option that is activated with the appropriate line in the Name File. Most packages are major options and require an input file, so, in general, each package is activated by at least one line in the Name File. For example, if a well is simulated, a file having a file type of "WEL" is required in the Name File, and an input file in the Well Package input format is required. Values of Ftype are character strings that can be a package acronym (for example, "BAS6", "UZF", or "SFR") or a functional description (for example, "LIST" or "DATA") and can be specified using uppercase or lowercase letters. Nunit is an integer value used to specify the Fortran unit number used to read from or write to the named file. Fname is a character string that identifies the file either as a full pathname or relative pathname (a relative pathname is specified from the user's current directory). For example, if a well is simulated, a line in the Name File must specify Ftype equal to "WEL", an integer value for Nunit that specifies a valid Fortran unit number, and a character string, Fname, identifying the name and location of the Well package input file.

A MODFLOW-2005 package is inactive if the Name File does not include a line having the option's Ftype. An option can be turned off either by removing the appropriate line from the Name File or by placing a pound sign ("#") at the beginning of a line. The MODFLOW-2005 Input Instructions section in appendix 1 defines the available values for Ftype and Nunit. The Name File must include definitions for two files required for each simulation involving MODFLOW-2005, the Basic Package File (BAS6) and the Discretization File (DIS). The Basic Package File specifies, among other values,

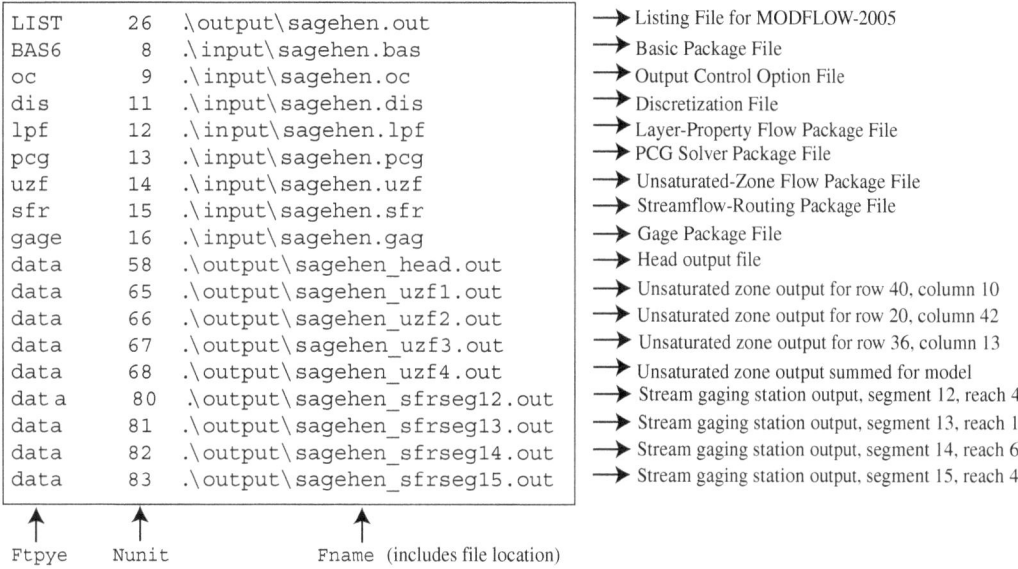

```
LIST      26    .\output\sagehen.out
BAS6       8    .\input\sagehen.bas
oc         9    .\input\sagehen.oc
dis       11    .\input\sagehen.dis
lpf       12    .\input\sagehen.lpf
pcg       13    .\input\sagehen.pcg
uzf       14    .\input\sagehen.uzf
sfr       15    .\input\sagehen.sfr
gage      16    .\input\sagehen.gag
data      58    .\output\sagehen_head.out
data      65    .\output\sagehen_uzf1.out
data      66    .\output\sagehen_uzf2.out
data      67    .\output\sagehen_uzf3.out
data      68    .\output\sagehen_uzf4.out
data      80    .\output\sagehen_sfrseg12.out
data      81    .\output\sagehen_sfrseg13.out
data      82    .\output\sagehen_sfrseg14.out
data      83    .\output\sagehen_sfrseg15.out
```

→ Listing File for MODFLOW-2005
→ Basic Package File
→ Output Control Option File
→ Discretization File
→ Layer-Property Flow Package File
→ PCG Solver Package File
→ Unsaturated-Zone Flow Package File
→ Streamflow-Routing Package File
→ Gage Package File
→ Head output file
→ Unsaturated zone output for row 40, column 10
→ Unsaturated zone output for row 20, column 42
→ Unsaturated zone output for row 36, column 13
→ Unsaturated zone output summed for model
→ Stream gaging station output, segment 12, reach 4
→ Stream gaging station output, segment 13, reach 11
→ Stream gaging station output, segment 14, reach 6
→ Stream gaging station output, segment 15, reach 4

Ftpye   Nunit   Fname (includes file location)

Figure 30. MODFLOW Name File used for the Sagehen Creek watershed example problem.

which finite-difference cells are variable-head in the model domain and the initial heads in each finite-difference cell. The Discretization File specifies data values such as the dimensions of the MODFLOW finite-difference grid (number of layers, rows, and columns), data units, and altitude of each cell in the top layer of the model. An example Name File for the Sagehen Creek watershed example problem described in section "Example Problem" is shown in figure 30. The MODFLOW-2005 Input Instructions in appendix 1 define all file types that can be used in a GSFLOW simulation.

In addition to specifying input data by way of the actual file, MODFLOW-2005 also allows model input to be entered from external formatted or unformatted files. Values of Ftype specified in the Name File as "DATA" and "DATA(BINARY)" (in either upper or lower case) define input or output files whose file unit numbers, Nunit, can be referenced by other MODFLOW-2005 packages. "DATA" file types are for formatted (text) files, whereas "DATA(BINARY)" file types are for binary (unformatted) files. The "DATA" and "DATA(BINARY)" file types can be specified for many files. An example is the input of an array of variables, such as hydraulic conductivity, by row and column grid indices. When using external files for model input, the model input values will not be read from the package input files themselves. Rather, the package file specifies that the input values must be read from a different file by specifying the Fortran unit number listed in the Name File for that input. For example, the values for the top altitude can be specified directly in the Discretization File (that is, in the file whose FTYPE is

"DIS"), or a file-unit flag can be set in the Discretization File to indicate that the values for the top altitude should be read from a file defined in the Name File with Ftype "DATA" or "DATA(BINARY)."

The full pathname of the MODFLOW Name File, meaning the directory location and name of the Name File, is specified in the GSFLOW Control File using parameter modflow_name. Similar to the original MODFLOW-2005 construction, the Name File pathname also can be specified in a MODFLOW "Batch File" that consists of a single line that specifies the Name File pathname beginning in column 1 and having no spaces (pathnames that include spaces must be quoted, for example "C:\Program Files\gsflow\data\sagehen.nam" or "C: \Documents and Setting\userid\gsflow\data\sagehen\sagehen.nam"). If a file named modflow.bf is found by GSFLOW in the user's current directory, it will be used as the MODFLOW Batch File. Otherwise, the value of the control parameter modflow_name is used; the default value is modflow.nam. For example, if a GSFLOW simulation is initiated from the user's current directory that contains a file named modflow.bf, GSFLOW will read the first line to define the pathname of the MODFLOW Name File. For this execution scenario and for the Sagehen Creek watershed example problem described in this report that uses the Name File named sagehen.nam, the MODFLOW Batch File consists of the following line:

./sagehen.nam

Output Files

Several output files can be generated during a GSFLOW simulation. The following output file descriptions are organized by model source—GSFLOW, PRMS, then MODFLOW.

GSFLOW

Two GSFLOW-specific output files can be optionally generated during a simulation. The first, the GSFLOW Water-Budget File, contains tabular summaries of the water budget computed by GSFLOW at user-specified frequency. The second, the GSFLOW Comma-Separated-Values (CSV) File, contains simulation results for each time step; these can be used to compute water budgets for each GSFLOW storage reservoir and a mass balance for the simulation. A third non-optional GSFLOW output file, named the GSFLOW Log File or gsflow.log, is generated in the user's current directory during a simulation. This Log File contains information that describes the GSFLOW model, any warning or error messages, simulation time period, execution time, and iteration statistics for each time step and simulation, and any information that was written to the user's computer display.

The GSFLOW Water-Budget File (fig. 29) is a text file that provides summary tables of flows into and out of the watershed and storage changes within the watershed by zones (land surface, soil, unsaturated, saturated). The name of the file is specified by parameter gsflow_output_file in the GSFLOW Control File. The frequency at which the tables are generated is specified in the GSFLOW Control File using parameter rpt_days, which defaults to every 7 days within the simulation time period beginning with the table generated for simulation day 7. The first line of each summary table in the output file provides the calendar date and corresponding time step of the simulation. The next line provides the MODFLOW stress period and time step within the stress period. The remainder of each summary table is divided into four columns. The last two columns list budget terms for the current time step in units of volumetric flow rate per time (that is, flow rates). The first two columns list cumulative flows (units of volume) for all time steps up to and including the current time step. All values are reported in the MODFLOW cubic length unit. Details of the GSFLOW water budget are described in section "Overall Water Budget."

The GSFLOW Comma-Separated-Values (CSV) File is a text file that provides GSFLOW simulation results in a format that can be easily imported to spreadsheet programs. Figure 31 shows the first three lines of the CSV File generated for the Sagehen Creek watershed example problem described in this report. Each line shown in figure 31 is wrapped over several lines to stay within the margins of the page. Values included in this file can be used to compute the budgets specified in

section "Water Budgets." The file is generated when parameter gsf_rpt is set to 1 in the GSFLOW Control File; the name of the file is specified by parameter csv_output_file in the GSFLOW Control File, which defaults to the name gsflow.csv. The first line of the file (the header) lists the names of each output variable in the sequence for which values are included in each subsequent line; these names can be used as spreadsheet column headers. The remaining lines of the file provide the simulation time-step date as a two-digit month, a two-digit day, and a four-digit year, each separated by a "/" or in abbreviated form as MM/DD/YYYY. The date is followed by model-calculated values for each variable defined in the header line. All values are separated by commas. One line of values is written for each simulation time step.

PRMS

Three PRMS output files can be generated as part of a GSFLOW simulation: the PRMS Water-Budget File of summary tables generated at a user-specified frequency, the PRMS Statistic Variables File of user-selected variable values for each time step, and the PRMS Animation Variables File of user-selected array variables for each time step.

The PRMS Water-Budget File (fig. 32) provides summary table(s) of the water budget for a PRMS simulation. The pathname of the Water-Budget File is specified by parameter model_output_file in the GSFLOW Control File. Three types of summary tables are available, depending on the value specified for parameter print_type in the PRMS Parameter File. The first is a listing of the observed and predicted flow, which is generated when print_type is set to 0. The second is a table of water-balance computations and includes the watershed-weighted averages for net precipitation, evapotranspiration from all sources, storage in all reservoirs, and the simulated (P-Runoff) and measured (O-Runoff) flows. This report is generated when print_type is set to 1. The third is a detailed summary of the rainfall, outflow, and state variables, and is generated when print_type is set to 2. Any of the summaries may be requested in any combination of the available time increments—daily, monthly, yearly, or total for the simulation. The frequency of output is specified by parameter print_freq in the PRMS Parameter File.

The PRMS Statistic Variables File (fig. 33) is a text file that provides PRMS model output that can be used with visualization and statistics programs. The file is generated when parameter statsON_OFF is set to 1 in the GSFLOW Control File. The name of the file is set by control parameter stat_var_file. The first line of the file is the number of variable values that are written in the file; this value is specified using control parameter nstatVars. The next group of lines (nstatVars in number) lists the names and array index of each output variable; the output variables

```
Date,basinppt,basinpervet,basinimpervevap,basinintcpevap,
basinsnowevap,basinstrmflow,basinsz2gw,basingw2sz,gw_inout,
stream_leakage,uzf_recharge,basinseepout,sat_stor,unsat_stor,
basinsoilmoist,basingravstor,basingwstor,basinintcpstor,
basinimpervstor,basinpweqv,basininterflow,basinsroff,strm_stor,
lake_stor,obs_strmflow,basinszreject,basinprefstor,uzf_et,
uzf_infil,uzf_del_stor,net_sz2gw,sat_change_stor,streambed_loss,
sfruz_change_stor,gwflow2strms,sfruz_tot_stor,lakebed_loss,
lake_change_stor,gwflow2lakes,basininfil,basindunnian,
basinhortonian,basinsm2gvr,basingvr2sm,basininfil_tot,
basininfil2pref,basindnflow,basinactet,basinsnowmelt,
basinhortonianlakes,basinlakeinsz,basinlakeevap,basinlakeprecip,
kkiter
    10/01/1980,  0.0000000E+00,   0.7730693E+04,   0.0000000E+00,
0.0000000E+00,  0.0000000E+00,   0.6287971E+04,   0.0000000E+00,
0.1590413E+05, -0.6488422E+03,  -0.6287971E+04,   0.2106252E+05,
0.1590659E+05,  0.1638903E+09,   0.9435054E+07,   0.7787383E+05,
0.1703915E+02,  0.0000000E+00,   0.0000000E+00,   0.0000000E+00,
0.0000000E+00,  0.0000000E+00,   0.0000000E+00,   0.7615586E+03,
0.0000000E+00,  0.5627124E+04,   0.0000000E+00,   0.0000000E+00,
0.0000000E+00,  0.0000000E+00,  -0.2102739E+05,   0.0000000E+00,
-0.1760261E+04,  0.6480000E+02,   0.0000000E+00,   0.6505124E+04,
0.5966707E+02,  0.0000000E+00,   0.0000000E+00,   0.0000000E+00,
0.0000000E+00,  0.0000000E+00,   0.0000000E+00,   0.0000000E+00,
0.1588709E+05,  0.0000000E+00,   0.0000000E+00,   0.0000000E+00,
0.7730693E+04,  0.0000000E+00,   0.0000000E+00,   0.0000000E+00,
0.0000000E+00,  0.0000000E+00,   7
    10/02/1980,  0.0000000E+00,   0.8015101E+04,   0.0000000E+00,
0.0000000E+00,  0.0000000E+00,   0.6194819E+04,   0.0000000E+00,
0.1584995E+05, -0.6488286E+03,  -0.6194818E+04,   0.2092071E+05,
0.1585241E+05,  0.1638886E+09,   0.9414168E+07,   0.8572385E+05,
0.1839345E+01,  0.0000000E+00,   0.0000000E+00,   0.0000000E+00,
0.0000000E+00,  0.0000000E+00,   0.0000000E+00,   0.7564084E+03,
0.0000000E+00,  0.5627124E+04,   0.0000000E+00,   0.0000000E+00,
0.0000000E+00,  0.0000000E+00,  -0.2088586E+05,   0.0000000E+00,
-0.1743168E+04,  0.6480000E+02,   0.0000000E+00,   0.6489369E+04,
0.5966872E+02,  0.0000000E+00,   0.0000000E+00,   0.0000000E+00,
0.0000000E+00,  0.0000000E+00,   0.0000000E+00,   0.0000000E+00,
0.1586515E+05,  0.0000000E+00,   0.0000000E+00,   0.0000000E+00,
0.8015101E+04,  0.0000000E+00,   0.0000000E+00,   0.0000000E+00,
0.0000000E+00,  0.0000000E+00,   6
```

Header line

First data line

Second data line

A data line is added for each one-day time step

Figure 31. Three lines of a GSFLOW Comma-Separated-Values (CSV) File that shows variables that can be used for water-budget calculations for each time step.

that are listed are specified using control parameter statVar_name. The array index for variables that are scalar, meaning a single value, is set to 1. The remaining lines provide the model-calculated values of each variable for each simulation time step. These data lines have the following order: model time-step number, year, month, day of month, hour, minute, second, and each variable value in the order specified by the list of variable names. Each value is separated by a space.

The PRMS Animation File (fig. 34) is a text file that provides PRMS model results as a time-series of spatial arrays that can be used with animation programs. The file is generated when parameter aniOutON_OFF is set to 1

in the GSFLOW Control File. The name of the file is set by parameter ani_output_file. The first group of lines in the file, starting with pound characters (#) describes the format of the file (that is, provide Meta data that define the file format and contents); these lines can be used by external programs to reformat the file. The first line beyond the Meta data is a tab-separated list of names of the output variables whose values are provided in a column in each data line. These output variables are specified using parameter aniOutVar_names. The next line is a tab-separated list of the field width and data type, defined as a single text string, of each output variable in the same sequence as the list of variable names. Each value in the list is single character appended to an integer value defining the field width.

```
Surface Water and Energy Budgets Simulated by PRMS Version 2.3730 2008-01-18
 Start time: 1980/10/01 00:00:00
 End time:  1996/09/30 00:00:00
 Sum of HRUs:     6782.50  Basin_area:     6782.50
1  Year Month Day    Precip      ET    Storage P-Runoff O-Runoff
                    (inches) (inches) (inches) (inches) (inches)
    1981.            27.189   22.711    0.794    9.266    6.181
    1982.            76.778   31.551    4.142   34.977   36.819
    1983.            64.524   33.072    2.415   31.771   38.374
    1984.            52.146   33.120    0.743   22.507   26.168
    1985.            29.538   24.228    1.103   11.245   11.260
    1986.            60.447   32.247    2.695   23.094   27.044
    1987.            17.887   19.568    0.295    7.768    6.438
    1988.            21.214   19.986    0.357    6.016    4.140
    1989.            43.851   27.848    2.042   12.659   12.683
    1990.            29.970   25.613    1.705    7.398    6.491
    1991.            25.841   24.300    0.529    6.032    5.574
    1992.            21.591   17.620    0.364    5.928    3.847
    1993.            55.061   29.138    0.368   18.250   18.886
    1994.            20.340   18.582    0.639    6.248    4.182
    1995.            69.736   31.238    0.544   28.132   29.493
    1996.            52.824   28.632    0.832   21.199   22.769
 *********************************************************************
 Total for run    668.938  419.456    0.832  252.492  260.352
```

Figure 32. PRMS Water-Budget File that shows annual and simulation-run summary tables (parameters print_type set to 1 and print_freq set to 3 in the PRMS Parameter File) for the Sagehen Creek watershed example problem.

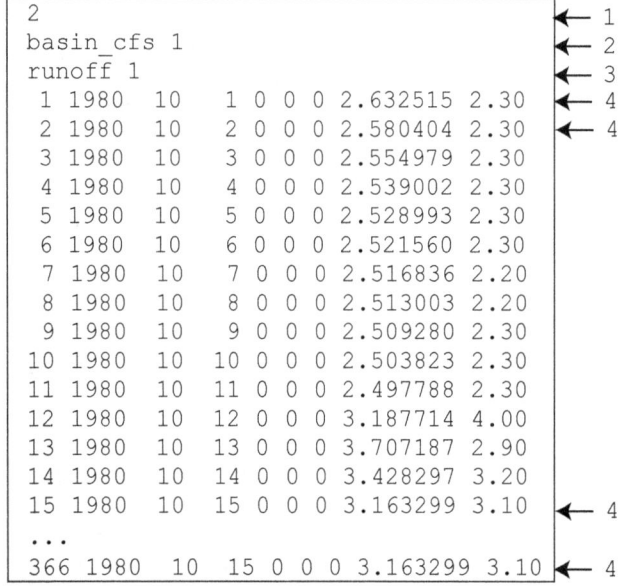

```
 2                                          ← 1
 basin_cfs 1                                ← 2
 runoff 1                                   ← 3
  1 1980   10    1 0 0 0 2.632515 2.30      ← 4
  2 1980   10    2 0 0 0 2.580404 2.30      ← 4
  3 1980   10    3 0 0 0 2.554979 2.30
  4 1980   10    4 0 0 0 2.539002 2.30
  5 1980   10    5 0 0 0 2.528993 2.30
  6 1980   10    6 0 0 0 2.521560 2.30
  7 1980   10    7 0 0 0 2.516836 2.20
  8 1980   10    8 0 0 0 2.513003 2.20
  9 1980   10    9 0 0 0 2.509280 2.30
 10 1980   10   10 0 0 0 2.503823 2.30
 11 1980   10   11 0 0 0 2.497788 2.30
 12 1980   10   12 0 0 0 3.187714 4.00
 13 1980   10   13 0 0 0 3.707187 2.90
 14 1980   10   14 0 0 0 3.428297 3.20
 15 1980   10   15 0 0 0 3.163299 3.10    ← 4
 ...
366 1980   10   15 0 0 0 3.163299 3.10    ← 4
```

[1] Two variables are included.

[2] Simulated streamflow and one column of numbers, respectively.

[3] Daily mean streamflow at outlet gage and one column of numbers, repsectively.

[4] Time step, year, month, day, time (HH MM SS), simulated streamflow, and daily mean streamflow, respectively. Line 4 is repeated for each daily time step of simulation.

Figure 33. Selected output from the PRMS Statistic Variables (statvar) File that lists simulated (basin_cfs) and measured (runoff) streamflow for the Sagehen Creek watershed example problem.

```
# Begin DBF
# timestamp,#FIELD_ISODATETIME,19,0
# nhru,#FIELD_DECIMAL,10,2
# pkwater_equiv,#FIELD_DECIMAL,10,2
# soil_moist_pct,#FIELD_DECIMAL,10,2
# hru_actet,#FIELD_DECIMAL,10,2
# hru_ppt,#FIELD_DECIMAL,10,2
# infil_tot,#FIELD_DECIMAL,10,2
# snowmelt,#FIELD_DECIMAL,10,2
# sroff,#FIELD_DECIMAL,10,2
# tavgf,#FIELD_DECIMAL,10,2
# End DBF
#
timestamp              nhru  pkwater_equiv  soil_moist_pct  hru_actet   hru_ppt   ...
19d                    10n        10n            10n           10n        10n     ...
1980-10-01:00:00:00      1    0.000e+000      2.446e-002     1.037e-002  0.000e+000 ...
1980-10-01:00:00:00      2    0.000e+000      3.593e-002     1.458e-002  0.000e+000 ...
1980-10-01:00:00:00      3    0.000e+000      2.959e-002     1.106e-002  0.000e+000 ...
1980-10-01:00:00:00      4    0.000e+000      2.480e-002     9.699e-003  0.000e+000 ...
1980-10-01:00:00:00      5    0.000e+000      4.660e-002     1.741e-002  0.000e+000 ...
1980-10-01:00:00:00      6    0.000e+000      2.442e-002     1.054e-002  0.000e+000 ...
1980-10-01:00:00:00      7    0.000e+000      2.453e-002     9.897e-003  0.000e+000 ...
1980-10-01:00:00:00      8    0.000e+000      2.593e-002     1.096e-002  0.000e+000 ...
1980-10-01:00:00:00      9    0.000e+000      4.020e-002     1.519e-002  0.000e+000 ...
1980-10-01:00:00:00     10    0.000e+000      2.766e-002     1.151e-002  0.000e+000 ...
1980-10-01:00:00:00     11    0.000e+000      2.484e-002     9.990e-003  0.000e+000 ...
1980-10-01:00:00:00     12    0.000e+000      2.607e-002     8.646e-003  0.000e+000 ...
1980-10-01:00:00:00     13    0.000e+000      2.518e-002     7.445e-003  0.000e+000 ...
1980-10-01:00:00:00     14    0.000e+000      2.519e-002     1.053e-002  0.000e+000 ...
1980-10-01:00:00:00     15    0.000e+000      4.414e-002     1.810e-002  0.000e+000 ...
...
```

Variables listed at beginning of file are printed for each time step until end of simulation.
Gaps in the output listing are indicated by an ellipsis

Figure 34. Selected output from a PRMS Animation File that lists snowpack-water equivalent, soil moisture, evapotranspiration, and precipitation for each HRU.

The single character designates the data type using the following scheme: d=date, n=number. The remaining lines contain the date and corresponding index number within the spatial feature dimension and variable values in the order specified by the list of variable names. A date value (or timestamp) is output as a 19-character string as a four-digit year, one- or two-digit month, two-digit day, two-digit hour, two-digit minute and two-digit second. Year, month and day are separated with a dash whereas hour, minute, and second are separated with a colon. The abbreviated form is: YYYY-MM-DD HH:MM:SS. The index number is an integer value using the next 5 characters. Data values are numbers written in a 10-character exponential format. All values are separated by tab characters.

MODFLOW-2005

MODFLOW-2005 output files are generated by specifying the file type (Ftype), unit number (Nunit), directory address, and name (Fname), in the MODFLOW Name File. Output-file types are categorized according to the MODFLOW-2005 process or package that writes to the file. Output file types are defined using the variable Ftype specified in the Name File, and output-file types common to the Ground-Water Flow process are "LIST", "DATA" "DATA(BINARY)", and "GAGE". The frequency and format for which output is written to all output files is specified by the user in the Output Control Option File of the Basic Package. The Output Control Option File has its own input file, which is specified using file type "OC" in the Name File (see fig. 30). The Output Control Option File can be excluded by omitting the file type "OC" from the Name File, which will result in output being written at the end of each stress period as 10-digit floating-point numbers with 3 digits following the decimal point to the Listing File.

The main output file for MODFLOW-2005 is called the Listing File, to which much of the MODFLOW input data and calculated results are written during a simulation. The primary model-calculated output for the Ground-Water Flow Process includes the ground-water head distribution, drawdown distribution, the distribution of variable-head cells, and the overall volumetric budget. The Listing File is used for all MODFLOW-2005 simulations and must be the first file that is specified within the Name File. In addition to calculated results, input data specified for any MODFLOW-2005 package can be written (echoed) to the Listing File to verify proper input data specification. Budget information for MODFLOW-2005 packages, such as the Lake, Streamflow-Routing, and Unsaturated-Zone Flow Packages, also can be written to the Listing File.

Output from the Ground-Water Flow Process also can be separated into output files other than the Listing File. Most packages can write calculated results as formatted text to files that are specified as type "DATA", or as binary (unformatted) data to files that are specified as type "DATA(BINARY)". Calculated results for the Lake and Streamflow-Routing Packages can be written to output files of type "DATA" using the Stream Gaging Station or Gage Package. Each MODFLOW-2005 package generates its own output, and the details regarding output data for the Lake, Streamflow-Routing, and Unsaturated-Zone Flow Packages supported by GSFLOW are briefly described here. A MODFLOW-2005 package can write detailed information using the output option flags specified in the input files for each package (appendix 1).

The Lake and Streamflow-Routing Packages can write calculated results to formatted files with type "DATA", or to unformatted files with type "DATA(BINARY)". Additionally, calculated results for a particular lake or stream can be written as formatted text to files with type "DATA", as defined in the Name File, using the Gage Package output options. There are four options that can be specified in the Gage Package File when using the Lake Package. These options can be used to write columns of values for stage, volume, runoff, inflows from and outflows to streams, and lakebed conductance for every time step. There are seven options that can be specified in the Gage Package File when using the Streamflow-Routing Package. These options can be used to write columns of values for stage, outflow, runoff, streambed conductance, hydraulic gradients beneath the streambed, and flow diversions every time step. If unsaturated flow beneath streams is active, then changes in unsaturated zone storage and recharge also can be written as well as profiles of water content. The Gage Package input instructions in appendix 1 describe these options in more detail.

The Unsaturated-Zone Flow Package can write recharge, evapotranspiration, and ground-water discharge to formatted files with type "DATA", or to unformatted files with type "DATA(BINARY)" as specified in the MODFLOW Name File using the values of output option flags. Another option is to write time-series output for infiltration, recharge, evapotranspiration, and ground-water discharge that are the sum of values for all finite-difference cells that contain the

water table. Finally, recharge, evapotranspiration, ground-water discharge, and water-content profiles for selected finite-difference cells can be written to files as formatted columns of values for every time step.

Example Problem – Sagehen Creek Watershed

This section describes an example GSFLOW model for simulation of the Sagehen Creek watershed. All files, including documentation, input and output data, source code, and the GSFLOW executable model necessary to run this example are available from the GSFLOW web page (Internet address is listed in the "Preface" to this report on page ix).

The derived and calibrated input files provided with the Sagehen Creek watershed example problem, graphs and tables, and other simulation results described in this section should not be interpreted for assessing water-resource assets in the Sagehen Creek watershed. Although the models described herein were calibrated with available data, few data are available regarding ground-water heads in the watershed; consequently, these data sets and simulation results are provided for illustrative purposes only.

Description of Sagehen Creek Watershed

Sagehen Creek is a USGS Hydrologic Benchmark Network Basin located on the east slope of the northern Sierra Nevada (fig. 35). The watershed drains an area of 27 km^2 and ranges in altitude from 1,935 to 2,653 m. Geology of the Sagehen Creek watershed consists of granodiorite bedrock overlain by andesitic, tertiary volcanics, which are overlain by till and alluvium composed of granodiorite and andesite clasts (Burnett and Jennings, 1965; Rademacher and others, 2005). Quaternary gravels also are present on the northwestern side of the watershed (Burnett and Jennings, 1965). Very little is known regarding the depths and thickness of these different geologic formations. However, the volcanics were assumed to make up the principal component of the watershed aquifer and the volcanics were assumed to range in thickness between 50 and 300 m. Alluvium is thin or nonexistent in the upper parts of the watershed and thickens in the lower parts and near stream channels (Burnett and Jennings, 1965). Alluvium was assumed to range in thickness between 0 and 10 m.

Fourteen perennial springs are known in the watershed. The chemical and isotopic composition of these springs were analyzed by Rademacher and others (2005) to estimate the age of ground water seeping into Sagehen Creek. Ground-water discharge into the creek from springs and leakage through

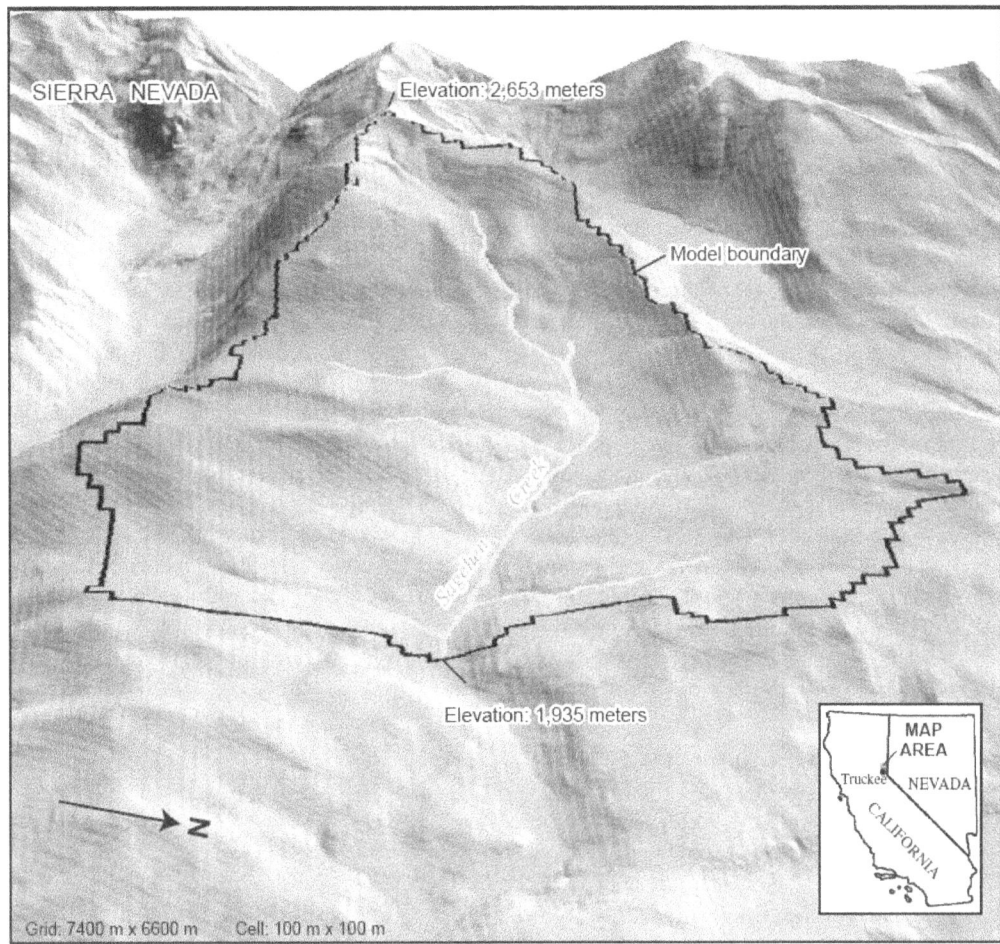

Shaded relief base from USGS 10-meter National Elevation
Data, illumination from the northwest at 45°. Universal
Transverse Mercator projection, Zone 11, North American
Datum of 1983. Perspective Information: Altitude is 6,300
meters above land surface, viewing angle is 24 degrees,
vertical exaggeration = 2x.

Figure 35. Sagehen Creek watershed located on the east slope of the northern Sierra Nevada near
Truckee, California.

the streambed was estimated to range in age between 26 and
33 years old (Rademacher and others, 2005). These ages and
the quantity of perennial base flow in the creek indicate that
ground-water discharge to streams is an important component
of the water budget in the Sagehen Creek watershed. Springs
present in the upper altitudes of the watershed also indicate
that the main ground-water flow direction generally follows
the slope of land surface (west to east) and leaves the
watershed as streamflow or ground-water flow beneath the
stream. Ground water is assumed not to flow through the
boundaries of the watershed other than beneath the stream at
the watershed outlet.

The model of the Sagehen Creek watershed relied on
precipitation and air temperature data that were collected from
four weather stations within the vicinity of the watershed.
Three of these weather stations are provided by the U.S.
Department of Agriculture and are named Independence Camp
(Western Region Climate Center, 2006a), Independence Lake
(Western Region Climate Center, 2006b), and Independence
Creek (Western Region Climate Center, 2006c). Data from
Sagehen Creek co-operative station were provided by
University of California, Berkeley (2005). Precipitation in
the Sagehen Creek watershed is highly variable in form and
intensity and generally increases with altitude. The areally
averaged annual precipitation from 1960 to 1991 was 970 mm

(Daly and others, 1994; G.H. Taylor, Oregon Climate Service, Oregon State University, written commun., 1997). Average annual precipitation ranged from 865 mm at land-surface altitudes less than 2,135 m to a maximum of 1,168 mm at a land-surface altitude of 2,576 m. Air temperature generally decreases with land-surface altitude; however, temperature inversions are common. Mean annual temperature near Sagehen Creek from 1980 to 2002 was 4°C at an altitude of 2,545 m (Western Region Climate Center, 2006b).

Daily mean streamflow values were obtained for Sagehen Creek at the streamflow gage near the outlet of the watershed. The gage is a U.S. Geological Survey hydrologic benchmark station near Truckee, California (Mast and Clow, 2000; station identification number–10343500). Mean daily streamflow was approximately 1 m^3/s during snowmelt in May and June for 1953–2003. The maximum daily mean discharge on record was approximately 23 m^3/s on January, 1, 1997. Minimum flows usually occur in late September when surface runoff and interflow are low and ET is still important. Average base flow was estimated using daily mean streamflow during September and October and was approximately 0.2 m^3s^{-1} for 1953–2003.

PRMS Input

A PRMS Data File was assembled for the PRMS Sagehen Creek watershed model for October 1, 1980 through April 25, 2004. This PRMS Data File included daily streamflow at the Sagehen Creek streamflow gage and climate data (precipitation and minimum and maximum air temperature) from the Independence Lake (Natural Resources Conservation Service snow-telemetry station 20K05S) and Sagehen Creek (National Weather Service co-operative station 047641). Climate data used for estimating precipitation and temperature for the PRMS component of the model were from the Sagehen Creek co-operative station, because this station provided a longer and more complete temperature record as compared with other weather stations near the Sagehen Creek watershed. Measured solar radiation data were not available, and were estimated by the model using the approach described in section "Computations of Flow" and by Leavesley and others (1983, p. 14-18).

The Sagehen Creek watershed was delineated into 128 HRUs (fig. 36), using the USGS GIS Weasel toolbox (Viger and Leavesley, 2007), determined by a combination of flow planes, climate and vegetation zones, and depth to the water table. This resulted in 4,691 gravity reservoirs when the HRUs were intersected with active cells in the top layer of the MODFLOW grid. The PRMS-only model of Sagehen Creek watershed had 128 ground-water reservoirs, one corresponding to each HRU. The ground-water reservoirs were not used in the integrated GSFLOW model simulations.

Measured data from the Independence Lake snow-telemetry and the Sagehen Creek co-operative stations for 1994–96 were used to develop the initial values of the temperature lapse rate parameters (`tmin_lapse` and `tmax_lapse`) and the precipitation adjustment factors (`snow_adj` and `rain_adj`) specified in the PRMS Parameter File. The years 1994–96 were chosen because of the completeness and quality of the measured data. Other parameters were estimated for each HRU using spatial data sets distributed with the GIS Weasel toolbox (Viger and Leavesley, 2007). Most parameters that do not vary in space were set to PRMS default values. Estimates of other PRMS parameters were from previous modeling studies of the Sagehen Creek watershed (Jeton, 1999) or were made by other methods that are described in table 13.

Table 13. Values and source of non-default PRMS parameters (appendix 1) used for the Sagehen Creek watershed example problem (modified from Koczot and others, 2005).

[**Abbreviations:** C, parameters that cannot be estimated from available data and are adjusted during calibration; CG, parameters that are initially computed in GIS and are adjusted, preserving relative spatial variation during calibration; L, parameters obtained from the literature as estimated or empirical estimates]

Parameter	Minimum value	Maximum value	Source
adjmix_rain	0.0055	2.32	L
covden_sum	.683	1.0	CG
covden_win	.683	1.0	CG
dday_intcp	-36.0	-10.0	L
dday_slope	.31	.65	L
fastcoef_lin	.4	.4	C
gwflow_coef	.00365	.00365	C
jh_coef	.0163	.0267	C
jh_coef_hru	13.6	15.7	CG
potet_sublim	.75	.75	C
pref_flow_den	.1	.1	C
rad_trncf	.233	.558	CG
sat_threshold	3.34	5.16	C
smidx_coef	.00037	.00037	C
snarea_curve	.05	1.0	L
snarea_thresh	.0	106.3	CG
snowinfil_max	2.75	2.75	C
soil_moist_max	2.39	3.68	CG
soil_rechr_max	1.48	2.09	CG
srain_intcp	.05	.05	L
ssr2gw_exp	.75	.75	C
ssr2gw_rate	.0378	.0557	C
tmax_allrain	60.0	70.0	L
tmax_allsnow	38.2	38.2	L
transp_beg	3	3	C
transp_end	11	11	C
tstorm_mo	0	1	C
wrain_intcp	.05	.05	L

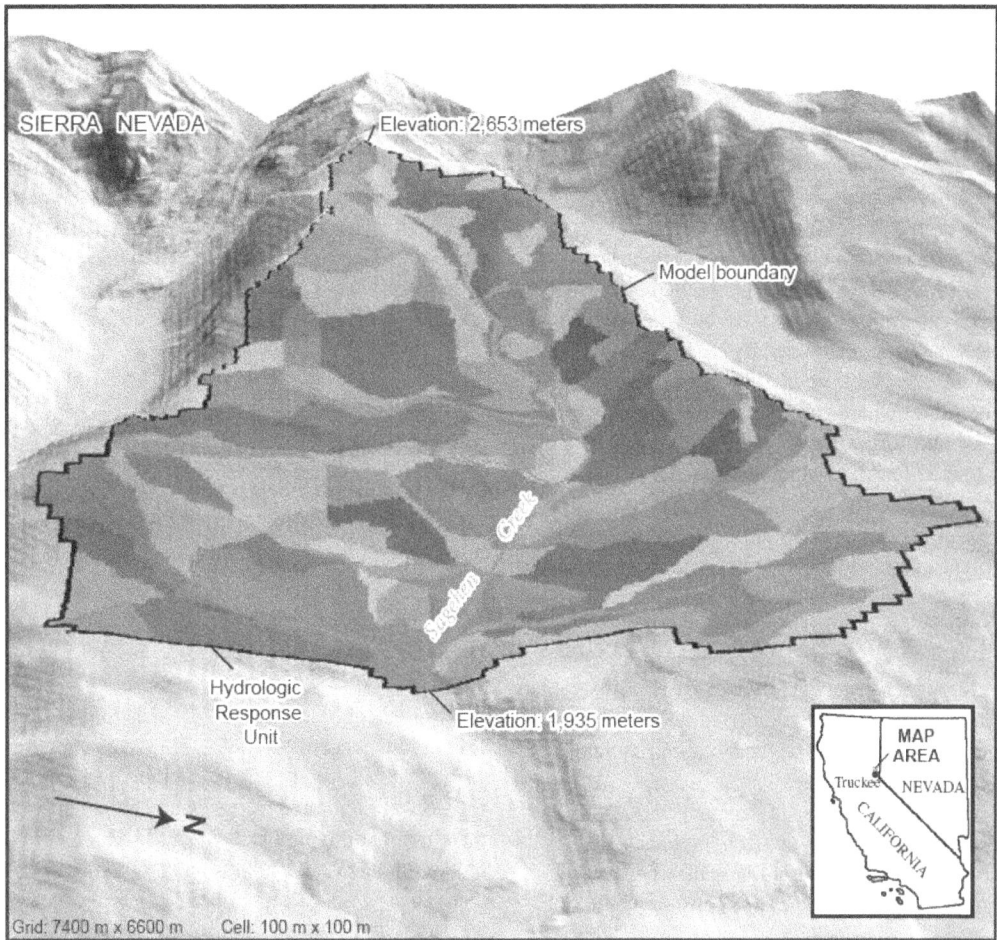

Shaded relief base from USGS 10-meter National Elevation
Data, illumination from the northwest at 45°. Universal
Transverse Mercator projection, Zone 11, North American
Datum of 1983. Perspective Information: Altitude is 6,300
meters above land surface, viewing angle is 24 degrees,
vertical exaggeration = 2x.

Figure 36.   Hydrologic response units discretized for Sagehen Creek watershed near Truckee, California.

MODFLOW-2005 Input

The MODFLOW-2005 model of Sagehen Creek
watershed consisted of two model layers, each layer with
73 rows and 81 columns in which all cells had a constant
width and length equal to 90 m. The top altitudes of finite-
difference cells were set equal to estimated land-surface
altitudes for each cell and ranged from 1,935 to 2,653 m; the
thickness of the top model layer ranged between 20 and 70 m
and the thickness of the bottom layer ranged between 10 and
115 m. The upper layer was created to represent the alluvium
in valley lowlands; however, it was made thicker than the
assumed thickness of the alluvium to avoid drying of cells.
Consequently, the hydraulic conductivity of the upper layer

was decreased, especially in the upland areas, to represent
a composite of alluvial and volcanic materials. No-flow
conditions were simulated across the bottom and sides of
the model except for six cells (three for each layer) beneath
and adjacent to the stream at the watershed outlet, which
were specified as constant-head cells. The simulation period
was divided into one steady-state stress period followed by
four transient stress periods, three had 182 1-day time steps,
and the last stress period had 5,500 1-day time steps. Five
stress periods were created in this example in order to test
the effects of pumping wells in the simulation; however, the
Well Package was not included in the version of the example
problem presented herein.

Several packages and optional files were used in the simulation of the Sagehen Creek watershed (table 14). The Layer-Property Flow Package (Harbaugh, 2005, p. 5-17) was used and both horizontal and vertical hydraulic conductivity values were specified. The upper model layer represented the shallow alluvial material, whereas the lower layer represented the volcanic material. All cells in the upper layer were specified as convertible (unconfined). The second layer was specified as confined because the water table was contained within the upper layer. The horizontal hydraulic conductivity (K) for the upper layer ranged from 0.026 m/d on the ridges to 0.39 m/d in the valleys as determined by calibration to match mean base flow of Sagehen Creek (table 15). A lower K was specified on the ridges where volcanic rocks are near land surface (fig. 37). The horizontal K for the lower layer ranged from 0.00045 to 0.027 m/d, in which the lower K values were specified beneath ridges. The hydraulic conductivity within each cell was assumed isotropic. Specific storage was set to 2 $\times10^{-6}$ m^{-1} for both layers, and the specific yield was specified as 0.08 on the ridges and 0.15 in the valleys near streams. The vertical hydraulic conductivity values of the unsaturated zone were specified within the Unsaturated-Zone Flow Package File and ranged between 0.018 and 0.27 m/d (table 15). A constant Brooks-Corey exponent of 4 and a saturated water content ranging from 0.15 to 0.25 were assigned to all unsaturated-zone cells.

Streams were represented by 15 segments made up of 201 reaches using the Streamflow-Routing Package and included the simulation of unsaturated flow beneath streams. All stream reaches were set to a constant width of 3 m and ranged in length from 30 to 120 m. The streambed hydraulic conductivity was set equal to 5.0 m/d (table 15). The streambed hydraulic conductivity was set to a large value such that ground-water exchange with the stream was a function of hydraulic conductivity of layer 1 and the unsaturated zone. Unsaturated-zone properties beneath all stream reaches were constant with a vertical hydraulic conductivity of 0.3 m/d, a saturated water content of 0.30 and a Brooks-Corey exponent of 3.5 (table 15). The initial water content beneath streams was set to 0.2. This was set as a starting value only because the initial water content beneath streams was calculated during the steady-state simulation for stream reaches where the water table in layer 1 was beneath the streambed.

Calibration

The GSFLOW model of the Sagehen Creek watershed was calibrated using a stepwise procedure: first, independent PRMS and MODFLOW-2005 models were calibrated by running these models separately (PRMS-only and MODFLOW-only simulations), followed by calibration of the integrated model. These steps are described in the following sections.

PRMS Calibration

Initial calibration of PRMS was done by running PRMS independently of MODFLOW-2005 (parameter `model_mode` set to "PRMS" in the GSFLOW Control File). PRMS was calibrated using a split sample approach such that three non-overlapping periods of the data record were used for model initialization, calibration, and verification. Data from October 1, 1980, to September 30, 1981 (water year 1981) were used to develop initial storages in the PRMS model. Data from October 1, 1981, to September 30, 1988 (water years 1982–88) were used for calibrating the PRMS model, and data from October 1, 1988, to September 30, 1995 (water years 1989–95) were used for verifying the calibrated PRMS model. The calibration procedure, described in detail below, consisted of first adjusting parameters that affect the average flow of water through the watershed until the model provided a reasonable match between the simulated and measured annual water balance. Parameters that affect the timing and magnitude of the simulated streamflow were then adjusted until the model provided a good fit between the simulated and measured daily streamflow.

The PRMS precipitation lapse rate factors (`snow_adj` and `rain_adj` in the PRMS Parameter File) were adjusted until the mean of the simulated annual precipitation approximately matched the mean of the measured values for the period when the lapse rates were computed. Then, evapotranspiration parameters were adjusted until the simulated mean annual evapotranspiration approximately matched a value of 27 in/yr (686 mm/yr) that was estimated from Kahrl (1979, p. 13).

Table 14. MODFLOW-2005 packages and files used for the Sagehen Creek watershed example problem.

MODFLOW-2005 packages and files
Basic Package File
Output Control Option File
Discretization File
Layer-Property Flow Package File
Preconditioned Conjugate Gradient Solver Package File
Unsaturated-Zone Flow Package File
Streamflow-Routing Package File
Gage Package File

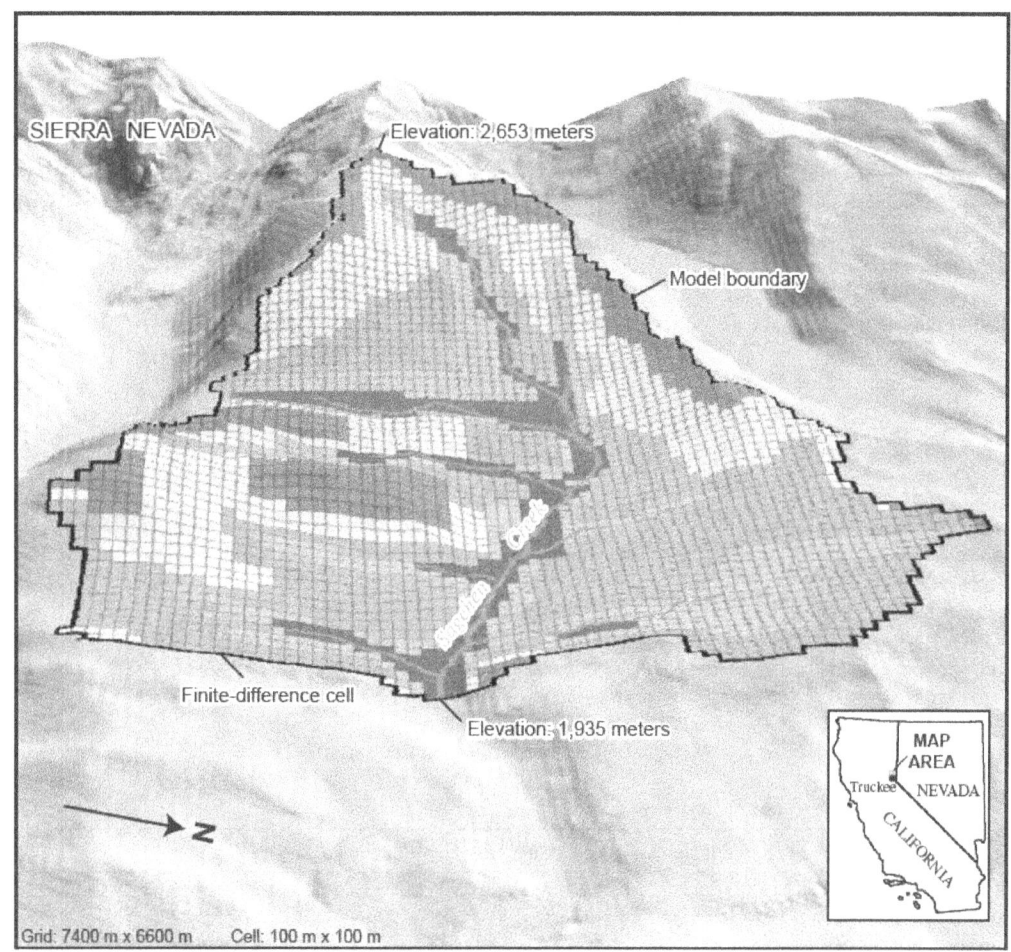

Shaded relief base from USGS 10-meter National Elevation
Data, illumination from the northwest at 45°. Universal
Transverse Mercator projection, Zone 11, North American
Datum of 1983. Perspective Information: Altitude is 6,300
meters above land surface, viewing angle is 24 degrees,
vertical exaggeration = 2x.

EXPLANATION

Hydraulic conductivity value—Meters per day

	Layer 1	Layer 2
	0.026	0.00045
	0.052	0.0009
	0.065	0.0027
	0.13	0.009
	0.39	0.027

Figure 37.   Hydraulic conductivity values used in ground-water model, Sagehen Creek watershed near
Truckee, California.

Table 15. Hydraulic properties and other selected variables used in the Layer-Property Flow, Unsaturated-Zone Flow, and Streamflow-Routing Packages for the Sagehen Creek watershed example problem.

[Note—Initial water content values must be specified if the first stress period is transient]

Variable	Units	Minimum value	Maximum value
Variables assigned to the Layer-Property Flow Package			
Horizontal hydraulic conductivity for upper layer	meters per day	0.026	0.39
Vertical hydraulic conductivity for upper layer	meters per day	.026	.39
Horizontal hydraulic conductivity for lower layer	meters per day	.00045	.027
Vertical hydraulic conductivity for lower layer	meters per day	.00045	.027
Specific storage for both layers	per meter of aquifer	2×10^{-6}	2×10^{-6}
Specific yield	cubic meter of water per cubic meter of aquifer	.08	.15
Variables assigned to the Unsaturated-Zone Flow Package			
Vertical hydraulic conductivity of the unsaturated zone below land surface	meters per day	0.018	0.27
Brooks-Corey exponent for unsaturated zone below land surface	dimensionless	4	4
Saturated water content of unsaturated zone below land surface	cubic meter of water per cubic meter of unsaturated zone	.15	.25
Initial water content of unsaturated zone below land surface	cubic meter of water per cubic meter of unsaturated zone	Not required for steady-state stress period [see note]	Not required for steady-state stress period [see note]
Steady-state infiltration rate	meters per day	.0007	.0019
Variables assigned to the Streamflow-Routing Package			
Hydraulic conductivity of streambed	meters per day	5.0	5.0
Streambed thickness	meters	1.0	1.0
Vertical hydraulic conductivity of unsaturated zone beneath streams	meters per day	.3	.3
Brooks-Corey exponent for unsaturated zone beneath streams	dimensionless	3.5	3.5
Saturated water content of unsaturated zone beneath streams	cubic meter of water per cubic meter of unsaturated zone	.3	.3
Initial water content of unsaturated zone beneath streams (see note)	cubic meter of water per cubic meter of unsaturated zone	.2	.2

All adjusted parameters were kept within a reasonable range so as not to violate any physical constraints and also to provide a close match of simulated and measured annual streamflow. The annual water budgets are summarized in table 16. The measured and simulated hydrographs for the calibration and verification periods are shown in figure 38. The goodness of fit between the simulated and measured hydrographs was assessed with the Nash-Sutcliffe efficiency coefficient (Nash and Sutcliffe, 1970). This coefficient is a relative measure of model predictive power. A value of 1.0 indicates a perfect simulation; a value of 0.0 indicates the mean measured value is equivalent to the simulated hydrograph in predictive power;

and a negative value indicates that the mean measured value is a better predictor than the simulated hydrograph. The Nash-Sutcliffe coefficient for the calibrated period is 0.79 and for the verification period is 0.71.

Visual inspection of the hydrographs in figure 38 indicate that the model generally performs well; however, there are several instances during winter months, where measured high streamflows are not simulated by the model, or the model simulates high streamflows that were not measured. This indicates that when air temperature in the watershed is near 0°C, the model cannot accurately determine the form of the precipitation.

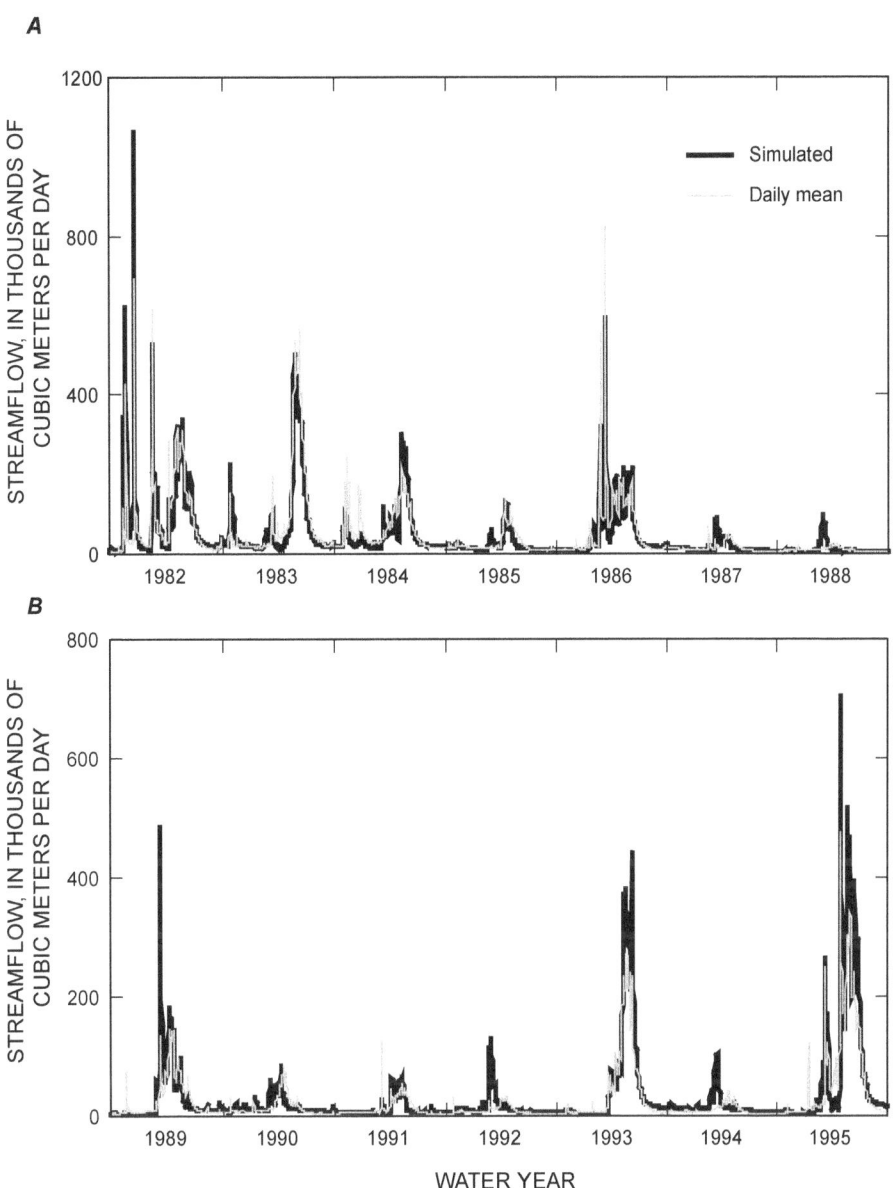

Figure 38. PRMS simulated and daily-mean streamflow at Sagehen Creek gage for (*A*) calibration period (water years 1982–88) and (*B*) verification period (water years 1989–95). Daily streamflow values are from a GSFLOW Comma-Separated Values (CSV) File using variables `basinstrmflow` and `obs_strmflow`.

Table 16. Annual water budget for PRMS calibration and verification periods for the Sagehen Creek watershed example problem.

[–, not determined]

Water year	Annual precipitation per drainage area (meters)	Annual evapotranspiration per drainage area (meters)	Annual storage change per drainage area (meters)	Simulated annual streamflow per drainage area (meters)	Measured annual streamflow per drainage area (meters)	Percent difference between simulated and measured annual streamflow
Calibration period						
1982	1.950	0.772	0.180	0.999	0.935	-6.6
1983	1.639	.829	-.042	.854	.975	13.2
1984	1.325	.786	-.059	.599	.665	10.4
1985	.750	.551	-.056	.257	.286	10.7
1986	1.535	.769	.090	.678	.687	1.3
1987	.454	.403	-.126	.179	.164	-8.7
1988	.539	.426	-.017	.130	.105	-21.3
Average	1.170	.648	–	.528	.545	3.2
Verification period						
1989	1.114	0.663	0.071	0.380	0.322	-16.5
1990	.761	.590	-.019	.191	.165	-14.6
1991	.656	.562	-.042	.137	.142	3.6
1992	.548	.385	-.001	.166	.098	-51.5
1993	1.399	.728	.065	.607	.480	-23.4
1994	.517	.401	-.053	.170	.106	-46.4
1995	1.771	.782	.090	.901	.749	-18.4
Average	.967	.587	–	.364	.294	-21.3

MODFLOW-2005 Steady-State Calibration

Initial calibration of ground-water flow was done by running MODFLOW-2005 independently of PRMS (parameter model_mode set to "MODFLOW" in GSFLOW Control File). Measurements of precipitation, streamflow, and the location of springs in the watershed were used to calibrate the ground-water flow model. These data were useful in this example for describing an approach for model calibration and to provide a realistic simulation. Calibration of the Sagehen Creek ground-water flow model was done using a trial-and-error approach; however, automated procedures can be used for complicated models when sufficient data are available.

The distribution of horizontal hydraulic conductivity (K) was created initially on the basis of the surface geology. During the steady-state calibration, spatially varying ground-water recharge was assumed proportional to the spatial distribution of mean annual precipitation with altitude (fig. 39). Calibration of the ground-water model was done by adjusting the steady-state infiltration-rate (Unsaturated-Zone Flow Package input variable FINF; appendix 1) and

the K (Array Reading Utility Subroutines variable CNSTNT; Harbaugh, 2005; p. 8–57). The infiltration-rate factor was adjusted until the steady-state streamflow at the watershed outlet was approximately equal to the mean base flow of 0.2 m³/s for 1953–2003. K factors specified in the Layer-Property Flow Package File were adjusted until the steady-state water table was above or at the soil-zone base near streams and where springs are located in the watershed.

Although no wells have been drilled on the ridges in the watershed, the maximum water-table depth was assumed to be less than 75 m in the watershed. This depth is based on depths measured in wells elsewhere in the northern Sierra Nevada. Some ground-water levels were measured in shallow (<5 m) wells near the Sagehen Creek watershed outlet close to the creek. These wells indicate that the water table remains close to land surface in this area. Simulated water-table depths were 60 m beneath the soil-zone base along the ridges and at, or slightly above, the soil-zone base in some of the upland areas where springs are located. The water-table depth in the lowland areas and valleys generally were at or above the soil-zone base (fig. 40).

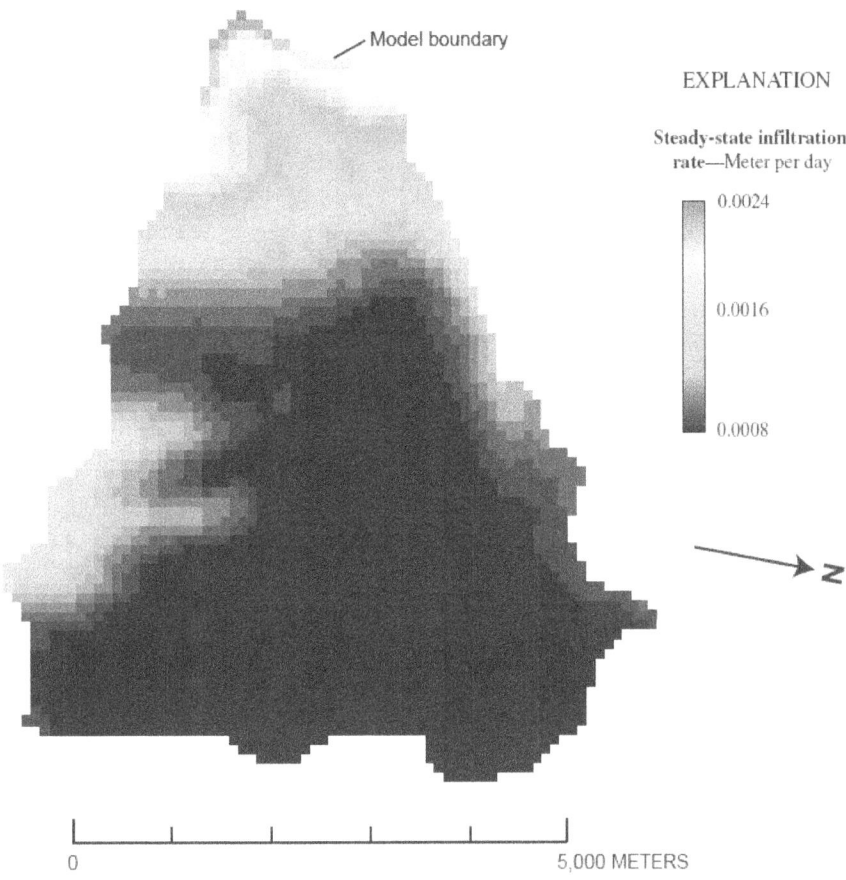

Figure 39. Steady-state infiltration rate for the ground-water model of Sagehen Creek watershed near Truckee, California.

GSFLOW Calibration

Parameters describing the overlay of PRMS HRUs and MODFLOW finite-difference cells (**fig. 41**) were developed using the USGS GIS Weasel toolbox, and the calibrated parameter sets from the PRMS and MODFLOW-2005 models of the Sagehen Creek watershed were initially used as input parameters for the GSFLOW model. GSFLOW could be calibrated using most of the parameters from these models unchanged, while focusing on just a few of the parameters that influence the flows between PRMS and MODFLOW-2005. The measured and simulated hydrographs for the calibration and verification periods are presented in **figure 42**. The Nash-Sutcliffe coefficient for the calibrated period is 0.81 and for the verification period is 0.85. Changes to parameters made during calibration of the integrated GSFLOW model are described in the following two sections.

Surface-Water Calibration

Initial evaluation of the GSFLOW model before calibration indicated two important water-balance problems. The first was that the gravity drainage rate from the soil zone to the unsaturated zone was insufficient to keep the transient ground-water head levels in balance with the steady-state heads. The second was that actual evapotranspiration rates were too high during the summer months. To address these problems, two PRMS calibration parameters were manually modified for the GSFLOW model of the Sagehen Creek watershed (**table 17**).

Table 17. Values of PRMS parameters (**appendix 1**) modified for the GSFLOW Sagehen Creek watershed example problem.

Parameter	PRMS values	GSFLOW values
jh_coef	0.0163 to 0.0267	0.0148 to 0.0267
ssr2gw_rate	0.0378 to 0.05571	0.189 to 0.278

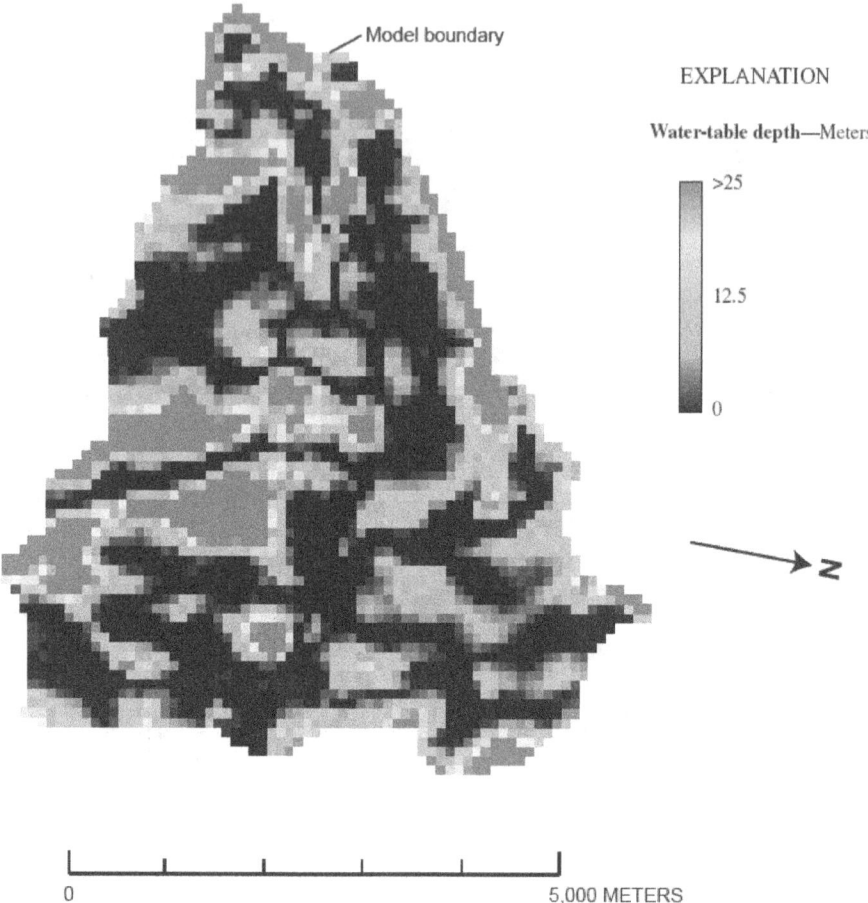

Figure 40.   Simulated steady-state water-table depth below land surface in the Sagehen Creek watershed near Truckee, California. Water-table depth below land surface calculated from a MODFLOW formatted head output file.

Transient Ground-Water Calibration

Generally, the ground-water component of GSFLOW (MODFLOW-2005) was calibrated to streamflow measured during September and October (autumn) from 1981 to 1995. The transient ground-water model calibration was done by further adjusting the K distribution from what was estimated using the steady-state calibration. Accordingly, the transient ground-water model calibration required an iterative process where the K values were adjusted for the transient calibration followed by further adjustment of the recharge factor for the steady-state simulation.

The calibrated model was able to simulate decadal oscillations in base flow caused by years of above and below normal precipitation (**fig. 43**), although there were some discrepancies between simulated and estimated base flow. Base flow (simulated and estimated) for this example is defined as the annual minimum daily mean discharge at the Sagehen Creek gage, which usually occurred between September and October. For example, estimated base flow was 0.175 m^3/s, whereas simulated base flow was 0.105 m^3/s during 1983 (**fig. 43**). These discrepancies occurred

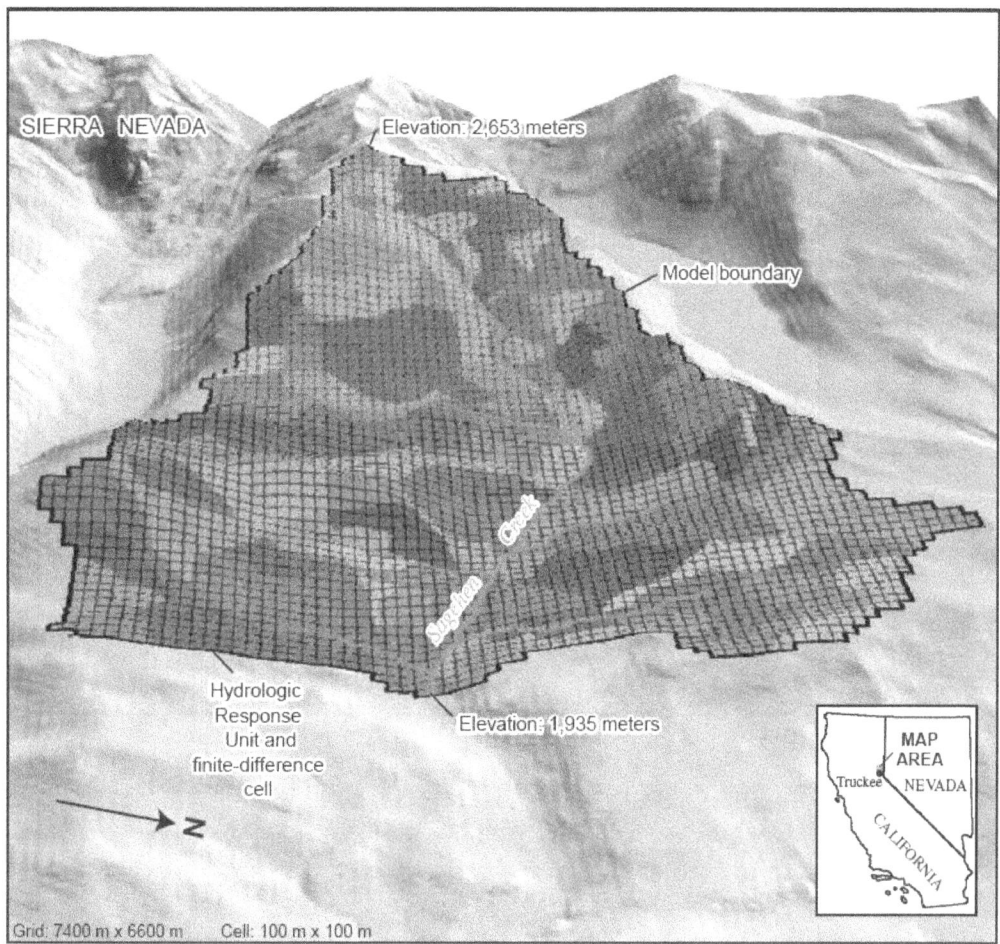

Shaded relief base from USGS 10-meter National Elevation
Data, illumination from the northwest at 45°. Universal
Transverse Mercator projection, Zone 11, North American
Datum of 1983. Perspective Information: Altitude is 6,300
meters above land surface, viewing angle is 24 degrees,
vertical exaggeration = 2x.

Figure 41.   Finite-difference cells in relation to hydrologic response units in Sagehen Creek watershed near
Truckee, California.

during years of above-average precipitation, which caused
the snowmelt period to extend into the autumn months. One
explanation for this discrepancy is that runoff and interflow
occurred throughout the autumn months into the following
water year. Thus, the discrepancy may be due to imperfect

calibration of either the surface-water and/or ground-water
components of GSFLOW. Decadal variations in estimated
base flow are caused by consecutive years of above-average
precipitation and the resulting increases in unsaturated-zone
and ground-water storage.

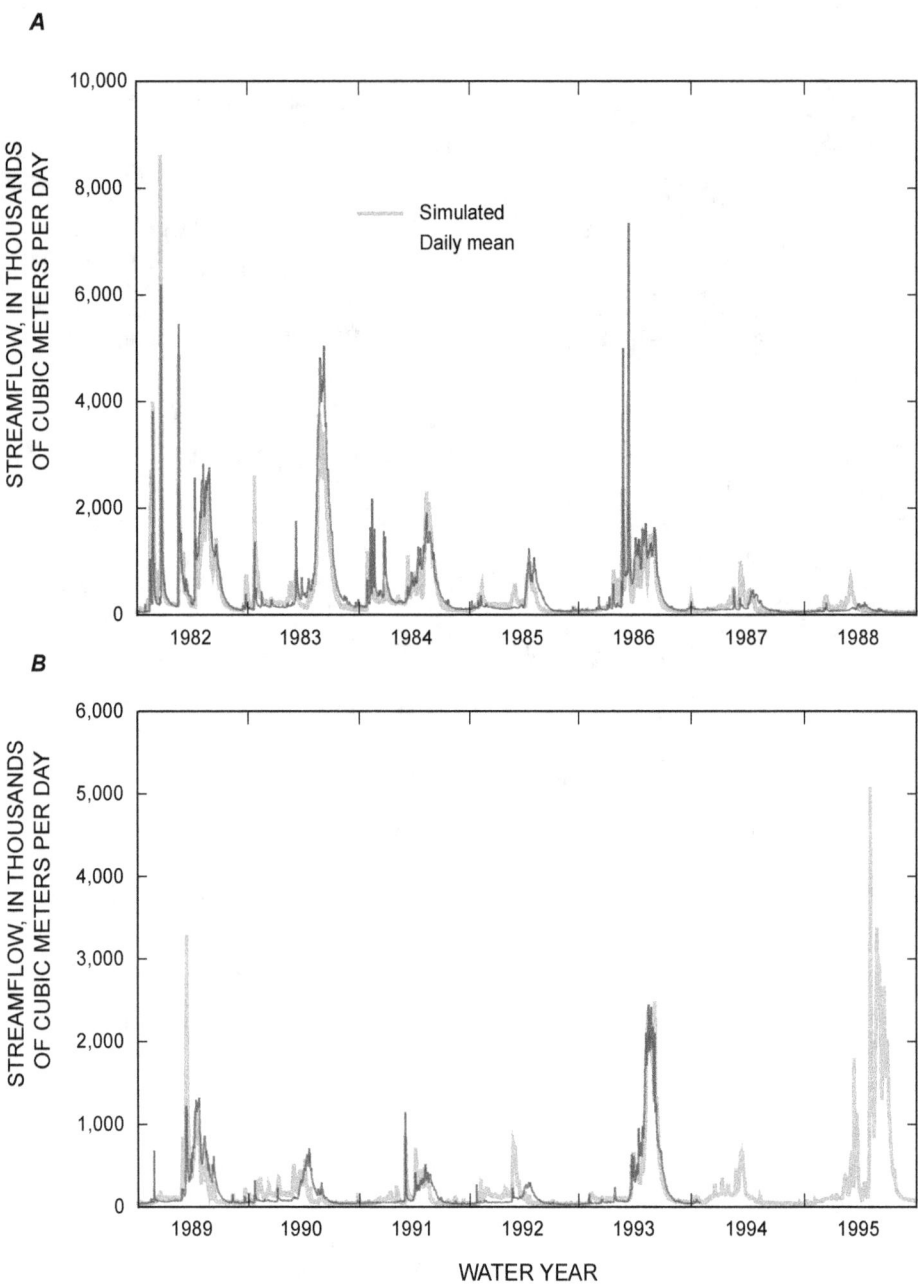

Figure 42.   GSFLOW simulated and daily-mean streamflow at Sagehen Creek gage for (*A*) calibration period (water years 1982–88) and (*B*) verification period (water years 1989–95). Daily streamflow values are from a GSFLOW Comma-Separated Values (CSV) File using variables `basinstrmflow` and `obs_strmflow`.

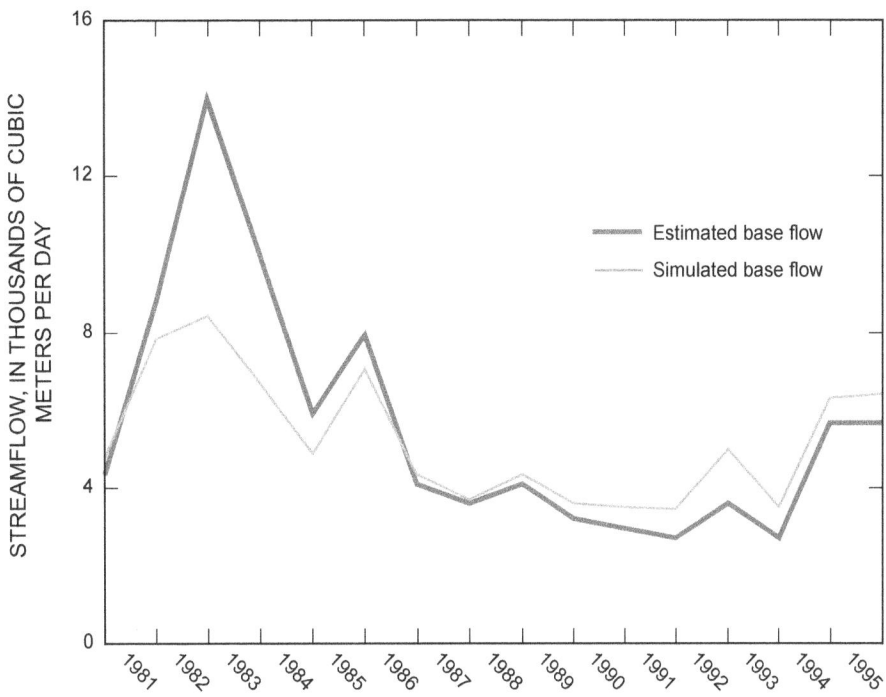

Figure 43. Comparison between simulated and estimated base flow for water years 1981–95. Base flow was determined from daily values of streamflow from a GSFLOW

Model Results

This section presents selected output from the calibrated GSFLOW simulation of the Sagehen Creek watershed. The model results described herein were simulated on a laptop personal computer, running the Microsoft© Windows XP Professional operating system, with a 2.0 gigahertz AMD processor, and 2.0 gigabytes of random access memory. No additional software was required. Run time was less than 7 minutes per year (365 days) simulated, with an average of 13 iterations per day. The number of iterations for each simulated time step from 1981 to 1994 is shown in figure 44.

Storages

Storage is divided into four major compartments in the GSFLOW model of the Sagehen Creek watershed. These are snowpack storage (principal surface storage), soil-zone storage, unsaturated-zone storage, and saturated-zone storage (fig. 45). The movement of water into and out of each compartment was dominated by the annual snowpack cycle as illustrated for the calibration period from 1981 to 1989. Three periods in the annual snowpack cycle generally can be identified: (1) the accumulation period begins October 1 and peaks toward the end of April; (2) the snowmelt period occurs from May through July; and (3) the no-snow period

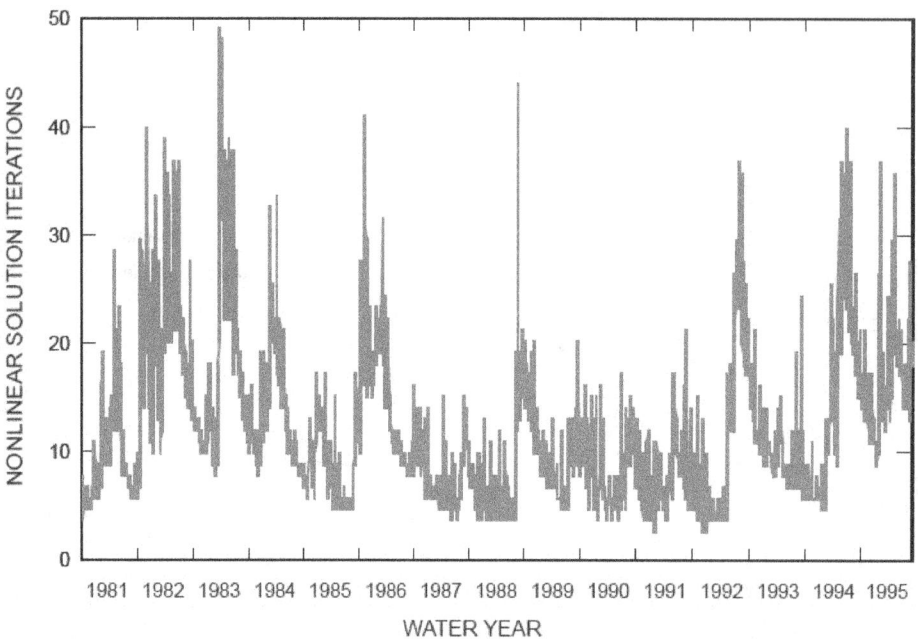

Figure 44.   Number of iterations per simulation time step for GSFLOW to converge during water years 1981–94. Iteration values are from a GSFLOW Comma-Separated Values (CSV) File using variable `kkiter`.

occurs from August through October. Soil-zone storage generally is higher through periods 1 and 2 because of reduced evapotranspiration and replenishment from snowmelt, and then decreases in period 3. The annual cycle is superimposed on a longer cycle of wet and dry years. Minimal snowpack was simulated during water years 1981 and 1987 and maximum snowpack was simulated during water year 1983. Storages shown in **figure 45** are summed over the entire watershed; simulated storage change can occur at different rates throughout the watershed.

Unsaturated-zone storage generally peaks in July after all snowmelt has occurred (**fig. 45***B*). Saturated-zone or groundwater storage generally peaks in August, decreasing until April or May (**fig. 45***B*). Variations in the unsaturated-zone storage also are affected by changes in the water-table depth and thus, generally trend opposite to changes in the saturated-zone storage. The saturated-zone storage increased to a maximum during water years 1983–84 and subsequently decreased during the drier years 1985–89, whereas the unsaturated-zone storage decreased during water years 1983–84 because the unsaturated-zone thickness decreased and then increased during the drier years 1985–89.

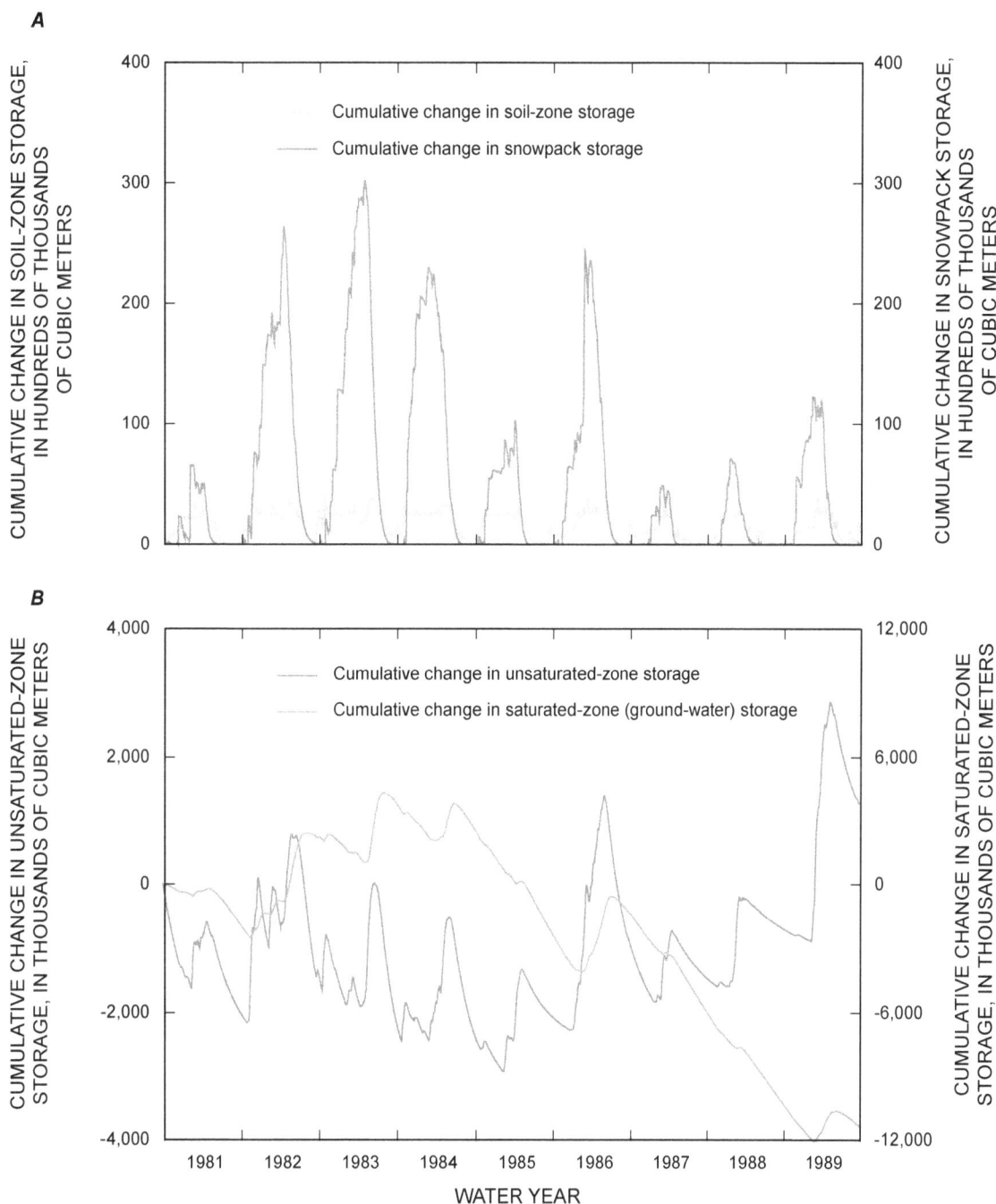

Figure 45. Cumulative change of (*A*) snowpack and soil-zone storage; and (*B*) unsaturated-zone and saturated-zone (ground-water) storage during water years 1981–89. Water year begins October 1 of each year. Storage values are from a GSFLOW Comma-Separated Values (CSV) File. Cumulative change in snowpack storage was calculated from variable `basinpweqv`, and cumulative change in soil-zone storage was calculated from variables `basinsoilmoist`, `basingravstor`, and `basinprefstor`. Cumulative change in unsaturated-zone and saturated-zone storage was calculated from variables `uzf_del_stor` and `sat_change_stor`, respectively.

Fluxes

The main fluxes of water across land surface for the Sagehen Creek watershed included precipitation, snowmelt, and to a lesser extent, evapotranspiration, as illustrated for water years 1981–89 (fig. 46). Precipitation occurred as rain, snow, or a mixture of both. Snowmelt occurred as the snowpack ablates during the snowmelt period. Evapotranspiration is the total evaporation from the plant canopy, sublimation, and evapotranspiration from the soil zone. Evapotranspiration generally occurred during the warmer periods of the year when sufficient water was available in the soil zone. Fluxes shown in figure 46 are areally averaged values over the entire watershed; simulated precipitation, snowmelt, and evapotranspiration can occur at different rates throughout the watershed depending on lapse rates, geographic characteristics, and subsurface properties.

Two major fluxes of water were simulated between the soil zone and the unsaturated and saturated zones in the Sagehen Creek watershed. The first was gravity drainage of water from the soil zone down to the unsaturated zone, whereas the second was ground-water discharge from the saturated zone into the soil zone, as illustrated for the calibration period from 1981 to1989 (fig. 47). Gravity drainage exceeded ground-water discharge during the annual snow accumulation and snowmelt period even during the drier years, whereas ground-water discharge exceeded gravity drainage following snowmelt and before snow accumulation began in late autumn. Fluxes shown in figure 47 also are areally averaged values over the entire watershed; simulated gravity drainage and ground-water discharge can occur at different rates throughout the watershed. In fact, at most time steps, both occurred simultaneously at different locations in the watershed.

Three major fluxes of water in the watershed add streamflow to Sagehen Creek. These are surface runoff, interflow, and ground-water discharge directly to streams, which is commonly called base flow (fig. 48). Surface runoff occurred sporadically from 1982–89 due to rainfall on saturated soils. Large fluxes of surface runoff were simulated when sufficient rain fell on the snowpack or on saturated or frozen soils. These events are difficult to simulate. Interflow was the predominant source of simulated streamflow during most years except for those years of large surface runoff. Net ground-water discharge directly to streams typically was the least of the three fluxes that generate streamflow in Sagehen Creek annually but was the predominant source of streamflow following the snowmelt period. As in the figures 45-47, the fluxes shown in figure 48 are areally averaged values over the entire watershed.

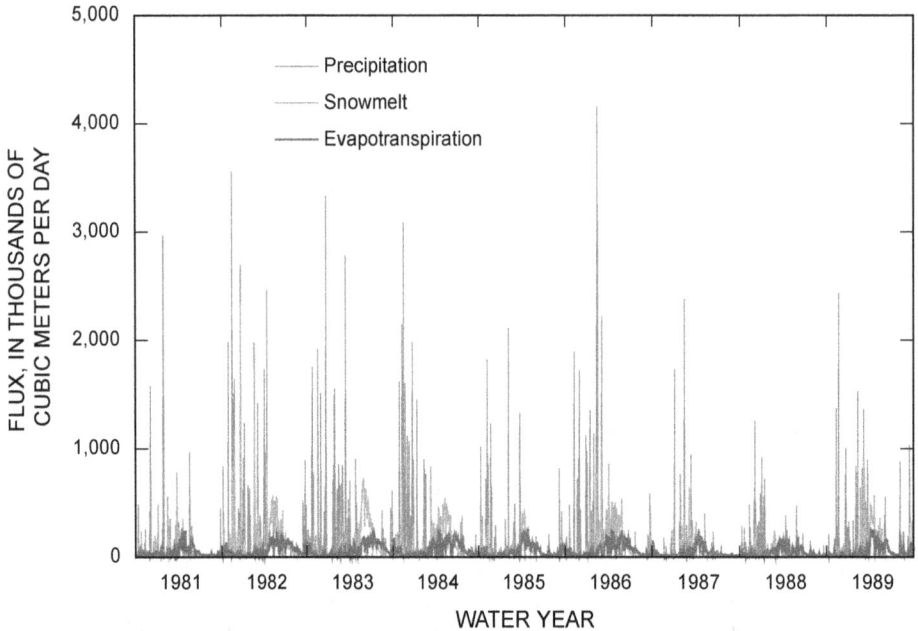

Figure 46.   Major fluxes across land surface during water years 1981–89 that include precipitation, snowmelt, and evapotranspiration. Flux values are from a PRMS Animation File using variables `basin_ppt`,

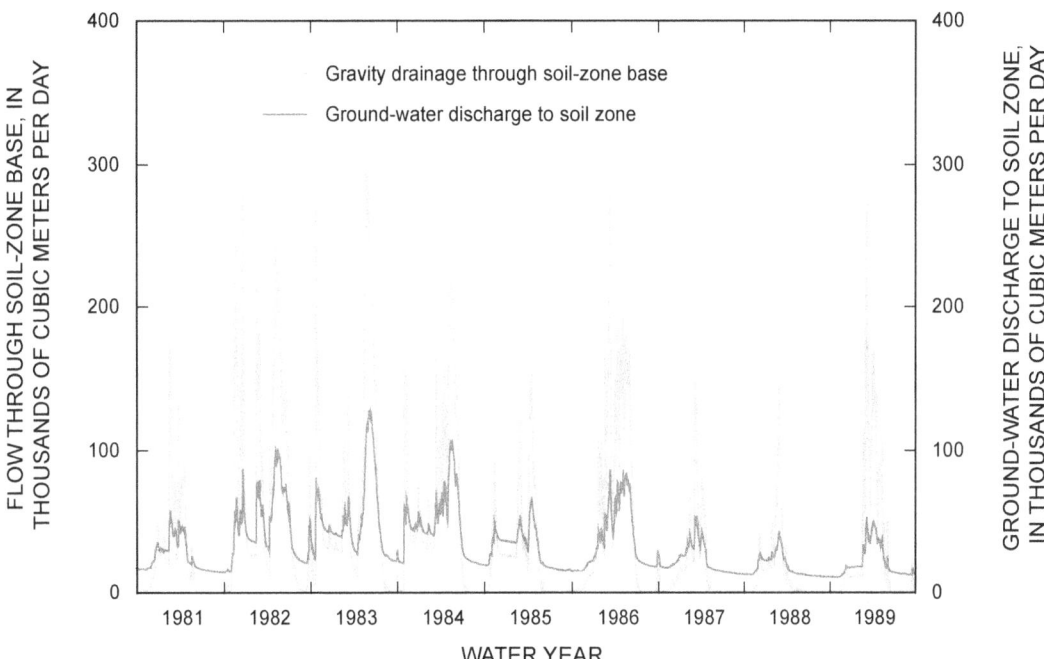

Figure 47. Fluxes of gravity drainage from soil zone down to unsaturated zone and ground-water discharge from saturated zone into soil zone during water years 1981–89. Flux values are from a GSFLOW Comma-Separated Values (CSV) File using variables `basinsz2gw` and `basingw2sz`.

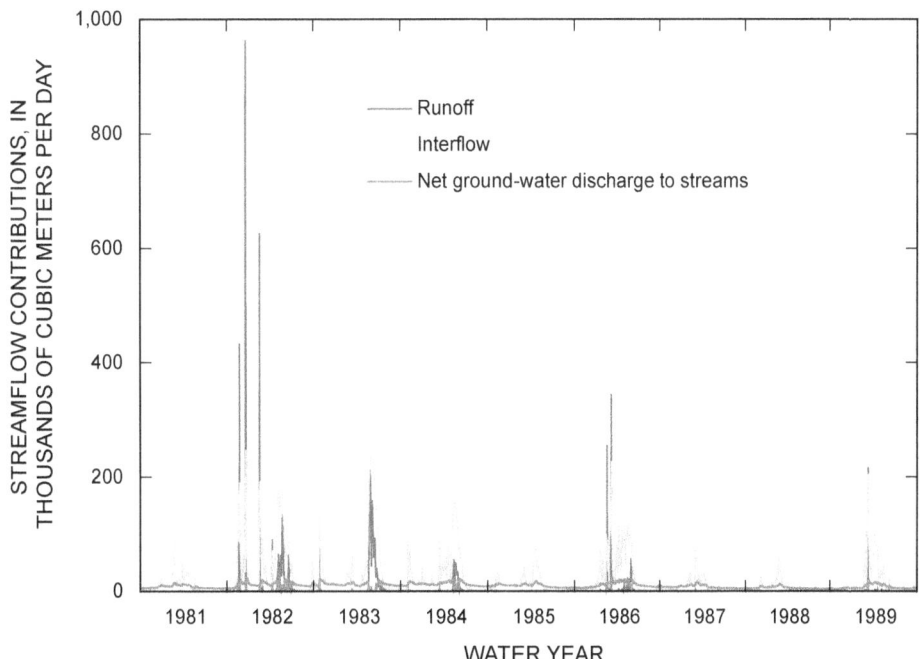

Figure 48. Components of streamflow during water years 1981–89. Runoff and interflow are from a GSFLOW Comma-Separated Values (CSV) File using variables `basinsroff`, and `basininterflow`; streamflow is from the same file using variable `basinstrmflow`; and net ground-water discharge to streams was calculated by subtracting runoff and interflow from streamflow.

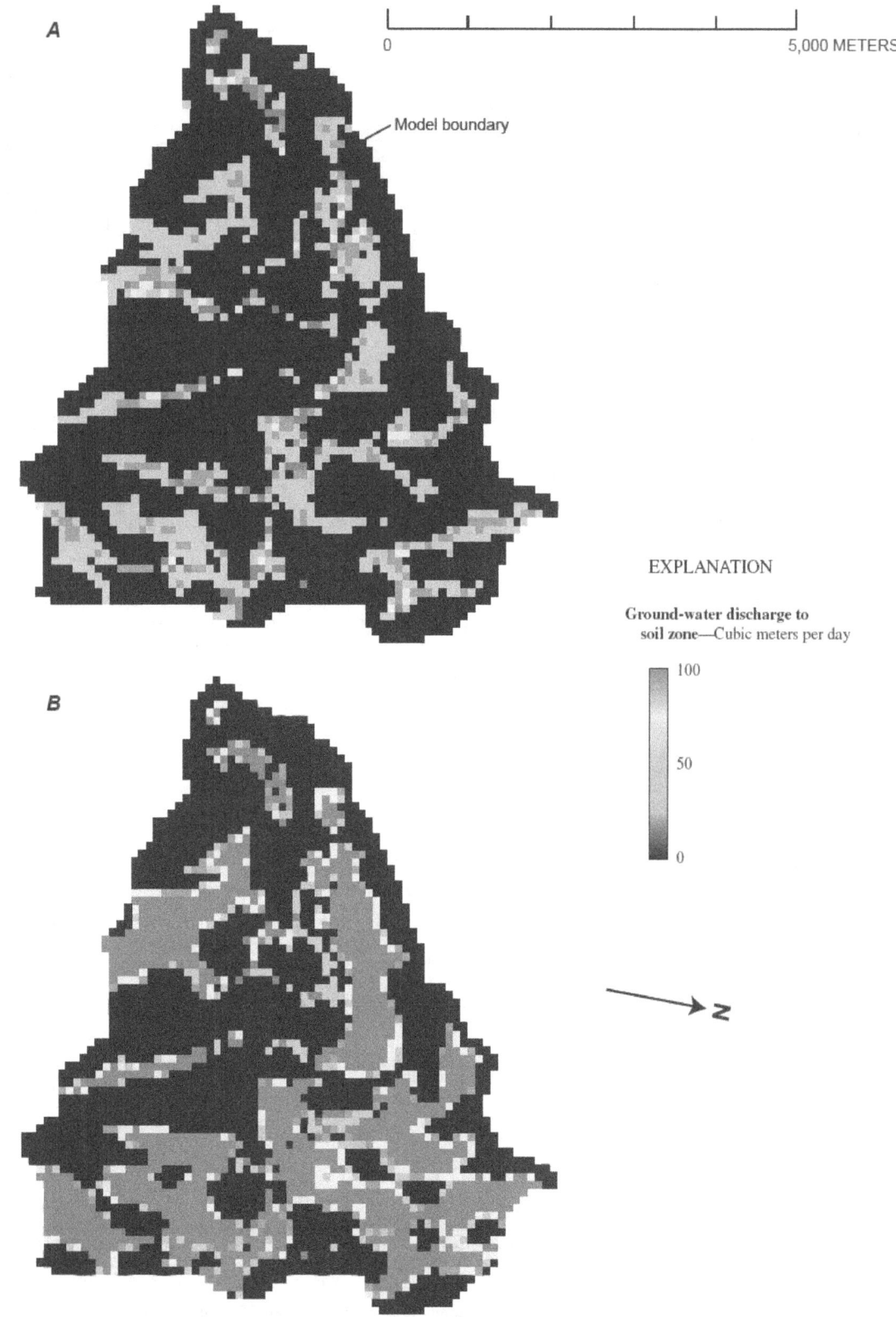

Figure 49.   Distribution of ground-water discharge to the soil zone for (*A*) steady-state simulation of the autumn low precipitation period, and (*B*) July 8, 1983. Cell-by-cell values of ground-water discharge are from an Unsaturated-Zone Flow Package unformatted output file.

Simulated ground-water discharge to the soil zone showed seasonal variations, as illustrated by the difference in the steady-state distribution of ground-water discharge (simulation of mean flow during the autumn low-precipitation period) and ground-water discharge during the peak snowmelt in July 1983 (fig. 49). The total ground-water discharge to the soil zone was about 0.19 m^3/s during October (corresponding to the steady-state discharge in fig. 49) and 0.28 m^3/s during June 1983. Ground-water discharge to the soil zone can contribute interflow and surface runoff and thus, ground water can contribute flow to streams through surface runoff and

interflow. Seasonal variation in ground-water discharge to the soil zone likely is realistic given the large seasonal variation in precipitation and streamflow in the watershed.

Simulated water-content profiles beneath the soil-zone base varied seasonally and generally were greatest during the spring snowmelt period. Water-content profiles are shown in figure 50 for a MODFLOW finite-difference cell located about midway between the peak of the watershed and the watershed outlet that had a land-surface altitude 82 m greater than the altitude of the watershed outlet (layer 1, row 45, and column 30). Maximum saturations in the upland

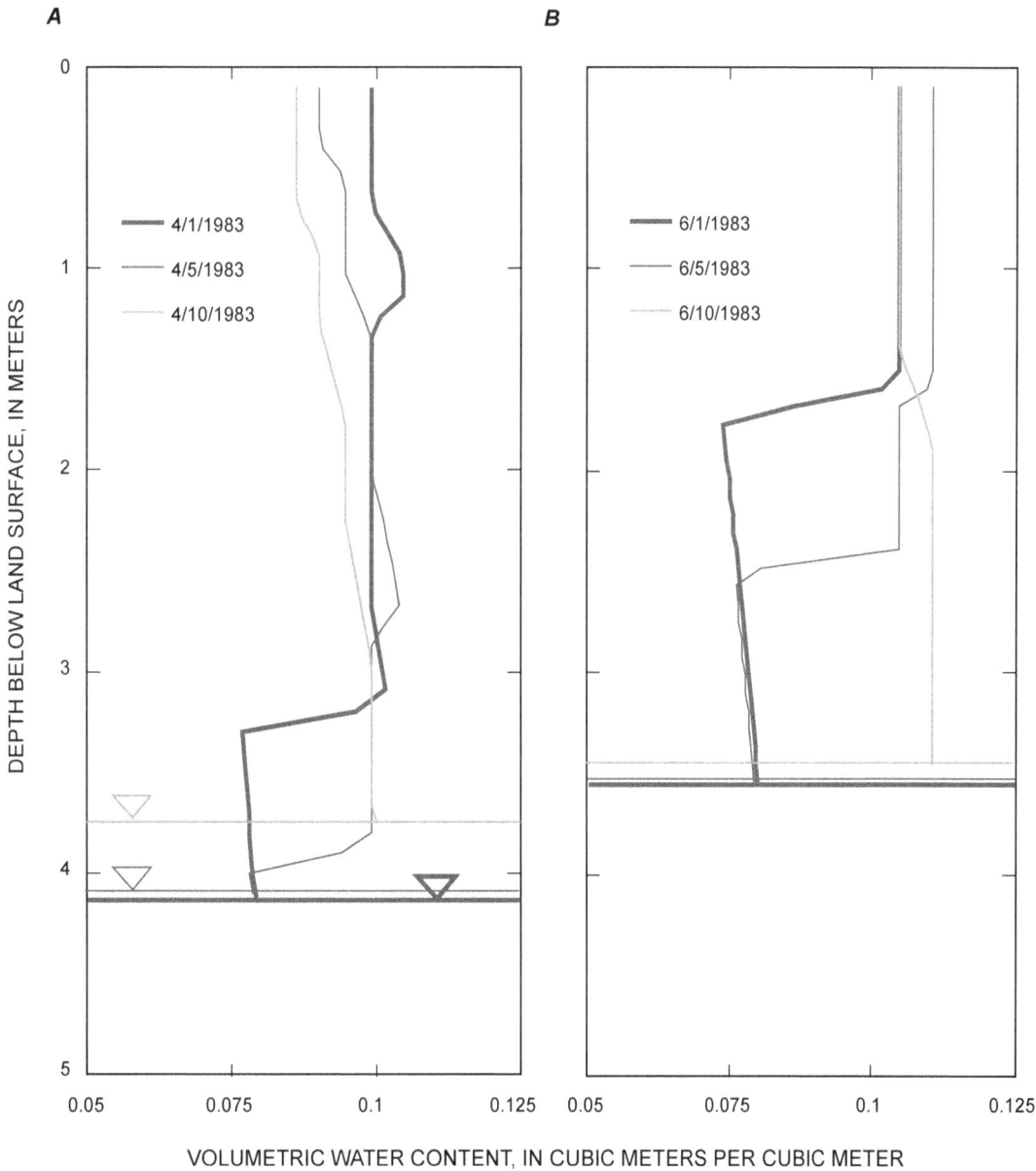

Figure 50. Water-content profiles below land surface (finite-difference cell layer 1, row 45, and column 30)—(A) during April 1983, and (B) during June 1983. Water-content values are from an Unsaturated-Zone Package formatted output file.

areas were approximately 67 percent of the saturated water content, which ranged between 0.15 and 0.25. The water table fluctuated seasonally by more than 4 m, resulting in seasonal disappearance of the unsaturated zone beneath the soil-zone base in places where the water-table depth was less than 4 m.

Annual Water Budgets and Errors

Volumetric budgets and errors, on an annual basis, are presented in table 18 for water years 1982–95. These results were obtained from the GSFLOW Water-Budget File (fig. 29), which is described in detail in section "Water Budget." The percent discrepancy values for each year are all very good.

Sensitivity to Unsaturated-Zone and Aquifer Hydraulic Conductivity

Model sensitivity to unsaturated-zone and aquifer hydraulic conductivity (K) was analyzed by adjusting the value of CNSTNT (Harbaugh, 2005; p. 8–57), which is used to equally scale K values for all cells. Net gravity drainage into the unsaturated zone increased proportionally to the increase in unsaturated zone and aquifer K, except during periods when soil-zone storage was depleted. The amount of water available for gravity drainage into the unsaturated zone beneath the soil zone during spring snowmelt was limited by the vertical K in the unsaturated zone for lower K values (fig. 51A). The decrease in gravity drainage for the lowest K values was enhanced by the increase in area where ground-water levels were at or above land surface; saturation excess caused a

Table 18. Annual water budget for GSFLOW simulation of the Sagehen Creek watershed example problem, 1982–95.

[All values are reported in units of cubic meters, except percent discrepancy is dimensionless]

| Water year | In | Out | | | In–Out |
	Precipitation	Evapotranspiration	Streamflow	Boundary flow	
1982	7.2483E+07	3.7929E+07	3.0458E+07	478,282.41	3,617,919.50
1983	1.1747E+08	6.1108E+07	5.2342E+07	720,214.69	3,297,289.25
1984	1.5372E+08	8.4367E+07	6.7816E+07	962,476.62	576,307.38
1985	1.7429E+08	1.0141E+08	7.5457E+07	1,201,432.75	-3,773,800.75
1986	2.1656E+08	1.2401E+08	9.1433E+07	1,441,176.00	-329,792.00
1987	2.2903E+08	1.3783E+08	9.6804E+07	1,678,463.38	-7,284,239.50
1988	2.4382E+08	1.5183E+08	1.0093E+08	1,913,408.12	-1.0854E+07
1989	2.7405E+08	1.7128E+08	1.0964E+08	2,148,526.75	-9,014,063.00
1990	2.9528E+08	1.8905E+08	1.1472E+08	2,383,076.75	-1.0870E+07
1991	3.1330E+08	2.0617E+08	1.1894E+08	2,616,651.50	-1.4427E+07
1992	3.2835E+08	2.1855E+08	1.2299E+08	2,849,880.75	-1.6040E+07
1993	3.6674E+08	2.3905E+08	1.3559E+08	3,084,391.75	-1.0983E+07
1994	3.8064E+08	2.5197E+08	1.3991E+08	3,318,625.50	-1.4556E+07
1995	4.2954E+08	2.7376E+08	1.5948E+08	3,555,781.75	-7,263,078.00

| Water year | Storage change | | | | | Error | Percent discrepancy |
	Surface	Soil zone	Unsaturated	Saturated	Total		
1982	120,517.80	2,660,699.00	-1,377,943.75	2,208,153.25	3,611,416.75	6,502.75	0.01
1983	7,197.33	1,636,156.88	-2,109,868.75	3,746,286.50	3,279,756.50	17,532.75	.01
1984	-1.00	385,343.59	-2,337,493.50	2,503,117.00	550,946.69	25,360.69	.02
1985	31,096.60	627,998.62	-2,079,286.75	-2,372,487.50	-3,792,694.50	18,893.75	.01
1986	35,995.13	1,863,144.12	-873,155.62	-1,377,784.75	-351,815.53	22,023.53	.01
1987	-.50	136,199.17	-1,461,251.12	-5,969,752.00	-7,294,821.00	10,581.50	.00
1988	4,490.44	185,758.05	-706,397.12	-1.0329E+07	-1.0845E+07	-9,349.00	.00
1989	2,953.11	1,075,756.12	1,275,628.12	-1.1354E+07	-8,999,396.00	-14,667.00	-.01
1990	-1.13	1,326,907.25	1,500,342.38	-1.3677E+07	-1.0850E+07	-20,031.00	-.01
1991	-.62	316,719.00	1,608,705.62	-1.6317E+07	-1.4392E+07	-35,615.00	-.01
1992	-.53	206,177.45	2,149,828.75	-1.8342E+07	-1.5986E+07	-53,920.00	-.02
1993	2.50	196,883.77	4,265,108.50	-1.5384E+07	-1.0922E+07	-60,893.00	-.02
1994	2.47	193,511.73	2,192,072.00	-1.6876E+07	-1.4490E+07	-66,133.00	-.02
1995	39,441.46	264,386.19	3,940,972.50	-1.1445E+07	-7199875.50	-63,202.50	-.01

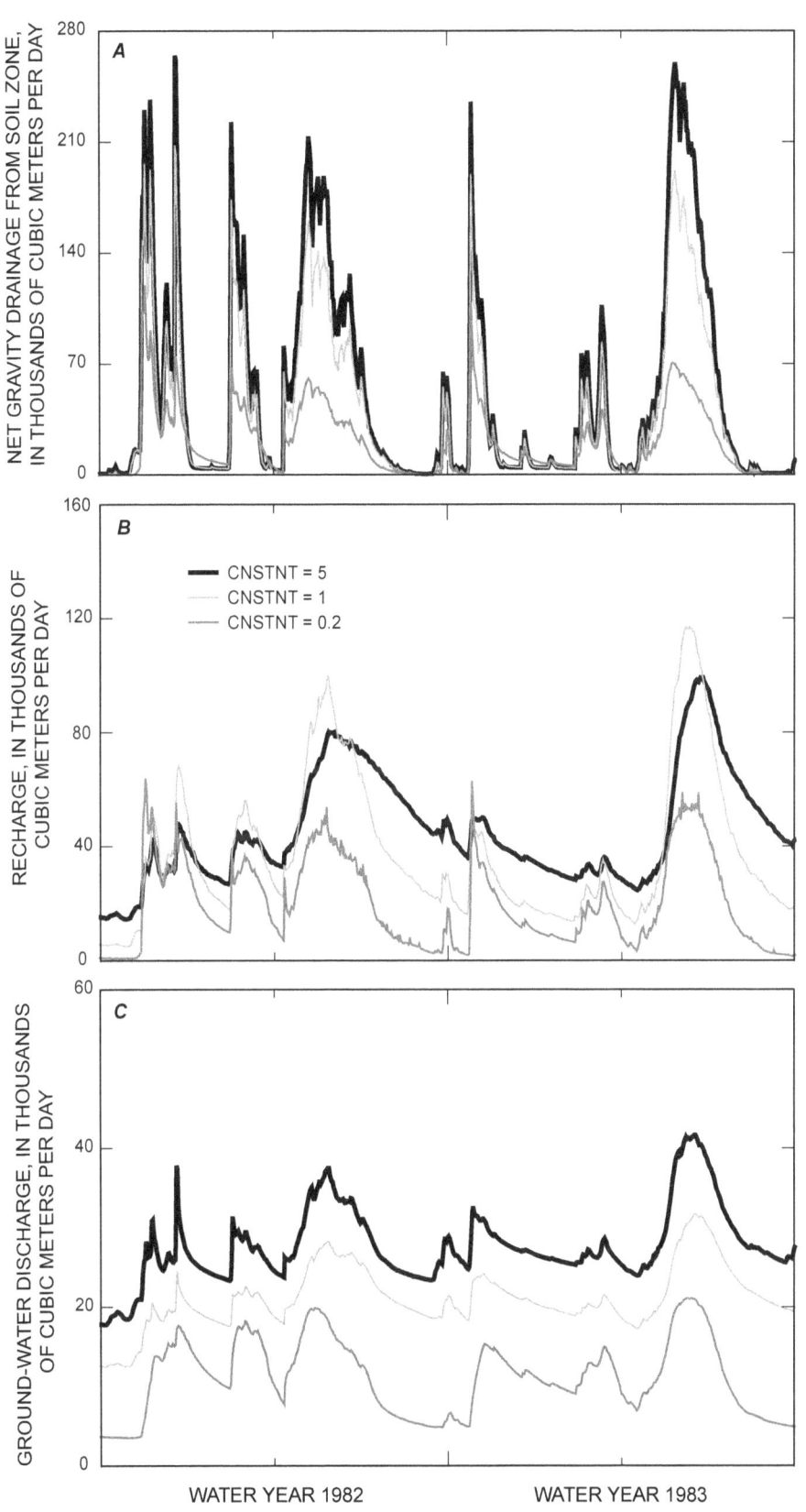

Figure 51. Sensitivity analysis for daily volumetric flow of (*A*) net gravity drainage from the soil zone, (*B*) recharge to the saturated zone, and (C) ground-water discharge to land surface summed over the Sagehen Creek watershed for water years 1982–83. The variable CNSTNT is multiplied by the hydraulic conductivity of all finite-difference cells (Harbaugh, 2005; p. 8–57). Values are from an Unsaturated-Zone Flow Package formatted output file.

non-linear relation between K and net gravity drainage for the Sagehen Creek watershed (fig. 51). Generally, several factors affect the net gravity drainage beneath the soil zone, including the amount of soil-zone storage, the area where ground-water levels are at or near land surface, and the vertical K in the unsaturated zone. The extent to which each of these factors affects net gravity drainage is dependent on the climate, land surface topography, and subsurface properties in a watershed or basin.

Total daily ground-water recharge rates did not increase proportionately to the increase in K. The medium-K distribution resulted in the greatest ground-water recharge rates during the snowmelt period (fig. 51B). Much of the recharge occurred where the water table was slightly beneath the soil-zone base, whereas in areas with a thick unsaturated zone, the total daily ground-water recharge rates were attenuated due to coalescence of high unsaturated-zone flow rates with lower rates. Attenuation occurs to a lesser degree in thin unsaturated zones because wetting fronts reach the water table before being attenuated. These results are attributed to the relation between the horizontal K and the thickness of the unsaturated zone. Generally, high horizontal K values allow ground water to drain out of the watershed, which thickens the unsaturated zone beneath much of the watershed. The high recharge rates for the medium-K distribution as compared with the low-K distribution are caused by much greater gravity drainage from the soil zone. The high-K distribution allowed for more water to be stored in the unsaturated zone, which recharged ground water much longer after the spring snowmelt than either the low– or medium–K distributions (fig. 51B). The upland areas comprise a larger fraction of the total watershed area compared with the valley lowland areas at Sagehen Creek; consequently, a change in K values in these areas can greatly change the timing and magnitude of recharge in the watershed.

Ground-water discharge to the soil zone increased with increases in K for the range in K values used in this sensitivity analysis (fig. 51C). However, yearly maximum ground-water discharge rates may not increase with higher values of K. The yearly maximum ground-water discharge rates decreased when K was increased by a factor of 100 for a simulation with one model layer (Niswonger and others, 2006b).

Discussion of Results

Runoff and ground-water recharge at Sagehen Creek watershed was dominated by water made available by snowmelt. Peak evapotranspiration rates generally occurred slightly later than peak snowmelt and runoff. Most of the precipitation at Sagehen Creek watershed occurred as snow and accumulated in snowpack. Water became available to plants as melt water during the warmer months of the year when plants reach their highest transpiration rates. This is dissimilar to lower altitude basins where winter precipitation occurs as rain and becomes runoff prior to the period of maximum evapotranspiration rates.

The simulated streamflow at Sagehen Creek received contributions from surface runoff, interflow, and net ground-water discharge directly to streams. Net ground-water discharge is the total ground-water discharge to streams minus total stream leakage to ground water upstream from the Sagehen Creek gage. Surface runoff and interflow provided most of the annual streamflow from 1981 to 1995. Some of the interflow (and surface runoff) simulated in GSFLOW was from ground-water discharge into the soil zone in lowland areas where the water table was greater than the soil-zone base. During years of above-average precipitation, interflow lasted throughout the water year and remained greater than the net ground-water discharge to stream reaches into the summer; for 1983, interflow began to decrease about July 1 (fig. 52A). However, during years of below-average precipitation, there were several month-long periods during which net ground-water discharge was greater than interflow and surface runoff. Net ground-water discharge decreased during peak streamflow because stream stage exceeded the ground-water heads in many stream reaches. However, ground-water head subsided slower than the streamflow due to bank storage effects and diffuse recharge from nearby cells. This resulted in net ground-water discharge back to the stream reaches to be greatest following high streamflow (fig. 52).

The integrated model resulted in a modest improvement of the Nash-Sutcliffe coefficient compared with the PRMS-only simulation from 0.79 to 0.81 for the calibration period (water years 1982–88) but the integrated model resulted in a marked increase from 0.71 to 0.85 for the verification period (water years 1989–95). The less rigorous approach for simulating subsurface flow and storage with PRMS-only was not able to simulate the observed decadal variations in base flow. The integrated GSFLOW simulation with MODFLOW-2005 was required to approximate the observed decadal variations in base flow, which improved the simulation of streamflow for the drier verification period. Decadal variations in base flow may not be important for flood predictions, or for assessing volumes of streamflow from snowmelt runoff. Integrated models may not be practical for such applications. However, integrated models are necessary whenever ground-water interactions with surface water are of concern. Examples include (1) effects of ground-water withdrawals on surface-water supplies and stream ecology, (2) effects caused by stream diversions and reservoirs on ground water and associated wetland systems, and (3) effects of conjunctive use of surface- and ground-water supplies. Thus, integrated models should not be assessed solely by how well the model predicts streamflow at the outlet of a basin using a single measure of goodness of fit. Rather, integrated models should be evaluated on the basis of a variety of different measures to evaluate the importance of changes in surface and subsurface flows and storages within the modeled domain. Such measures might include changes in snowpack depth and water content, unsaturated zone moisture profiles, and ground-water heads.

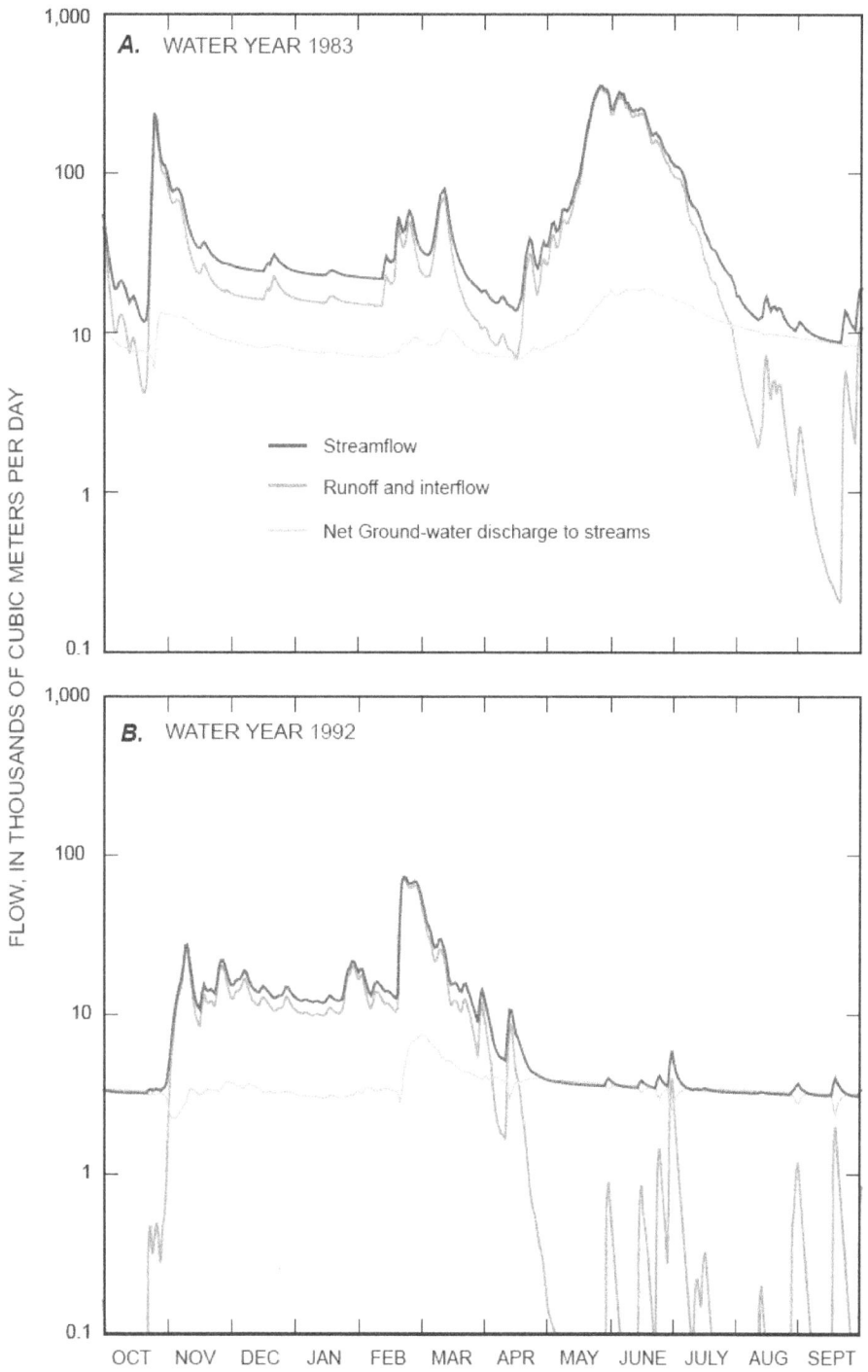

Figure 52. *Streamflow components for (A)* 1983, a year of above-average precipitation, and *(B)* 1992, a year of below-average precipitation. Runoff and interflow are from a GSFLOW Comma-Separated Values (CSV) File using variables `basinsroff` and `basininterflow`; streamflow is from same file using variable `basinstrmflow`; and net ground-water discharge to stream reaches was calculated by subtracting runoff and interflow from streamflow.

Acknowledgments

The development of GSFLOW was supported through the USGS Ground-Water Resources Program, National Streamflow Information Program, the Watershed and River System Management Program, and National Research Program; the Pennsylvania Department of Environmental Protection; and the Clearwater Conservancy of State College, Pennsylvania. The authors thank George H. Leavesley (retired USGS National Research Program, Lakewood, Colorado), Arlen Harbaugh (retired, USGS Office of Ground Water, Reston, Virginia), Wolfgang-Albert Fluegel (Friedrich-Schiller-Universitat, Jena, Germany), Arlen Feldman (retired, U.S. Army Corps of Engineers, Davis, California), Jon P. Fenske (U.S. Army Corps of Engineers, Davis, California), and Douglas Boyle (Desert Research Institute, University of Nevada, Reno, Nevada) for their support and encouragement throughout the development of GSFLOW. The Hydrologic Research Center, U.S. Army Corps of Engineers provided initial support in designing a method for connecting rainfall-runoff models with MODFLOW, which led to the development of the Unsaturated-Zone Flow Package (Niswonger and others, 2006a). Discussions with Greg Pohll (Desert Research Institute, Reno, Nevada) led to important changes to the conceptualization of GSFLOW. Roland Viger (USGS National Research Program, Lakewood, Colorado) was instrumental in developing programs using Geographic Information System (GIS) software to create input for the Sagehen Creek watershed example problem. Rose Medina (USGS, Carson City, Nevada) developed the three-dimensional diagrams of model input data and results used to illustrate the Sagehen Creek watershed example.

The authors also thank Randall J. Hunt and Steven M. Westenbroek (USGS, Middleton, Wisconsin) and John Fulton (USGS, Pittsburgh, Pennsylvania) and Dennis Risser (USGS, New Cumberland, Pennsylvania) for their ideas and testing of initial versions of GSFLOW using sample data sets. We greatly appreciate the thorough reviews and comments that greatly improved the quality of this report. These technical reviewers were Arlen Harbaugh (retired, USGS Office of Ground Water, Reston, Virginia), David M. Bjerklie (USGS, East Hartford, Connecticut), Randall J. Hunt, John F. Walker, and Steven M. Westenbroek (USGS, Middleton, Wisconsin), and Kathleen M. Flynn (USGS Office of Surface Water, Reston, Virginia).

References Cited

Ahlfeld, D.P., Barlow, P.M., and Mulligan, A.E., 2005, GWM-a ground-water management process for the U.S. Geological Survey modular ground-water model (MODFLOW-2000): U.S. Geological Survey Open-File Report 2005-1072, 124 p.

American National Standards Institute, 1992, American national standard for programming language—Fortran—Extended: X3.198-1992, 369 p.

Anderman, E.R., and Hill, M.C., 2000, MODFLOW-2000, the U.S. Geological Survey modular ground-water model—Documentation of the Hydrogeologic-Unit Flow (HUF) Package: U.S. Geological Survey Open-File Report 2000-342, 89 p.

Anderman, E.R., and Hill, M.C., 2003, MODFLOW-2000, the U.S. Geological Survey modular ground-water model—Three additions to the Hydrogeologic-Unit Flow (HUF) Package: Alternative storage for the uppermost active cells, flows in hydrogeologic units, and the hydraulic-conductivity depth-dependence (KDEP) capability: U.S. Geological Survey Open-File Report 2003-347, 36 p.

Anderman, E.R., Kipp, K.L., Hill, M.C., Valstar, J., and Neupauer, R.M., 2002, MODFLOW-2000, the U.S. Geological Survey modular ground-water model—Documentation of the model-layer variable-direction horizontal anisotropy (LVDA) capability of the Hydrogeologic-Unit Flow (HUF) Package: U.S. Geological Survey Open-File Report 2002-409, 61 p.

Anderson, E.A., 1968, Development and testing of snow pack energy balance equations: Water Resources Research, v. 4, no. 1, p. 19-38.

Anderson, E.A., 1973, National Weather Service river forecast system—Snow accumulation and ablation model: NOAA Tech. Memorandum NWS Hydro-17, U.S. Dept. of Commerce, Silver Spring, Maryland, p. 3-7.

Anderson, E.I., 2005, Modeling groundwater–surface water interactions using the Dupuit approximation: Advances in Water Resources, v. 28, no. 4, p. 315-327.

Bauer H.H., and Vaccaro, J.J., 1987, Documentation of a deep percolation model for estimating ground-water recharge: U.S. Geological Survey Open-File Report 86-536, 180 p.

Bear, Jacob, 1972, Dynamics of fluids in porous media: New York, American Elsevier Publishing Co., Inc., 764 p.

Berris, S.N., Hess, G.W., and Bohman, L.R., 2001, River and reservoir operation model, Truckee River basin, California and Nevada, 1998: U.S. Geological Survey Water-Resources Investigations Report 01-4017, 138 p.

Beven, K.J., Kirkby, M.J., Schoffield, N., and Tagg, H., 1984, Testing a physically-based flood forecasting model (TOPMODEL) for three UK catchments: Journal of Hydrology, v. 69, p. 119–143.

Bredehoeft, J.D., and Pinder, G.F., 1970, Digital analysis of areal flow in multiaquifer groundwater systems: a quasi three-dimensional model: Water Resources Research, v. 6, no. 3, p. 883-888.

Brooks, R.H., and Corey, A.T., 1966, Properties of porous media affecting fluid flow: Journal of Irrigation and Drainage, v. 101, p. 85-92.

Burden, R.L., and Faires, J.D., 1997, Numerical Analysis: Pacific Grove, Calif., Brooks/Cole Publishing Company, 811 p.

Burnett, J.L., and Jennings, C.W., 1965, Chico Quadrangle, scale 1:250,000: State of California, Division of Mines and Geology.

Charbeneau, R.J., 1984, Kinematic models for soil moisture and solute transport: Water Resources Research, v. 20, no. 6, p. 699-706.

Chen, Z., Govindaraju, R.S., and Kavvas, M.L., 1994, Spatial averaging of unsaturated flow equations under infiltration conditions over areally heterogeneous fields-1. Development of models: Water Resources Research, v. 30, no. 2, p. 523-533.

Cheng, X., and Anderson, M.P., 1993, Numerical simulations of ground-water interaction with lakes allowing for fluctuating lake levels: Ground Water, v. 31, no. 6, p. 929-933.

Chow, V.T., Maidment, D.R., and Mays, L.W., 1988, Applied Hydrology: New York, New York, McGraw-Hill, Inc., 572 p.

Clark, M.P., and Hay, L.E., 2004, Use of medium-range numerical weather prediction model output to produce forecasts of streamflow: Journal of Hydrometeorology, v. 5, no. 1, p. 15-32.

Colbeck, S.C., 1972, A theory of water percolation in snow: Journal of Glaciology, v. 2, no. 63, p. 369-385.

Cosner, O.J., and Harsh, J.F., 1978, Digital-model simulation of the glacial-outwash aquifer, Otter Creek –Dry Creek basin, Cortland County, New York: U.S. Geological Survey Water-Resources Investigations Report 78-71, 34 p.

Council, G.W., 1998, A lake package for MODFLOW, in Poeter, E., Zheng, C., and Hill, M.C., MODFLOW '98: Colorado School of Mines, Golden, Colo., v. 2, p. 675-682.

Daly, C., Neilson, R.P., and Phillips, D.L., 1994, A statistical-topographic model for mapping climatological precipitation over mountainous terrain: Journal of Applied Meteorology, v. 33, p. 140-158.

Danskin, W.R., 1998, Evaluation of the hydrologic system and selected water-management alternatives in the Owens Valley, California: U.S. Geological Survey Water-Supply Paper 2370-H, 175 p.

Dean, J.D., and Snyder, W.M., 1977, Temporarily and areally distributed rainfall: Proceedings of the American Society of Civil Engineers, Journal of the Irrigation and Drainage Division, v. 103, no. IR2, p. 221-229.

Dickinson, W.T., and Whiteley, H.Q., 1970, Watershed areas contributing to runoff: International Association of Hydrologic Sciences Publication 96, 1.12-11.28 p.

Dunne, T., and Black, R.G., 1970, An experimental investigation of runoff production in permeable soils: Water Resources Research, v. 6, p. 478-490.

Ely, D.M., 2006, Analysis of sensitivity of simulated recharge to selected parameters for seven watersheds modeled using the precipitation-runoff modeling system: U.S. Geological Survey Scientific Investigations Report 2006-5041, 21 p.

Emerson, D.G., 1991, Documentation of a heat and water transfer model for seasonally frozen soils with application to a precipitation-runoff model: U.S. Geological Survey Open-File Report 91-462, 92 p.

Federer, A.C., and Lash, D., 1978, Brook: A hydrologic simulation model for eastern forests: University of New Hampshire, Water Resources Research Center, Research Report No. 19, 84 p.

Fenske, J.P., Leake, S.A., and Prudic, D.E., 1996, Documentation of a computer program (RES1) to simulate leakage from reservoirs using the modular finite-difference ground-water flow model (MODFLOW): U.S. Geological Survey Open-File Report 96-364, 51 p.

Fenske, J.P., and Prudic, D.E., 1998, Development of HMS/MODFLOW for simulation of surface and groundwater flow, in Poeter, E., Zheng, C., and Hill, M., MODFLOW '98: Golden, Colo., Colorado School of Mines, v. 1, p. 463-470.

Ford, L.R., and Fulkerson, D.R., 1956, Maximal flow through a network: Canadian Journal of Mathematics, p. 399-404.

Frank, E.C., and Lee, R., 1966, Potential solar beam irradiation on slopes: U. S. Department of Agriculture, Forest Service Research Paper RM-18, 116 p.

Fread, D.L., 1993, Flow routing, *in* Maidment, D.R., ed., Handbook of Hydrology: New York, McGraw-Hill, Inc., p. 36.

Freeze, R.A., 1971, Three-dimensional, transient, saturated-unsaturated flow in a groundwater basin: Water Resources Research, v. 7, no. 2, p. 347-366.

Freeze, R.A., 1972, Role of subsurface flow in generating overland runoff, 1—Baseflow contributions to channel flow: Water Resources Research, v. 8, no. 3, p. 609-623.

Freeze, R.A., and Cherry, J.A., 1979, Groundwater: Englewood Cliffs, New Jersey, Prentice-Hall, Inc., 604 p.

Fulp, T.J., Vickers, W.B., Williams, B., and King, D.L., 1995, Decision support for water resources management in the Colorado River region, *in* Ahuja, L., Leppert, J., Rojas, K., and Seely, E., Workshop on computer applications in water management: Fort Collins, Colo., p. 24-27.

Goode, D.J., and Appel, C.A., 1992, Finite-difference interblock transmissivity for unconfined aquifers and for aquifers having smoothly varying transmissivity: U.S. Geological Survey Water-Resources Investigations Report 92-4124, 79 p.

Green, W.H., and Ampt, G.A., 1911, Studies on soil physics, part I, the flow of air and water through soils: Journal of Agricultural Science, v. 4, no. 1, p. 1-24.

Guo, W., and Langevin, C.D., 2002, User's Guide to SEAWAT: A computer program for simulation of three-dimensional variable-density ground-water flow: U.S. Geological Survey Techniques of Water-Resources Investigations, Book 6, Chapter A7, 77 p.

Halford, K.J., and Hanson, R.T., 2002, User guide for the drawdown-limited, multi-node well (MNW) package for the U.S. Geological Survey's modular three-dimensional finite-difference ground-water flow model, versions MODFLOW-96 and MODFLOW-2000: U.S. Geological Survey Open-File Report 2002-293, 33 p.

Hamon, W.R., 1961, Estimating Potential evapotranspiration: Proceedings of the American Society of Civil Engineers, Journal of the Hydraulic Division, v. 87, no. HY3, p. 107-120.

Harbaugh, A.W., 2005, MODFLOW-2005, the U.S. Geological Survey modular ground-water model-the Ground-Water Flow Process: U.S. Geological Survey Techniques and Methods 6-A16, variously paginated.

Harbaugh, A.W., Banta, E.R., Hill, M.C., and McDonald, M.G., 2000, MODFLOW-2000, the U.S. Geological Survey modular ground-water model-modularization concepts and the ground-water flow process: U.S. Geological Survey Open-File Report 00-92, 121 p.

Harbaugh, A.W., and McDonald, M.G., 1996, User's documentation for MODFLOW-96, and update to the U.S. Geological Survey modular finite-difference ground-water flow model: U.S. Geological Survey Open-File Report 96-485, 56 p.

Harter, T., and Hopmans, J.W., 2004, Role of vadose-zone flow processes in regional scale hydrology-review, opportunities and challenges, *in* Feddes, R.A., De Rooij, G.H., and Van Dam, J.C., eds., Unsaturated Zone Modeling—Progress, Challenges and Applications: Dordrecht, The Netherlands, Kluwer Academic Publisher, Wageningen Frontis Series, p. 179-208.

Hay, L.E., and Clark, M.P., 2003, Use of statistically and dynamically downscaled atmospheric model output for hydrologic simulations in three mountainous basins in the western United States: Journal of Hydrology, v. 282, p. 56-75.

Hay, L.E., Clark, M.P., Wilby, R.L., Gutowski, W.J., Leavesley, G.H., Pan, Z., Arritt, R.W., and Takle, E.S., 2002, Use of regional climate model output for hydrologic simulations: Journal of Hydrometeorology, v. 3, p. 571-590.

Hay, L.E., Wilby, R.L., and Leavesley, G.H., 2000, A comparison of delta change and downscaled GCM scenarios for three mountainous basins in the United States: Journal of the American Water Resources Association, v. 36, p. 387-397.

Hewlett, J.D., and Nutter, W.L., 1970, The varying source area of streamflow from upland basins, *in* Symposium on Interdisciplinary Aspects of Watershed Management, Montana State University, Bozeman, Mont., 1970 [Proceedings], p. 65-83.

Hill, M.C., 1990, Preconditioned Conjugate Gradient 2 (PCG2), A computer program for solving ground-water flow equations: U.S. Geological Survey Water-Resources Investigations Report 90-4048, 43 p.

Hill, M.C., Banta, E.R., and Harbaugh, A.W., 2000, MODFLOW-2000, the U.S. Geological Survey Modular Ground-Water Model—User Guide to the observations, sensitivity, and parameter-estimation processes and three post-processing programs: U.S. Geological Survey Open-File Report 00-184, 210 p.

Hoffmann, John, Leake, S.A., Galloway, D.L., and Wilson, A.M., 2003, MODFLOW-2000 Ground-Water Model— User Guide to the Subsidence and Aquifer-System Compaction (SUB) Package: U.S. Geological Survey Open-File Report 2003-233, 44 p.

Horton, R.E., 1933, The role of infiltration in the hydrological cycle: American Geophysical Union Transactions, v. 23, p. 479-482.

Hsieh, P.A., and Freckleton, J.R., 1993, Documentation of a computer program to simulate horizontal-flow barriers using the U.S. Geological Survey's modular three-dimensional finite-difference ground-water flow model: U.S. Geological Survey Open-File Report 92-477, 32 p.

Hunt, R.J., and Steuer, J.J., 2000, Simulation of the recharge area for Frederick Springs, Dane County, Wisconsin : U.S. Geological Survey Water-Resources Investigations Report 00-4172, 33 p.

Jensen, M.E., Rob, D.C.N., and Franzoy, C.E., 1969, Scheduling irrigations using climate-crop-soil data, National Conference on Water Resources Engineering of the American Society of Civil Engineers: New Orleans, La. [Proceedings], p. 20.

Jeton, A.E., 1999, Precipitation-runoff simulations for the upper part of the Truckee River Basin, California and Nevada: U.S. Geological Survey Water-Resources Investigations Report 99-4282, 41 p.

Jobson, H.E., and Harbaugh, A.W., 1999, Modifications to the diffusion analogy surface-water flow model (DAFLOW) for coupling to the modlular finite-difference ground-water flow model (MODFLOW): U.S. Geological Survey Open-File Report 99-217, 107 p.

Kahrl, W.L. (ed.), 1979, The California Water Atlas: North Highlands, Calif., Governer's Office of Planning and Research, State of California, 118 p.

Koczot, K.M., Jeton, A.E., McGurk, B.J., and Dettinger, M.D., 2005, Precipitation-runoff processes in the Feather River Basin, northeastern California, with prospects for streamflow predictability, water years 1971-97: U.S. Geological Survey Scientific Investigations Report 2004-5202, 82 p.

Konikow, L.F., Goode, D.J., and Hornberger, G.Z., 1996, A three-dimensional method-of-characteristics solute-transport model (MOC3D): U.S. Geological Survey Water-Resources Investigations Report 96-4267, 87 p.

Langevin, C.D., Shoemaker, W.B., and Guo, W., 2003, MODFLOW-2000, the U.S. Geological Survey modular ground-water model-documentation of the SEAWAT-2000 version with the variable-density flow process (VDF) and the integrated MT3DMS transport process (IMT): U.S. Geological Survey Open-File Report 03-426, 43 p.

Leaf, C.F., and Brink, G.E., 1973, Hydrologic simulation model of Colorado subalpine forest: U.S. Department of Agriculture, Forest Service Research Paper RM-107, 23 p.

Leake, S.A., and Lilly, M.R., 1997, Documentation of a computer program (FHB1) for assignment of transient specified-flow and specified-head boundaries, in Applications of the Modular Finite-Difference Ground-Water Flow Model (MODFLOW): U.S. Geological Survey Water-Resources Investigations Report 97-571, 50 p.

Leake, S.A., and Prudic, D.E., 1991, Documentation of a computer program to simulate aquifer-system compaction using the modular finite-difference ground-water flow model: U.S. Geological Survey Techniques of Water-Resources Investigations, Book 6, Chapter A2, 68 p.

Leavesley, G.H., Lichty, R.W., Troutman, B.M., and Saindon, L.G., 1983, Precipitation-runoff modeling system—User's manual: U.S. Geological Survey Water-Resources Investigations Report 83-4238, 207 p.

Leavesley, G.H., Markstrom, S.L., Brewer, M.S., and Viger, R.J., 1996a, The Modular Modeling System (MMS)—The physical process modeling component of a database-centered decision support system for water and power management: Water, Air, and Soil Pollution, v. 90, p. 303-311.

Leavesley, G.H., Markstrom, S.L., Viger, R.J., and Hay, L.E., 2005, USGS Modular Modeling System (MMS)— Precipitation-Runoff Modeling System (PRMS) MMS-PRMS, in Singh, V., and Frevert, D., eds., Watershed Models: Boca Raton, Fla., CRC Press, p. 159-177.

Leavesley, G.H., Restrepo, P.J., Markstrom, S.L., Dixon, M., and Stannard, L.G., 1996b, The Modular Modeling System (MMS): User's manual: U.S. Geological Survey Open-File Report 96-151, 142 p.

Leavesley, G.H., and Stannard, L.G., 1995, The precipitation-runoff modeling system—PRMS, in Singh, V.P., ed., Computer Models of Watershed Hydrology: Highlands Ranch, Colo., Water Resources Publications, p. 281-310.

Lee, R., 1963, Evaluation of solar beam irradiation as a climatic parameter of mountain watersheds: Colorado State University Hydrology Papers, no. 2, 50 p.

Leopold, L.B., Wolman, M.G., and Miller, J.P., 1992, Fluvial Processes in Geomorphology (2d ed.): Mineola, New York, Dover Publications, Inc., 522 p.

Lighthill, M.J., and Whitham, G.B., 1955, On kinematic floods—flood movements in long rivers: Proceedings, R. Soc. London, v. A220, p. 281-316.

Lindgren, R.J., and Landon, M.K., 1999, Effects of ground-water withdrawals on the Rock River and associated valley aquifer, eastern Rock County, Minnesota: U.S. Geological Survey Water-Resources Investigations Report 98-4157, 103 p.

Linsley, R.K., Kohler, M.A., and Paulhus, J.L.H., 1975, Hydrology for Engineers: New York, McGraw-Hill, Inc., 482 p.

Lohman, S.W., and others, 1972, Definitions of selected ground-water terms-revisions and conceptual refinements: U.S. Geological Survey Water-Supply Paper, 21 p.

Male, D.H., and Gray, D.M., 1981, Snowcover ablation and runoff, in Gray, D.M., and Male, D.H., eds., Handbook of snow: Ontario, Pergamon Press Canada Ltd., p. 360-430.

Mantoglou, A., 1992, A theoretical approach for modeling unsaturated flow in spatially variable soils—Effective flow models in finite domains and nonstationarity: Water Resources Research, v. 28, no. 1, p. 251-267.

Mast, M.A., and Clow, D.W., 2000, Environmental characteristics and water-quality of Hydrologic Benchmark Network stations in the Western United States: U.S. Geological Survey Circular 1173-D, 114 p.

Mastin, M.C., and Vaccaro, J.J., 2002, Documentation of Precipitation Runoff Modeling System modules for the Modular Modeling System modified for the Watershed and River Systems Management Program: U.S. Geological Survey Open-File Report 2002-362, 5 p.

McDonald, M.G., and Harbaugh, A.W., 1988, A modular three-dimensional finite-difference ground-water flow model: U.S. Geological Survey Techniques of Water-Resources Investigations, Book 6, Chapter A-1, 586 p.

Meeus, J., 1999, Astronomical Algorithms: Richmond, Va., Willmann-Bell, Inc., 477 p.

Mehl, S.W., and Hill, M.C., 2001, MODFLOW-2000, the U.S. Geological Survey modular ground-water model — User Guide to the Link-AMG (LMG) Package for solving matrix equations using an algebraic multigrid solver: U.S. Geological Survey Open-File Report 01-177, 33 p.

Mehl, S.W., and Hill, M.C., 2006, MODFLOW-2005, the U.S. Geological Survey modular ground-water model — Documentation of shared node local grid refinement (LGR) and the boundary flow and head (BFH) package: U.S. Geological Survey Techniques and Methods 6-A12, 78 p.

Meinzer, O.E., 1923, Outline of ground-water hydrology, with definitions: U.S. Geological Survey Water-Supply Paper 494, 71 p.

Merritt, M.L., and Konikow, L.F., 2000, Documentation of a computer program to simulate lake-aquifer interaction using the MODFLOW ground-water flow model and the MOC3D solute-transport model: U.S. Geological Survey Water-Resources Investigations Report 00-4167, 146 p.

Morgan, D.S., 1988, Geohydrology and numerical model analysis of ground-water flow in the Goose Lake Basin, Oregon and California: U.S. Geological Survey Water-Resources Investigations Report 87-4058, 92 p.

Murray, F.W., 1967, On the computation of saturation vapor pressure: Journal of Applied Meteorology, v. 6, p. 203-204.

Nash, J.E., and Sutcliffe, J.V., 1970, River flow forecasting through conceptual models part I—A discussion of principles: Journal of Hydrology, v. 10, no. 3, p. 282-290.

Nishikawa, T., Izbicki, J.A., Hevesi, J.A., Stamos, C.L., and Martin, P., 2005, Evaluation of geohydrologic framework, recharge estimates, and ground-water flow of the Joshua Tree area, San Bernardino County, California: U.S. Geological Survey Scientific Investigations Report 2004–5267, 127 p.

Niswonger, R.G., and Prudic, D.E., 2005, Documentation of the Streamflow-Routing (SFR2) Package to include unsaturated flow beneath streams—A modification to SFR1: U.S. Geological Survey Techniques and Methods 6-A13, 62 p.

Niswonger, R.G., Prudic, D.E., and Regan, R.S., 2006a, Documentation of the Unsaturated-Zone Flow (UZF1) Package for modeling unsaturated flow between the land surface and the water table with MODFLOW-2005: U.S. Geological Survey Techniques and Methods 6-A19, 74 p.

Niswonger, R.G., Prudic, D.E., Markstrom, L.S., Regan, R.S., and Viger, R.J., 2006b, GSFLOW—A basin-scale model for coupled simulation of ground-water and surface-water flow—Part B. Concepts for modeling saturated and unsaturated subsurface flow with the U.S. Geological Survey modular ground-water model: Joint Federal Interagency Conference, Reno, Nev., 8 p.

Obled, C., and Rosse, B.B., 1977, Mathematical models of a melting snowpack at an index plot: Journal of Hydrology, v. 32, p. 139-163.

Olson, S.A., 2002, Flow-frequency characteristics of Vermont streams: U.S. Geological Survey Water-Resources Investigations Report 2002-4238, 47 p.

Panday, S., and Huyakorn, P.S., 2004, A fully coupled physically-based spatially-distributed model for evaluating surface/subsurface flow: Advances in Water Resources, v. 27, no. 4, p. 361-382.

Phillip, J.R., 1957, The theory of infiltration: 2. The profile at infinity: Soil Science, v. 83, p. 435-448.

Pikul, M.F., Street, R.L., and Remson, I., 1974, A numerical model based on coupled one-dimensional Richards and Boussinesq equations: Water Resources Research, v. 10, no. 2, p. 295-302.

Prudic, D.E., 1989, Documentation of a computer program to simulate stream-aquifer relations using a modular, finite-difference, ground-water flow model: U.S. Geological Survey Open-File Report 88-729 113 p.

Prudic, D.E., Konikow, L.F., and Banta, E.R., 2004, A new Streamflow-Routing (SFR1) Package to simulate stream-aquifer interaction with MODFLOW-2000: U.S. Geological Survey Open-File Report 2004-1042, 95 p.

Rademacher, L.K., Clark, J.F., Clow, D.W., and Hudson, G.B., 2005, Old groundwater influence on stream hydrochemistry and catchment response times in a small Sierra Nevada catchment: Sagehen Creek, California: Water Resources Research, v. 41, no. W02004, doi:10.1029/2003WR002805, 10 p.

Rankl, J.G., 1987, Analysis of sediment production from two small semiarid basins in Wyoming: U.S. Geological Survey Water-Resources Investigations Report 85-4314, 27 p.

Refsgaard, J.C., and Storm, B., 1995, MIKE SHE, in Singh, V.P., ed., Computer Models of Watershed Hydrology: Highlands Ranch, Colo., Water Resources Publications, p. 809-846.

Riley, J.P., Israelsen, E.K., and Eggleston, K.O., 1973, Some approaches to snowmelt prediction, in The role of snow and ice in hydrology: International Association of Hydrological Sciences Publication 107, p. 956-971.

Rorabaugh, M.I., 1953, Graphical and theoretical analysis of step-drawdown test of artesian well: Proceedings of the American Society of Civil Engineers, v. 79, no. 362, 23 p.

Ross, M.A., Tara, R.D., Geurink, J.S., and Stewart, M.T., 1997, FIPR Hydrologic Model Users Manual and Technical Documentation: Florida Institute of Phosphate Research, FIPR-OFR-88-03-085, Barton, Fla.

Said, A., Stevens, O.K., and Sehlke, G., 2005, Estimating water budget in a regional aquifer using HSFP-MODFLOW integrated model: Journal of the American Water Resources Association, v. 41, no. 1, p. 55-66.

Schmid, W., Hanson, R.T., Maddock, T.I., and Leake, S.A., 2006, User guide for the farm process (FMP1) for the U.S. Geological Survey's modular three-dimensional finite-difference ground-water flow model, MODFLOW-2000: U.S. Geological Survey Techniques and Methods 6-A17, 127 p.

Selby, S.M., 1970, Standard mathematical tables (15th ed.): Cleveland, Ohio, The Chemical Ruber Co., 664 p.

Selker, J..S., Keller, C.K., and McCord, J.T., 1999, Vadose zone processes: Boca Raton, Fla., CRC Press LLC, 339 p.

Shen, H.W., and Julien, P.Y., 1993, Erosion and sediment transport, in Maidment, D.R., ed., Handbook of hydrology: New York, McGraw-Hill, Inc., p. 61.

Smith, R.E., 1983, Approximate sediment water movement by kinematic characteristics: Soil Science Society of America Journal, v. 47, p. 3-8.

Smith, R.E., and Hebbert, R.H.B., 1983, Mathematical simulation of interdependent surface and subsurface hydrologic processes: Water Resources Research, v. 19, no. 4, p. 987-1001.

Sophocleous, M., and Perkins, S.P., 2000, Methodology and application of combined watershed and groundwater model in Kansas: Journal of Hydrology, v. 236, p. 185-201.

Steuer, J.J., and Hunt, R.J., 2001, Use of a watershed-modeling approach to assess hydrologic effects of urbanization, North Fork Pheasant Branch basin near Middleton, Wisconsin: U.S. Geological Survey Water-Resources Investigations Report 2001-4113, 49 p.

Streeter, V.L., and Wylie, E.B., 1985, Fluid Mechanics (8th ed.): New York, McGraw-Hill, Inc., 586 p.

Sun, R.J., ed., 1986, Regional Aquifer-System Analysis Program of the U.S. Geological Survey—Summary of projects, 1978-84: U.S. Geological Survey Circular 1002, 264 p.

Swain, E.D., and Wexler, E.J., 1996, A coupled surface-water and ground-water flow model (MODBRANCH) for simulation of stream-aquifer interaction: U.S. Geological Survey Techniques of Water-Resources Investigations, Book 6, Chapter A6, 125 p.

Swift, L.W., Jr., 1976, Algorithm for solar radiation on mountain slopes: Water Resources Research, v. 12, no. 1, p. 108-112.

Tangborn, W.V., 1978, A model to predict short-term snowmelt runoff using synoptic observations of streamflow, temperature, and precipitation *in* Colbeck, S.C., and Ray, M., eds., Modeling of snow cover runoff, Hanover, New Hampshire, 1978 [Proceedings]: U.S. Army Corps of Engineers, Cold Region Research and Engineering Laboratory, p. 414–426.

Thompson, E.S., 1976, Computation of solar radiation from sky cover: Water Resources Research, v. 12, no. 5, p. 859-865.

Thoms, R.B., Johnson, R.L., and Healy, R.W., 2006, User's guide to the Variably Saturated Flow (VSF) process for MODFLOW: U.S. Geological Survey Techniques and Methods 6-A18, 58 p.

University of California Berkeley, 2005, Sagehen Creek Field Station, UC Berkeley, Historical Weather Data, accessed June 15, 2005: at http://www.wrcc.dri.edu/cgi-bin/rawMAIN.pl?casagh%20: Sagehen Creek, California.

U.S. Geological Survey, 2000, U.S. GeoData Digital Elevation Models: U.S. Geological Survey Fact Sheet 040-00, accessed January 15, 2006 at http://pubs.er.usgs.gov/usgspubs/fs/fs04000, 4 p.

U.S. Soil Conservation Service, 1971, SCS national engineering handbook, Section 4—Hydrology: Washington, D.C., U.S. Government Printing Office, 654 p.

Vaccaro, J.J., 1992, Sensitivity of groundwater recharge estimates to climate variability and change, Columbia Plateau, Washington: Journal of Geophysical Research, v. 97, no. D3, p. 2821–2833.

Vaccaro, J.J., 2007, A deep percolation model for estimating ground-water recharge—Documentation of modules for the modular modeling system of the U.S. Geological Survey: U.S. Geological Survey Scientific Investigations Report 2006-5318, 30 p.

VanderKwaak, J.E., 1999, Numerical simulation of flow and chemical transport in integrated surface-subsurface hydrologic systems: Ontario, Canada, University of Waterloo, Department of Earth Sciences, Ph.D. Dissertation, 217 p.

Viger, R.J., and Leavesley, G.H., 2007, The GIS Weasel Users Manual: U.S. Geological Survey Techniques and Methods 6-B4, 201 p.

Vining, K.C., 2002, Simulation of streamflow and wetland storage, Starkweather Coulee subbasin, North Dakota, water years 1981-98: U.S. Geological Survey Water-Resources Investigations Report 02-4113, 28 p.

Wang, H.F., and Anderson, M.P., 1982, Introduction to groundwater modeling: San Francisco, W.H. Freeman and Company, 237 p.

Western Region Climate Center, 2006a, Western U.S. Climate Historical Summaries, accessed January 15, 2006, at http://www.wcc.nrcs.usda.gov/snotel/snotel.pl?sitenum=539&state=ca.

Western Region Climate Center, 2006b, Western U.S. Climate Historical Summaries, accessed January 15, 2006, at http://www.wcc.nrcs.usda.gov/snotel/snotel.pl?sitenum=541&state=ca.

Western Region Climate Center, 2006c, Western U.S. Climate Historical Summaries, accessed January 15, 2006, at http://www.wcc.nrcs.usda.gov/snotel/snotel.pl?sitenum=540&state=ca.

Wilby, R.L., Hay, L.E., and Leavesley, G.H., 1999, A comparison of downscaled and raw GCM output: implications for climate change scenarios in the San Juan River basin, Colorado: Journal of Hydrology, v. 225, p. 67-91.

Wilson, J.D., and Naff, R.L., 2004, MODFLOW-2000, the U.S. Geological Survey modular ground-water model—GMG Linear Equation Solving Package: U.S. Geological Survey Open-File Report 2004-1261, 47 p.

Yates, D.N., Warner, T.T., and Leavesley, G.H., 2000, Prediction of a flash flood in complex terrain: Part II—A comparison of flood discharge simulations using rainfall input from radar, a dynamic model, and an automated algorithmic system: Journal of Applied Meteorology, v. 39, no. 6, p. 815-825.

Zahner, R., 1967, Refinement in empirical functions for realistic soil-moisture regimes under forest cover, *in* Sopper, W.E., and Lull, H.W., eds., International Symposium of Forest Hydrology: New York, Pergamon Press, p. 261-274.

Zarriello, P.J., and Ries, K.G., III, 2000, A precipitation-runoff model for analysis of the effects of water withdrawals on streamflow, Ipswich River basin, Massachusetts: U.S. Geological Survey Water- Resources Investigations Report 2000-4029, 99 p.

Appendix 1. Input Instructions for GSFLOW.

Contents

Introduction

This appendix describes the input data requirements for a GSFLOW simulation. Because GSFLOW is based on different models, the format and contents of data in each file depends on whether the genesis of the file was for the PRMS or MODFLOW model. The GSFLOW Control File, PRMS Data File(s), and PRMS Parameter File are based on the Modular Modeling System input-data file formats (Leavesley and others, 1996b). All other files described in this section are based on MODFLOW input file formats as documented for each MODFLOW package. The MODFLOW Name File is described separately from the MODFLOW Packages because the file is used to identify the names of MODFLOW-based files that are read from and(or) written to during a GSFLOW simulation.

Input Terminology

GSFLOW input is described on the basis of terminology used for PRMS and MODFLOW. The input terminology of each program was adapted, where possible, for the purpose of consistency within GSFLOW. Thus, input terminology used in this appendix may differ from that used in previous reports documenting PRMS, MODFLOW, and MODFLOW packages. The terminology used to define data input for PRMS and MODFLOW differ because the models were developed by different disciplines within the U.S. Geological Survey.

Styles and Formats

Consistent styles and formats for PRMS and MODFLOW input instructions are used for the following:

- Variables and parameters that are directly related to user-supplied values are identified using 10-point, **bold Courier** font.

- Variables from model code not directly related to user-supplied values are identified using 10-point, Times font.

- Example input variables and parameters are identified using 10-point, `Courier` font.

- Optional input is enclosed between brackets ([ ]).

The following presents descriptions of the terms variable, parameter, dimension, and item in the context of input for PRMS and MODFLOW, respectively.

Modular Modeling System Based Files

The Modular Modeling System input structure is used to define input data for the GSFLOW Control File, PRMS Data File(s), and PRMS Parameter File. This input structure uses the following definitions.

- **Variables** (or states and fluxes) may vary from one time step to another, and are either specified in the PRMS Data File(s) (such as the daily maximum air temperature at a temperature station) or calculated (such as the soil-infiltration rate for a HRU). Variables may have a single value or they may include multiple values (one- or two-dimensional arrays).

- **Parameters** are user-specified data that do not change during a simulation, such as the name of the PRMS Parameter File, the number of columns of maximum air temperature values in a PRMS Data File, and the area of a hydrologic response unit (HRU) as specified in the GSFLOW Control File, PRMS Data File, and PRMS Parameter File, respectively. Parameters, like variables, may have a single value or they may include multiple values (one- or two-dimensional arrays).

- **Dimensions** are a type of parameter that define the number of spatial features and constants, such as the number of HRUs, number of months in a year, and the number of temperature stations for which time-series data are specified in the PRMS Data File(s). Dimensions are specified in the PRMS Parameter File.

- **Items** are groups of input values used to define a parameter, dimension, or variable. Items may be specified on a single line or on multiple lines in a file.

MODFLOW Based Input Files

MODFLOW-2005 input structure is used to define input data for ground-water flow, unsaturated flow, and flow in streams and lakes. The input structure of MODFLOW uses the following definitions.

- **Variables** (or states and fluxes) may include a single value or multiple values. One-dimensional variables are multi-valued variables in which the number of values is indicated by a single number in parentheses after the name. Two-dimensional variables are multi-valued variables in which the number of values is indicated by two numbers in parentheses after the name. The total number of values represented is the product of the two numbers.

- **Parameters** are single values that can be used to determine data values for multiple finite-difference cells (Harbaugh, 2005, p. 8-2). Input instructions for defining parameters are not included in this appendix. Input instructions for parameters are included in Harbaugh (2005).

- **Dimensions** are a not used to refer to MODFLOW input data.

- **Items** are numbered groups of input variables used to define a parameter, dimension, or variable. Items may be specified on a single line or on multiple lines in a file.

GSFLOW Control File

The GSFLOW Control File consists of a header followed by a sequence of control-parameter items specified in the Modular Modeling System Data File format. Each line in the Control File can include descriptive text to allow annotation of any data, such as to describe the source of a parameter value. The descriptive text is added to lines following the required data and preceded by at least one space and two forward slashes (//). Each line in the Control File can be up to 256 characters. The header is a single line of text that can be used to identify the file. For example, the header used for Sagehen Creek watershed example problem is:

```
GSFLOW Control File for the Sagehen Creek Watershed
```

Control-parameter items consist of four lines followed by parameter values, one per line, that have the following general structure and order:

```
####        [ // descriptive_text]
NAME        [ // descriptive_text]
N_VALUES    [ // descriptive_text]
DATA_TYPE   [ // descriptive_text]
VALUE(S)    [ // descriptive_text]
```

The first line is used as a delimiter signaling the start of a control-parameter item and must specify a string of four pound signs (####) beginning in column 1. The second line specifies the name of the control parameter (NAME). The third line specifies the number (N_VALUES) of parameter values that are specified. The fourth line identifies the data type of the control parameter using an integer flag (DATA_TYPE). The DATA_TYPE options are:

1 is for integer;
2 is for real (single-precision, floating decimal point);
3 is for double (double-precision, floating decimal point); and
4 is for character string.

Note, no double-precision real (option 3) control parameters are required in GSFLOW.

The next N_VALUES lines specify data values (VALUES(S)), one value per line. Thus, each control-parameter item must consist of at least five lines. Table A1-1 is a list of the GSFLOW control parameters, and provides parameter names, definitions, number of values, data-type flag, and whether the parameter is optional. Control-parameter names are case-sensitive and must be specified as defined in table A1-1.

Table A1-1.   Control parameters specified in the GSFLOW Control File.

[Data Type: 1, integer; 2, single precision floating point (real); 3, double precision floating point (real); 4, character string. **Abbreviation:** HRU, hydrologic response unit]

Parameter name (NAME)	Definition	Number of values (N_value)	Data type (Data_type)	Default value or optional
Parameters related to model execution				
model_mode	Model to run (GSFLOW, PRMS, MODFLOW)	1	4	GSFLOW
start_time[1]	Simulation start time specified in order as: year, month, day, hour, minute, second	6	1	2000, 10, 1, 0, 0, 0
end_time[1]	Simulation end time specified in order as: year, month, day, hour, minute, second	6	1	2001, 9, 30, 0, 0, 0
Parameters related to model input				
data_file[2,3]	Pathname(s) for PRMS Data File(s); typically, a single Data File is specified for a GSFLOW simulation	Equal to the number of data files	4	prms.data
param_file[3]	Pathname for PRMS Parameter File	1	4	prms.params
modflow_name[3]	Pathname for MODFLOW Name File	1	4	modflow.nam
precip_module[1]	Module name for precipitation-distribution method (precip_prms, precip_laps_prms, xyz_dist, or precip_dist2_prms)	1	4	precip_prms
temp_module[1]	Module name for temperature-distribution method (temp_1sta_prms, temp_2sta_prms, xyz_dist, or temp_dist2.prms)	1	4	temp_1sta_prms
solrad_module[1]	Module name for solar-radiation-distribution method (ccsolrad_hru_prms or ddsolrad_hru_prms)	1	4	ddsolrad_hru_prms
et_module[1]	Module name for potential-evapotranspiration computation method (potet_hamon_prms, potet_jh_prms, or potet.pan.prms)	1	4	potet_jh_prms
srunoff_module[1]	Module name for surface-runoff/infiltration computation method (srunoff_carea_prms or srunoff_smidx_prms)	1	4	srunoff_smidx_prms
Parameters related to model output				
gsflow_output_file[3]	Pathname for GSFLOW Water-Budget File of summaries of each component of GSFLOW water budget	1	4	gsflow.out
model_output_file[3]	Pathname for PRMS Water-Budget File of summaries of each component of PRMS water budget	1	4	prms.out
csv_output_file[3]	Pathname for GSFLOW Comma-Separated-Values (CSV) File of GSFLOW water budget and mass balance results for each time step	1	4	gsflow.csv
gsf_rpt	Switch to specify whether or not the GSFLOW Comma-Separated-Values (CSV) File is generated (0=no; 1=yes)	1	1	1
rpt_days	Frequency that summary tables are written to GSFLOW Water-Budget File (0=none, >0 frequency in days, e.g., 1=daily, 7=every 7th day)	1	1	7
statsON_OFF[1]	Switch to specify whether or not PRMS Statistic Variables (statvar) File of selected time-series values is generated (0=no; 1=yes)	1	1	0
stat_var_file[3]	Pathname for PRMS Statistic Variables (statvar) File of time-series values; required only when statsON_OFF = 1	1	4	optional
nstatVars[1]	Number of variables to include in PRMS Statistic Variables File and names specified in statVar_names; required only when statsON_OFF = 1	1	1	optional

Table A1-1.   Control parameters specified in the GSFLOW Control File.—Continued

[Data Type: 1, integer; 2, single precision floating point (real); 3, double precision floating point (real); 4, character string. **Abbreviation:** HRU, hydrologic response unit]

Parameter name (NAME)	Definition	Number of values (N_value)	Data type (Data_type)	Default value or optional
	Parameters related to model output—Continued			
statVar_names[1]	List of variable names for which output is written to PRMS Statistic Variables File; required only when statsON_OFF = 1	nstatVars	4	optional
statVar_element	List of identification numbers corresponding to variables specified in statVar_names file (1 to variable's dimension size); required only when statsON_OFF = 1	nstatVars	1	optional
aniOutON_OFF[1]	Switch to specify whether or not PRMS Animation Variables File(s) of spatially-distributed values is generated (0=no; 1=yes)	1	1	0
ani_output_file[1,3]	Root pathname for PRMS Animation Variables File(s) to which a filename suffix based on dimension names associated with selected variables is appended; required only when gisOutOn_Off = 1	1	4	optional
naniOutVars[1]	Number of output variables specified in the aniOutVar_names list; required only when aniOutON_OFF = 1.	1	1	optional
aniOutVar_names[1]	List of variable names for which all values of the variable (that is, the entire dimension size) for each time step are written to PRMS Animation Variables File(s), use only for aniOutON_OFF = 1	naniOutVars	4	optional
	Parameters related to PRMS model initial conditions			
init_vars_from_file[1]	Flag to determine if a PRMS Initial Conditions File is specified as an input file (0=no; 1=yes)	1	1	0
var_init_file[1,3]	Pathname for the PRMS Initial Conditions File; only required when init_vars_from_file = 1	1	4	optional
save_vars_to_file[1]	Flag to determine if a PRMS Initial Conditions File (var_save_file) will be generated at the end of simulation (0=no; 1=yes)	1	1	0
var_save_file[1,3]	Pathname for the PRMS Initial Conditions File to be generated at end of simulation; only required when save_vars_to_file = 1	1	4	optional

[1] Additional description of parameter provided in appendix 1 section "Control Parameters Related to Model Input."

[2] Multiple PRMS Data Files can be specified, although typically, only one is used with GSFLOW.

[3] Pathnames can be 1 to 256 characters and must be specified as a valid pathname for the operating system.

An example control-parameter item follows. The example is for the control parameter with the NAME set to data_file that is used to specify the pathname(s) of the PRMS Data File(s). The parameter consists of a single value (N_VALUES set to 1) with a value specified as a character string (DATA_TYPE set to 4).

```
####  // PRMS Data File of time-series values
data_file
1
4
D:/gsflow/data/sagehen/input/sagehen.data
```

Although this example specifies that the data_file parameter is a single value, multiple PRMS Data Files can be specified for a GSFLOW simulation by specifying N_VALUES greater than one and including that number of pathnames.

Control-parameter items can be specified in any order in the GSFLOW Control File; table A1-1 provides a suggested order for specifying control-parameter items grouped by whether the parameter controls functions related to (a) model execution, (b) input, (c) output, or (d) initial conditions. Any control-parameter item not specified in the Control File that is required by a GSFLOW simulation is assigned default values. Any control-parameter item specified in the Control File that is not required by a GSFLOW simulation is ignored. Additional information for selected control parameters beyond the definitions provided in table A1-1 follows.

Control Parameters Related to Model Input

start_time and end_time—are used to specify the starting and ending times of the simulation. They each are specified using six lines of integer values in the following order: year, month, day, hour, minute, and second. The specified times must have corresponding time-series data items of the exact same date and time in the PRMS Data File(s). GSFLOW only allows a daily time step.

data_file—is used to specify a list of PRMS Data File(s) to use in a simulation. Typically, a single PRMS Data File is used for GSFLOW simulations (N_VALUES set to 1). Use of multiple PRMS Data Files is described in section, "PRMS Data File" in this appendix.

precip_module, temp_module, solrad_module, et_module, and srunoff_module—are used to select a PRMS module for simulation processes to distribute precipitation, temperature, and solar radiation and to compute potential evapotranspiration and surface runoff, respectively. The computations for each process are described in section, "Computations of Flow" in the body of this report and input requirements for each module are described below in this appendix. If the module xyz_dist is selected for the temperature-distribution method (temp_module), this module also must be selected as the precipitation-distribution module (precip_module), and vice versa.

Control Parameters Related to Model Output

statVar_names and aniOutVar_names—are used to specify lists of names of output variables (states and fluxes) to include in the PRMS Statistic Variables (statvar) File and PRMS Animation Variables File(s), respectively. A description of these files is given in section, "Output Files" in the body of this report. A list with definitions of the 35 most commonly used GSFLOW output variables is given in table A1-2. Many additional output variables are available for inclusion in the PRMS Statistic Variables and Animation Variables Files for specialized analysis and debugging purposes. The complete list of the 270 output variables with definitions can be found in the file *gsflow.var_name* that is provided in the *gsflow/bin* directory included in the GSFLOW distribution. Users can select as many output variables as desired to include in both files using control parameters nstatVars and naniOutVars, respectively. If control parameter statsON_OFF is set to 0, the Statistic Variables File is not created and values of statVar_names and nstatVars are ignored. Likewise, if control parameter aniOutON_OFF is set to 0, PRMS Animation Variables File(s) are not created and values of aniOutVar_names and naniOutVars are ignored.

ani_output_file—is used as the root filename for PRMS Animation Variables File(s). A separate output file is generated for each dimension size associated with the list of variables specified by aniOutVar_names when aniOutON_OFF is set to 1. Variables with the same dimension are included in a single file. The names of each file differ in the suffix appended to ani_output_file. The suffix for each file is set to the dimension name. For example, if ani_output_file is specified as *./output/sagehen_ani* and variables with PRMS dimension nhru and nreach are specified in the aniOutVar_names list of variable names, the files *./output/sagehen_ani.nhru* and *./output/sagehen_ani.nreach* are generated.

Table A1-2. Selected GSFLOW variables for which values can be written to the PRMS Statistic Variables File and PRMS Animation Variables File(s) for each simulation time step.

[**Dimension variable:** nhru, number of HRUs; nhrucell, number of intersections between HRUs and MODFLOW grid cells; one, dimension of one; **Abbreviations:** MF_L, MODFLOW length unit; HRU, hydrologic response unit; cfs, cubic feet per second; T, time unit]

Variable name	Definition	Units	Dimension variable
basin_cfs	Streamflow out of watershed	cfs	one
basin_et	Total evapotranspiration on watershed as sum for evaporation from snowpack, impervious areas, plant canopy, and soil zone and transpiration from soil zone	inches	one
basin_gwflow_cfs	Area-weighted average ground-water flow for watershed	cfs	one
basin_potet	Area-weighted average potential evapotranspiration for watershed	inches	one
basin_ppt	Area-weighted average precipitation for watershed	inches	one
basin_pweqv	Area-weighted average pack-water equivalent of snowpack for watershed	inches	one
basin_reach_latflow	Area-weighted average lateral flow into stream reaches for watershed	cfs	one
basin_sroff_cfs	Area-weighted average Hortonian and Dunnian surface runoff into stream reaches for watershed	cfs	one
basin_ssflow_cfs	Area-weighted average interflow into stream reaches for watershed	cfs	one
basinactet	Volumetric flow rate of evapotranspiration for watershed	$(MF\_L)^3T^{-1}$	one
basingravstor	Total volume of soil water in gravity reservoirs of soil zone for watershed	$(MF\_L)^3$	one
basingw2sz	Volumetric flow rate of ground-water discharge added to soil zone for watershed	$(MF\_L)^3T^{-1}$	one
basininfilprev	Volumetric flow rate of soil infiltration into preferential-flow reservoirs of soil zone including precipitation, snowmelt, and cascading Hortonian flow for watershed	$(MF\_L)^3T^{-1}$	one
basininfil_tot	Volumetric flow rate of soil infiltration into capillary reservoirs of soil zone including precipitation, snowmelt, and cascading Hortonian flow for watershed	$(MF\_L)^3T^{-1}$	one
basininterflow	Volumetric flow rate of slow interflow to stream reaches for watershed	$(MF\_L)^3T^{-1}$	one
basinprefstor	Total volume of soil water in preferential-flow reservoirs of soil zone for watershed	$(MF\_L)^3$	one
basinpweqv	Total volume of water in snowpack storage for watershed	$(MF\_L)^3$	one
basinsnowevap	Volumetric flow rate of snowpack sublimation for watershed	$(MF\_L)^3T^{-1}$	one
basinsnowmelt	Volumetric flow rate of snowmelt for watershed	$(MF\_L)^3T^{-1}$	one
basinsoilmoist	Total volume of soil water in capillary reservoirs of soil zone for watershed	$(MF\_L)^3$	one
basinsroff	Volumetric flow rate of Hortonian and Dunnian surface runoff for watershed	$(MF\_L)^3T^{-1}$	one
basinstrmflow	Volumetric flow rate of streamflow leaving the watershed	$(MF\_L)^3T^{-1}$	one
gw2sm	Average ground-water discharge to soil zone in an HRU	inches	nhru
gwc_head	Head at each MODFLOW ground-water cell	MF_L	ngwcell
gwflow2strms	Volumetric flow rate of ground-water discharge to stream reaches	$(MF\_L)^3T^{-1}$	one
hru_ppt	Adjusted precipitation on HRU	inches	nhru
kkiter	Current iteration in GSFLOW simulation	dimensionless	one
obsq_cfs	Streamflow at streamflow-gaging station	cfs	nobs
pkwater_equiv	Pack-water equivalent of snowpack	inches	nhru
reach_cfs	Streamflow leaving each stream reach	cfs	nreach
reach_latflow	Lateral flow (surface runoff and interflow) into each stream reach	cfs	nreach
reach_wse	Water-surface elevation in each stream reach	MF_L	nreach
sat_store	Total storage in saturated MODFLOW cells	$(MF\_L)^3$	one
sm2gw_grav	Gravity drainage from each gravity reservoir to each MODFLOW cell	inches	nhrucell
snowcov_area	Fraction of snow-covered area on HRU	dimensionless	nhru
snowmelt	Snowmelt from the snowpack on HRU	inches	nhru
soil_moist	Water content of capillary reservoir for HRU	inches	nhru
soil_moisture_pct	Decimal fraction of the saturation of capillary reservoir	dimensionless	nhru
sroff	Surface runoff to streams for HRU	inches	nhru
ssr_to_gw	Area-weighted average gravity drainage from soil zone for HRU	inches	nhru
ssres_flow	Interflow to streams for HRU	inches	nhru
ssres_stor	Average gravity reservoir storage for HRU	inches	nhru
stream_leakage	Total leakage from stream segments to associated MODFLOW cells	$(MF\_L)^3$	one
swrad	Computed shortwave radiation for HRU	langleys	nhru

Table A1-2. Selected GSFLOW variables for which values can be written to the PRMS Statistic Variables File and PRMS Animation Variables File(s) for each simulation time step.—Continued

[**Dimension variable:** nhru, number of HRUs; nhrucell, number of intersections between HRUs and MODFLOW grid cells; one, dimension of one. **Abbreviations:** MF_L, MODFLOW length unit; HRU, hydrologic response unit]

Variable name	Definition	Units	Dimension variable
tmaxf	Adjusted daily maximum temperature for HRU	degrees Fahrenheit	nhru
tminf	Adjusted daily minimum temperature for HRU	degrees Fahrenheit	nhru
unsat_store	Total storage in unsaturated MODFLOW cells as simulated by the Unsaturated-Zone Flow Package	$(MF\_L)^3$	one
uzf_infil	Net gravity drainage to the unsaturated zone as simulated by the Unsaturated-Zone Flow Package	$(MF\_L)^3$	one

Control Parameters Related to Initial Conditions for PRMS

init_vars_from_file, var_init_file, save_vars_to_file, var_save_file—are used to specify whether or not the PRMS initial conditions for all computation variables (or states) will be specified and(or) saved for use with a GSFLOW simulation. Most initial condition values are set to 0.0 internally by PRMS modules, with some initial conditions set to values specified in the PRMS Parameter File, such as the initial water content of the soil zone (parameters soil_moist_init, soil_rechr_init, and ssstor_init; see section, "PRMS Parameter File" in this appendix for an explanation of these and other parameters that are required for each GSFLOW and PRMS module). Typically, the control parameters init_vars_from_file and save_vars_to_file are specified as 0 for GSFLOW simulations.

If the user has a valid PRMS Initial Conditions File that was generated during a prior GSFLOW simulation, it can be selected for use by setting init_vars_from_file to 1 and specifying the pathname of the PRMS Initial Conditions File using control parameter var_init_file. To generate the PRMS Initial Conditions File, the user would set save_vars_to_file to 1 and the pathname using control parameter var_save_file.

One use of the PRMS Initial Conditions File is for forecasting hydrologic conditions. A series of predictive simulations can be generated using the results from a single simulation driven by climate data up to a selected date. GSFLOW could use these results to set the internal states for new simulations that are driven by different sets of predicted climate data. The predicted climate data, for example, results from a climate model or historical climate data, can be specified in additional PRMS Data File(s).

The PRMS Initial Conditions File created from a GSFLOW simulation only includes the variables (states) for the date and time at the end of the simulation. Thus, the file should only be used as a continuation from that date and time.

PRMS Data File

A PRMS Data File consists of three items: a header, input-variable declaration items, and time-series data items (fig. A1-1 illustrates part of a PRMS Data File), and is specified in the Modular Modeling System Data File format. The number of Data Files and file names of each PRMS Data File is specified using the data_file control-parameter item in the GSFLOW Control File. If multiple PRMS Data Files are specified for use in a single GSFLOW simulation the files will be combined according to the specified date and time of each time-series data item. If data overlap, values for the smallest time increment between consecutive values during the overlap period are used, with the last PRMS Data File having precedence. Multiple PRMS Data Files can be used to separate data by years or decades. Multiple PRMS Data Files must have identical input-variable declaration items in sequence and number but may have different comment lines. The time increment between two consecutive time-series data items sets the GSFLOW time step. GSFLOW simulates on a daily time step, therefore, time-series data items must be specified for each day with exactly 24 hours between the date and time specified for consecutive values. Thus, PRMS Data Files with time increments other than 24 hours cannot be used in a GSFLOW simulation.

```
Sagehen Data File: Independence Lake and Sagehen Creek data stations

// tmax stations are:
//    INDEPENDENCE LAKE
//    SAGEHEN CREEK
tmax 2

// tmin stations are:
//    INDEPENDENCE LAKE
//    SAGEHEN CREEK
tmin 2

// precip stations are:
//    INDEPENDENCE LAKE
//    SAGEHEN CREEK
precip 2
form_data 0
solrad 0
pan_evap 0

// runoff value is:
//    10343500      SAGEHEN C NR TRUCKEE C
runoff 1

####################################################################
1980 10 1 0 0 0 -901.0 85.0 -901.0 30.0 0.00 0.00 2.3
1980 10 2 0 0 0 -901.0 81.0 -901.0 32.0 0.00 0.00 2.3
1980 10 3 0 0 0 -901.0 83.0 -901.0 30.0 0.00 0.00 2.3
```

remaining lines have same format as the three previous lines and are not shown

Figure A1-1.   Example of a PRMS Data File.

The set of PRMS Data Files determines the range of time of a GSFLOW simulation, but not the simulation time period, which is specified by control-parameters items start_time and end_time in the GSFLOW Control File. Both start_time and end_time must be specified as dates and times that occurred between the earliest beginning and latest ending time of the time-series data specified in the set of PRMS Data Files. Typically, all time-series data are contained in a single PRMS Data File.

The header item in the PRMS Data File is a single line of text, up to 256 characters in length, which can be used to identify the file. For example, in the Sagehen Creek watershed example problem, the following header is used:

```
Sagehen Data File: Independence Lake and Sagehen Creek data stations
```

Input-variable declaration items are used to specify the type of time-series data included in each PRMS Data File. One line is used to identify each time-series data item. Blank lines or comment lines can be included before or after an input-variable declaration item; comment lines begin with two backslashes (//) in columns 1 and 2 to add descriptive information about an input variable. Each item specifies two values: (1) a character string that is the name of the input variable; and (2) an integer value that is the number of values (or columns) specified in each time-series data item for the input variable. The number of values must equal the size of the dimension associated with the input variable as defined by the dimensions specified in the PRMS Parameter File. For example, the two input-variable declarations items

```
tmax 2
tmin 2
```

indicate that daily maximum (tmax) and minimum (tmin) air temperature data will be specified for two observation stations in each time-series data item (and dimension ntemp is set to 2 in the dimensions section of the PRMS Parameter File). Each time-series data item must contain two columns of tmax values followed by two columns of tmin values. If the number of values and the associated dimension specified in the PRMS Parameter File are different an error message is printed and model execution stops.

A delimiter that consists of a single line specifying at least four pound symbols (####; fig. A1) beginning in column 1 signals the end of the input-variable declaration items and that the next and following lines each specify a time-series data item for consecutive dates and times. Each time-series data item consists of columns of data separated by at least one blank space, with the number of columns equal to six plus the sum of the number of values for each input variable. The first six columns specify the date and time of each data item as integer values in the order: year, month, day, hour, minute, and second. For GSFLOW integrated simulations, a time-series data item must be specified for each day with exactly 24 hours between the date and time specified for consecutive lines for the full extent of the time-series data. The hour, minute, and second must be specified as zero (that is, columns 4 through 6 can be 0 0 0). The remaining columns for each time-series data item specify the data values, which must be specified in the sequence the input-variable declarations items were defined. Value(s), depending on the specified number of values for the variable, must be specified in the corresponding column for each input variable. Extra values on any time-series data item beyond the specified number of columns (6 + the sum of the number of values for each input variable) are ignored.

Table A1-3 is a list of valid time-series input variables for GSFLOW, and provides definitions, units, valid range, and the dimension name associated with each variable. The input-variable names are case-sensitive and must be specified as defined in table A1-3.

A portion of a PRMS Data File is shown in figure A1-1. The input-variable declarations items specify that values for the following input variables are included or are not included in each time-series data item:

- two air-temperature stations with the two maximum air-temperature values specified before the two minimum air-temperature values (input variables tmax and tmin are set to 2 and dimension ntemp is set to 2 in the PRMS Parameter File);

- two precipitation stations (input variable precip is set to 2 and dimension nrain is set to 2 in the PRMS Parameter File);

- no data are included to specify the form of precipitation (input variable form_data is set to 0 and dimension nform is set to 0 in the PRMS Parameter File);

- no data are included to specify solar radiation (input variable solrad is set to 0 and dimension nsol is set to 0 in the PRMS Parameter File);

- no data are included to specify pan evaporation (input variable pan_evap is set to 0 and dimension nevap is set to 0 in the PRMS Parameter File); and

- one streamflow station (input variable runoff is set to 1 and dimension nobs is set to 1 in the PRMS Parameter File).

Thus, each time-series data item (only three are shown in fig. A1-1) specifies 6 columns for the date and time, followed by two maximum air-temperature values, two minimum air-temperature values, two precipitation values, and one streamflow value, for a total of 13 columns of data values on each line. For the time period shown in figure A1-1, no data were available for the first air-temperature measurement station, thus a value of -901.0 was specified as the first maximum and first minimum air temperatures (columns 7 and 9).

Table A1-3. Time-series data that can be specified in a PRMS Data File.

[**Dimension variable:** nevap, number of measurement stations that measure pan evaporation; nobs, number of streamflow gaging stations; nrain, number of measurement stations that measure precipitation; nsol, number of measurement stations that measure solar radiation; ntemp, number of measurement stations that measure air temperature; nform, is either 0 or 1. **Abbreviation:** cfs: cubic feet per second]

Variable name	Definition	Units	Valid range	Dimension variable
pan_evap	Pan evaporation at each measurement station that measures pan evaporation	inches	greater than 0.0	nevap
runoff[1]	Streamflow at each streamflow-gaging station	cfs	greater than 0.0	nobs
precip[1]	Precipitation at each measurement station that measures precipitation	inches	greater than 0.0	nrain
solrad	Solar radiation at each measurement station that measures solar radiation	langleys	greater than 0.0	nsol
tmax[1]	Daily maximum air temperature at each measurement station that measures air temperature	degrees Celsius or Fahrenheit	-50 to 150	ntemp
tmin[1]	Daily minimum air temperature at each measurement station that measures air temperature	degrees Celsius or Fahrenheit	-50 to 150	ntemp
form_data[1]	Form of precipitation (0=not known; 1=snow; 2=rain)	dimensionless	0, 1, or 2	nform
rain_day	Day is treated as a rain day (0=no; 1=yes)	dimensionless	0 or 1	one

[1] Additional description of variable provided below.

Additional information for selected input variables beyond the definitions provided in table A1-3 follows.

runoff—is the measured streamflow data and can be specified in units of cubic feet per second or cubic meters per second. Set the parameter runoff_units to a value of 1 in the parameters section of the PRMS Parameter File to indicate that runoff values are specified in cubic meters per second. Measured streamflow is not used in simulation computations and are thus optional. They are written to the GSFLOW Comma-Separated Values File and optionally can be written to the PRMS Statistic Variables and Animation Variables Files for dimension nobs, if included in the output variables list for either file.

tmax and tmin—are the measured daily maximum and minimum air temperatures and can be specified in units of degrees Fahrenheit or Celsius. Set the parameter temp_units to a value of 1 in the parameters section of the PRMS Parameter File to indicate that air-temperature values are specified in degrees Celsius. At least one column of tmax and tmin values must be specified, which also means that dimension ntemp in the dimensions section of the PRMS Parameter File must be set to at least 1.

precip—is the measured daily precipitation and can be specified in units of inches or millimeters. Set the parameter precip_units to a value of 1 in the parameters section of the PRMS Parameter File to indicate that precipitation values are in millimeters. At least one column of precip values must be specified, which also means that dimension nrain in the dimensions section of the PRMS Parameter File must be set to at least 1.

form_data—and the dimension nform in the dimensions section of the PRMS Parameter File are used to indicate whether or not the form of the precipitation will be specified as an input column. Typically, these data are not available. If form_data and nform are set to 0, specified temperature data will be used to determine the form of the precipitation (rain, if temperature is greater than or equal to parameter tmax_allrain, or snow, if temperature is less than or equal to parameter tmax_allsnow, with both specified in the parameters section of the PRMS Parameter File). If form_data and nform are set to 1, the form of the precipitation must be specified in the appropriate column on each line of the time-series data item. The form of the precipitation can be specified as unknown (temperature data will be used to determine the form), snow, or rain, using the values 0, 1, and 2, respectively. The specified form of the precipitation is applied to precipitation values for all HRUs. Typically, form of precipitation data are not known or is not applicable for an entire watershed, thus form_data and nform are set to 0 for a GSFLOW simulation.

PRMS Parameter File

The PRMS Parameter File specifies dimensions and parameters required for PRMS and GSFLOW modules formatted according to the Modular Modeling System Parameter File format. The first part of this section describes the overall format of the PRMS Parameter File. The second part provides 22 subsections that describe the parameters that are specified for each PRMS and GSFLOW module included in GSFLOW. These subsections describe each parameter that is required for each PRMS and GSFLOW module. The modules that are included in GSFLOW, described in the order in which they are used computationally, are listed in table 1 in the body of this report.

A table is provided for each module that includes the following information to define each parameter: (1) name; (2) description; (3) dimension name; (4) measurement units (acres, degrees Fahrenheit, and so forth); (5) data type (integer, real, or character); (6) the range of values that can be specified; and (7) the default value used by GSFLOW if the parameter is not defined in the PRMS Parameter File. No parameters have the double-precision data type (TYPE set to 3) in GSFLOW.

For some modules, an eighth column is included that provides the equation number and variable name for parameters that are used directly in equations documented in section, "Computations of Flow" in the body of this report. Additional information for selected parameters is given in the text that accompanies each table.

Parameter File Format

The PRMS Parameter File consists of three sections: header, dimensions, and parameters. Figure A1-2 shows a portion of the Parameter File used in the Sagehen Creek watershed example problem that includes the header section, a partial dimensions section (8 of 24 dimensions), and two parameter-declaration items (out of 132). The number of parameters specified in a Parameter File equals the number of unique parameters declared for all PRMS and GSFLOW modules included in a simulation. Each module can have different numbers of parameters. Some parameters are declared in multiple modules, but need only to be specified once in the PRMS Parameter File. The user can select different PRMS modules to distribute air temperature, precipitation, and potential solar radiation, and to compute potential evapotranspiration and surface runoff using control parameters `precip_module`, `temp_module`, `solrad_module`, `et_module`, and `srunoff_module`, respectively.

The header section consists of two items:

```
TITLE
Version: 1.7
```

The first item is a single line that specifies a description (or title) of the PRMS Parameter File; the `TITLE` line can contain up to 256 characters. The second item is a single line that specifies the Modular Modeling System Parameter File format version, this must begin in column 1 and be specified as `Version: 1.7`.

The dimensions section is used to define the size of dimensions that are used to allocate memory for parameters and variables required by the GSFLOW and PRMS modules of a particular GSFLOW simulation. The dimensions section begins in line 3 of the PRMS Parameter File with the following identifier that begins in column 1:

```
** Dimensions **
```

The identifier is followed by a series of 3-line dimension declarations items that have the following format:

```
####
NAME
SIZE
```

The first line is used as a delimiter for each of the dimension declarations, specified as a string of four pound signs (`####`) that begins in column 1. The second line (`NAME`) is the name of the dimension, specified as a character string without spaces using lowercase letters. The third line (`SIZE`) is the dimension size, specified as an integer value. Table A1-4 lists the names, definitions, and default values of the 24 dimensions that can be specified in the dimensions section. If one or more of the dimensions listed in table A1-4 are not specified, the default value is used. Although dimension names are listed alphabetically within functional groups in table A1-4, dimension declarations items may be specified in any order in the dimensions section. An example dimension declaration item for dimension `nhru`, which specifies the number of hydrologic response units and has a size of `128` for this particular example, follows:

```
####
nhru
128
```

```
Sagehen Parameter File with 128 HRUs, 201 reaches, 15 segments, cascading flow
Version: 1.7
** Dimensions **
####
one
1
####
ncascade
317
####
nsegment
15
####
nreach
201
####
nmonths
12
####
nhru
128
####
ngwcell
5913
####
nhrucell
4691
** Parameters **
####
albset_snm 10
1
one
1
2
0.2000000029802
####
tmin_lapse 10
1
nmonths
12
2
0.0
-3.200000047684
-0.2000000029802
0.4000000059605
0.0
-2.5
-4.900000095367
-6.099999904633
-5.800000190735
-4.400000095367
-2.700000047684
-1.799999952316
```

Figure A1-2.   Example portion of a PRMS Parameter File.

Table A1-4.   Dimension-variables specified in the dimensions section of the PRMS Parameter File.

[HRU, hydrologic response unit]

Variable name	Definition	Default value
Spatial dimensions		
ngw	Number of PRMS ground-water reservoirs (used in PRMS-only simulations)	1
ngwcell	Number of MODFLOW finite-difference cells in a layer (includes active and inactive cells)	0
nhru	Number of HRUs	1
nhrucell	Number of unique intersections between gravity reservoirs in PRMS soil zone and MODFLOW finite-difference cells	0
nreach	Number of stream reaches on all stream segments	0
nsegment	Number of stream segments	0
nsfres	Number of on-channel detainment reservoirs (used in PRMS-only simulations)	0
nssr	Number of PRMS subsurface reservoirs (must be specified equal to nhru)	1
Time-series input data dimensions		
nevap	Number of measurement stations that measure pan evaporation	0
nform	Number of input columns in PRMS Data File used to specify form of precipitation (0 if no form data, 1 if form data)	0
nobs	Number of streamflow-gaging stations	0
nrain	Number of measurement stations that measure precipitation	1
nsol	Number of measurement stations that measure solar radiation	0
ntemp	Number of measurement stations that measure air temperature	1
Computation dimensions		
mxnsos	Maximum number of table values for computing storage in and flow from detention reservoirs using Puls routing (PRMS-only simulations)	0
ncascade	Number of cascade paths associated with HRUs	0
ncascdgw	Number of cascade paths associated with PRMS ground-water reservoirs	0
ndepl	Number of snow-depletion curves used for snowmelt calculations	1
ndeplval	Number of snow-depletion values for each snow-depletion curve	ndepl*11
Fixed dimensions		
ndays	Maximum number of days in a year	366
nlapse	Number of lapse rates in the x, y, and z directions (used by module xyz_dist)	3
nmonths	Number of months in a year	12
one	A constant	1

The parameters section begins with the following identifier line that begins in column 1:
```
** Parameters **
```

The identifier is followed by a series of 7 items to declare each parameter. Each parameter declaration item has the following structure:
```
####
NAME WIDTH
NO_DIMENSIONS
DIMENSION_NAMES(S) (Repeat this item for each DIMENSION_NAME)
N_VALUES
TYPE
VALUE(S) (Repeat this item N_VALUES times)
```

The first line is used as a delimiter for each parameter declarations item, specified as a string of four pound signs (####) that begins in column 1. The second line specifies two values: (1) the name of the parameter (NAME), specified as a character string without spaces; and (2) a column width (WIDTH), specified as an integer value that is used by a Modular Modeling System spreadsheet editor that is not implemented in GSFLOW. The WIDTH value is ignored by GSFLOW, but a value, such as 10, must be specified. The third line is used to set the number of dimensions (NO_DIMENSIONS) that defines the array used to store the parameter values. NO_DIMENSIONS is specified as an integer value (1 for scalars and one-dimensional arrays; 2 for two-dimensional arrays). Scalar parameters (that is, those with a single value) are stored in the Modular Modeling System as a one-dimensional array; therefore, the user must specify a 1 for scalar variables.

The next NO_DIMENSIONS lines specify the dimension name(s) (DIMENSION_NAME(S)), one value per line, over which the parameter is declared, as character strings without spaces (see table A1-4 for a list of valid dimension names). DIMENSION_NAME must be specified as one if the parameter is a scalar. Most of the parameters have NO_DIMENSIONS equal to 1, and a DIMENSION_NAME that is either one or the name of a one-dimensional array (such as nhru or nmonths). A few parameters consist of two-dimensional arrays; that is, an array of values that consists of rows and columns. An example parameter that consists of a two-dimensional array of values is rain_adj, which is used in the PRMS precipitation modules precip_prms and precip_laps_prms. This parameter is defined over the dimensions nhru by nmonths; that is, the parameter has nhru rows and nmonths columns.

The line following the line(s) specifying dimension name(s) specifies the number of values (N_VALUES) that are input for the parameter. N_VALUES is specified as an integer value. For parameters that are scalar or a one-dimensional array, N_VALUES is specified as the size of the associated dimension, as declared in the dimensions section of the file. The total number of values specified for each two-dimensional parameter is equal to the product of its two dimension sizes. For the rain_adj parameter mentioned above, N_VALUES is specified as the product of the sizes of nhru and nmonths.

The line following specification of N_VALUES specifies type (TYPE) of the parameter values as an integer value; options are:

1 for integer
2 for real (single-precision, floating decimal point)
3 for double (double-precision, floating decimal point)
4 for character string

Note, no double-precision real (option 3) parameters are required in GSFLOW.

The line(s) following specification of TYPE, N_VALUES in number, each specifies one parameter value [VALUE(S)]. Two-dimensional array values are read column by column. For example, for the rain_adj parameter mentioned above, a value for each HRU (nhru values) for January (nmonths set to 1) are specified first, followed by nhru values for nmonths set to 2, and so forth, until a total of the product of nhru and nmonths values are specified.

Two example parameter-declaration items follow. The first example is for the latitude of the watershed centroid (VALUE set to 39.42900085449), parameter basin_lat, which is a scalar parameter (NO_DIMENSIONS set to 1; DIMENSION_NAME set to one) with a single value (N_VALUE(S) set to 1) of real type (TYPE set to 2):

```
####
basin_lat 15
1
one
1
2
39.42900085449
```

The second example is for the area of each HRU, parameter hru_area, which is a one-dimensional array declared using dimension nhru (NO_DIMENSIONS set to 1; DIMENSION_NAME set to nhru). The parameter consists of 128 values (N_VALUE(S) set to 128), which equals the size of nhru specified in the dimensions section. The values are of real type (TYPE set to 2). Only the first five values are shown:

```
####
hru_area 15
1
nhru
128
2
75.16000366211
38.68999862671
13.11999988556
108.9700012207
26.23999977112
    (remaining 123 values not shown)
```

Parameter-declaration items can be listed in any order, although it may be convenient to group items alphabetically or by each of the PRMS and GSFLOW modules. However, as some parameters are used by multiple modules, care must be taken in grouping parameter-declaration items by module. All PRMS parameters and the modules they are used in are listed alphabetically in table A1-5. If multiple parameter-declaration items are specified for the same parameter, the values specified last in the Parameter File will be used. Any parameter not specified in the Parameter File that is required by a GSFLOW simulation is assigned a default value. Any parameter specified in the Parameter File that is not required by a GSFLOW simulation is ignored. Warning message(s) are printed in both cases.

Table A1-5.   Parameters in the PRMS Parameter File listed alphabetically and their associated modules.

Parameter name	Module or modules
adjmix_rain	precip_dist2_prms, precip_laps_prms, precip_prms, xyz_dist
adjust_rain	xyz_dist
adjust_snow	xyz_dist
albset_rna	snowcomp_prms
albset_rnm	snowcomp_prms
albset_sna	snowcomp_prms
albset_snm	snowcomp_prms
basin_area	basin_prms
basin_cfs_init	gsflow_budget, strmflow_prms
basin_lat	soltab_hru_prms
basin_solsta	ccsolrad_hru_prms, ddsolrad_hru_prms
basin_tsta	basin_sum_prms, ccsolrad_hru_prms, temp_1sta_prms, temp_dist2_prms, temp_laps_prms
basin_tsta_hru	basin_sum_prms, xyz_dist
carea_max	srunoff_carea_casc, srunoff_smidx_casc
carea_min	srunoff_carea_casc
cascade_flg	cascade_prms
cascade_tol	cascade_prms
ccov_intcp	ccsolrad_hru_prms
ccov_slope	ccsolrad_hru_prms
cecn_coef	snowcomp_prms
conv_flag	xyz_dist
cov_type	intcp_prms, snowcomp_prms, soilzone_gsflow
covden_sum	hru_sum_prms, intcp_prms, snowcomp_prms, soilzone_gsflow
covden_win	hru_sum_prms, intcp_prms, snowcomp_prms, soilzone_gsflow
crad_coef	ccsolrad_hru_prms
crad_exp	ccsolrad_hru_prms
dday_intcp	ddsolrad_hru_prms
dday_slope	ddsolrad_hru_prms
den_init	snowcomp_prms
den_max	snowcomp_prms
elev_units	basin_prms, xyz_dist
emis_noppt	snowcomp_prms
epan_coef	intcp_prms, potet_pan_prms
fastcoef_lin	soilzone_gsflow
fastcoef_sq	soilzone_gsflow
freeh2o_cap	snowcomp_prms
gvr_cell_id	gsflow_budget, gsflow_mf2prms, gsflow_prms2mf, gsflow_setconv
gvr_cell_pct	gsflow_prms2mf, gwflow_setconv
gvr_hru_id	gsflow_budget, gsflow_mf2prms, gsflow_prms2mf, soilzone_gsflow
gvr_hru_pct	gsflow_budget, gsflow_mf2prms, gsflow_prms2mf, soilzone_gsflow
gw_down_id	cascade_prms
gw_pct_up	cascade_prms
gw_strmseg_down_id	cascade_prms
gw_up_id	cascade_prms
gwflow_coef	gwflow_casc_prms

Table A1-5.   Parameters in the PRMS Parameter File listed alphabetically and their associated modules.—Continued

Parameter name	Module or modules
gwsink_coef	gwflow_casc_prms
gwstor_init	gwflow_casc_prms
hamon_coef	potet_hamon_hru_prms
hru_area	basin_prms, cascade_prms, ccsolrad_hru_prms, ddsolrad_hru_prms, gsflow_budget, gsflow_mf2prms, gsflow_prms2mf, gwflow_casc_prms, intcp_prms, potet_hamon_hru_prms, potet_jh_prms, potet_pan_prms, precip_dist2_prms, precip_laps_prms, precip_prms, snowcomp_prms, soilzone_gsflow, srunoff_carea_casc, srunoff_smidx_casc, strmflow_prms, temp_1sta_prms, temp_dist2_prms, temp_laps_prms, xyz_dist
hru_aspect	soltab_hru_prms
hru_deplcrv	snowcomp_prms
hru_down_id	cascade_prms
hru_elev	basin_prms, precip_laps_prms, temp_1sta_prms, temp_dist2_prms, temp_laps_prms, xyz_dist
hru_gwres	cascade_prms, gwflow_casc_prms, strmflow_prms
hru_lat	soltab_hru_prms
hru_pansta	potet_pan_prms
hru_pct_up	cascade_prms
hru_percent_imperv	basin_prms, hru_sum_prms, srunoff_carea_casc, srunoff_smidx_casc
hru_plaps	precip_laps_prms
hru_psta	precip_laps_prms, precip_prms
hru_segment	gsflow_prms2mf
hru_sfres	strmflow_prms
hru_slope	basin_prms, soltab_hru_prms
hru_strmseg_down_id	cascade_prms
hru_tlaps	temp_laps_prms
hru_tsta	temp_1sta_prms, temp_laps_prms
hru_type	basin_prms, cascade_prms, gsflow_budget, gsflow_mf2prms, gsflow_prms2mf, hru_sum_prms, intcp_prms, snowcomp_prms, soilzone_gsflow, srunoff_carea_casc, srunoff_smidx_casc
hru_up_id	cascade_prms
hru_x	xyz_dist
hru_xlong	precip_dist2_prms, temp_dist2_prms
hru_y	xyz_dist
hru_ylat	precip_dist2_prms, temp_dist2_prms
id_obsrunoff	gsflow_sum
imperv_stor_max	srunoff_carea_casc, srunoff_smidx_casc
jh_coef	potet_jh_prms
jh_coef_hru	potet_jh_prms
lake_hru_id	gsflow_budget, gsflow_mf2prms, gsflow_prms2mf
lapsemax_max	temp_dist2_prms
lapsemax_min	temp_dist2_prms
lapsemin_max	temp_dist2_prms
lapsemin_min	temp_dist2_prms
local_reachid	gsflow_prms2mf
max_lapse	xyz_dist
maxmon_prec	precip_dist2_prms
melt_force	snowcomp_prms
melt_look	snowcomp_prms
min_lapse	xyz_dist
monmax	temp_dist2_prms
monmin	temp_dist2_prms
moyrsum	hru_sum_prms
mxsziter	gsflow_modflow, gsflow_prms2mf
nsos	strmflow_prms
numreach_segment	gsflow_prms2mf
o2	strmflow_prms
objfunc_q	basin_sum_prms
padj_rn	precip_laps_prms
padj_sn	precip_laps_prms

Table A1-5. Parameters in the PRMS Parameter File listed alphabetically and their associated modules.—Continued

Parameter name	Module or modules
pmn_mo	precip_laps_prms
pmo	hru_sum_prms
potet_sublim	intcp_prms, snowcomp_prms
ppt_add	xyz_dist
ppt_div	xyz_dist
ppt_lapse	xyz_dist
ppt_rad_adj	ccsolrad_hru_prms, ddsolrad_hru_prms
precip_units	precip_dist2_prms, precip_laps_prms, precip_prms, xyz_dist
pref_flow_den	soilzone_gsflow
print_freq	basin_sum_prms
print_objfunc	basin_sum_prms
print_type	basin_sum_prms
psta_elev	precip_laps_prms, xyz_dist
psta_freq_nuse	xyz_dist
psta_mon	precip_dist2_prms
psta_month_ppt	xyz_dist
psta_nuse	xyz_dist
psta_x	xyz_dist
psta_xlong	precip_dist2_prms
psta_y	xyz_dist
psta_ylat	precip_dist2_prms
rad_conv	ccsolrad_hru_prms, ddsolrad_hru_prms
rad_trncf	snowcomp_prms
radadj_intcp	ddsolrad_hru_prms
radadj_slope	ddsolrad_hru_prms
radj_sppt	ccsolrad_hru_prms, ddsolrad_hru_prms
radj_wppt	ccsolrad_hru_prms, ddsolrad_hru_prms
radmax	ccsolrad_hru_prms, ddsolrad_hru_prms
rain_adj	precip_prms
rain_code	obs_prms, xyz_dist
rain_mon	precip_dist2_prms
reach_segment	gsflow_prms2mf
runoff_units	basin_sum_prms, gsflow_sum
s2	strmflow_prms
sat_threshold	soilzone_gsflow
segment_pct_area	gsflow_prms2mf
settle_const	snowcomp_prms
sfres_coef	strmflow_prms
sfres_din1	strmflow_prms
sfres_init	strmflow_prms
sfres_qro	strmflow_prms
sfres_type	strmflow_prms
slowcoef_lin	soilzone_gsflow
slowcoef_sq	soilzone_gsflow
smidx_coef	srunoff_smidx_casc
smidx_exp	srunoff_smidx_casc
snarea_curve	snowcomp_prms
snarea_thresh	snowcomp_prms
snow_adj	precip_prms
snow_intcp	intcp_prms
snow_mon	precip_dist2_prms
snowinfil_max	srunoff_carea_casc, srunoff_smidx_casc
soil_moist_init	soilzone_gsflow
soil_moist_max	soilzone_gsflow, srunoff_carea_casc, srunoff_smidx_casc
soil_rechr_init	soilzone_gsflow
soil_rechr_max	soilzone_gsflow, srunoff_carea_casc
soil_type	soilzone_gsflow

Table A1-5.   Parameters in the PRMS Parameter File listed alphabetically and their associated modules.—Continued

Parameter name	Module or modules
soil2gw_max	soilzone_gsflow
solrad_elev	xyz_dist
srain_intcp	intcp_prms
ssr2gw_exp	soilzone_gsflow
ssr2gw_rate	soilzone_gsflow
ssrmax_coef	soilzone_gsflow
ssstor_init	soilzone_gsflow
szconverge	gsflow_prms2mf
temp_units	potet_hamon_hru_prms, potet_jh_prms, precip_dist2_prms, precip_laps_prms, precip_prms, temp_1sta_prms, temp_dist2_prms, temp_laps_prms, xyz_dist
tmax_add	xyz_dist
tmax_adj	temp_1sta_prms, temp_dist2_prms, temp_laps_prms, xyz_dist
tmax_allrain	ddsolrad_hru_prms, precip_dist2_prms, precip_laps_prms, precip_prms, xyz_dist
tmax_allsnow	precip_dist2_prms, precip_laps_prms, precip_prms, snowcomp_prms, xyz_dist
tmax_div	xyz_dist
tmax_index	ddsolrad_hru_prms
tmax_lapse	temp_1sta_prms
tmin_add	xyz_dist
tmin_adj	temp_1sta_prms, temp_dist2_prms, temp_laps_prms, xyz_dist
tmin_div	xyz_dist
tmin_lapse	temp_1sta_prms
transp_beg	potet_hamon_hru_prms, potet_jh_prms, potet_pan_prms
transp_end	potet_hamon_hru_prms, potet_jh_prms, potet_pan_prms
transp_tmax	potet_hamon_hru_prms, potet_jh_prms
tsta_elev	temp_1sta_prms, temp_dist2_prms, temp_laps_prms, xyz_dist
tsta_month_max	xyz_dist
tsta_month_min	xyz_dist
tsta_nuse	xyz_dist
tsta_x	xyz_dist
tsta_xlong	temp_dist2_prms
tsta_y	xyz_dist
tsta_ylat	temp_dist2_prms
tstorm_mo	snowcomp_prms
upst_res1	strmflow_prms
upst_res2	strmflow_prms
upst_res3	strmflow_prms
wrain_intcp	intcp_prms
x_add	xyz_dist
x_div	xyz_dist
y_add	xyz_dist
y_div	xyz_dist
z_add	xyz_dist
z_div	xyz_dist

PRMS Modules

Input instructions for those modules that relate to a single process are grouped together, such as modules temp_1sta_prms, temp_laps_prms, and temp_dist2_prms, which relate to the distribution of temperature across the modeled area. The module xyz_dist is listed separately because it is used to distribute both temperature and precipitation across the modeled area. Several parameters are used by more than one module. Because the modules are defined independently of one another, the parameter description is repeated for each module. For example, the parameter hru_area is used by several modules, including modules basin_prms and temp_1sta_prms.

Basin Module

The PRMS Basin Module (basin_prms) computes shared watershed-wide variables. Shared variables include the area of each HRU that is pervious and impervious determined on the basis of the fraction in each HRU that is impervious (parameter hru_precent_imperv), the total area of the watershed determined as the sum of the area in each HRU (parameter hru_area), and the total area occupied by lake HRUs and land HRUs determined as the sum of the area in each lake and land HRU (parameter hru_type), respectively. Input parameters for the Basin Module are defined in table A1-6. Checks for consistency of watershed-wide variables are done with the input parameters.

Table A1-6.   Input parameters required for the PRMS Basin Module: basin_prms.

[HRU, hydrologic response unit; nhru, number of HRUs]

Parameter name	Description	Dimension variable	Units	Type	Range	Default value
basin_area	Total area of watershed	one	acres	real	0.1 to 1.0e9	1.0
elev_units	Units of altitude (0=feet; 1=meters)	one	dimensionless	integer	0 or 1	0
hru_area	Area of HRU	nhru	acres	real	0.1 to 1.0e9	1.0
hru_elev	Mean land-surface altitude of HRU	nhru	elev_units	real	-300.0 to 30,000.0	0.0
hru_percent_imperv	Decimal fraction of HRU area that is impervious	nhru	dimensionless	real	0.0 to 1.0	0.0
hru_slope	Slope of HRU, specified as change in vertical length divided by change in horizontal length	nhru	dimensionless	real	0.0 to 10.0	0.0
hru_type	Type of HRU (0=inactive; 1=land; 2=lake)	nhru	dimensionless	integer	0 to 2	1

Cascade Module

The PRMS Cascade Module (cascade_prms) determines the computational order of the HRUs and ground-water reservoirs for routing flow downslope in a cascading pattern. Input parameters for Cascade Module are defined in table A1-7; parameters are arranged by type of routing (HRU or ground-water reservoir) with those that are common to both routing patterns listed first. Each link in an HRU cascade path is specified by four parameters (hru_up_id, hru_down_id, hru_strmseg_down_id, and hru_pct_up). The number of HRU cascade links must equal dimension ncascade. Likewise, each ground-water reservoir cascade link is specified by four parameters (gw_up_id, gw_down_id, gw_strmseg_down_id, and gw_pct_up). The number of ground-water reservoir cascade links must equal dimension ncascdgw.

Additional information for selected parameters defined in table A1-7 follows:

cascade_flg—routing switch. A value of 0 allows for multiple routing paths of inflows and outflows among HRUs (many-to-many routing) and to stream segments. A value of 1 forces routing to be from an upslope HRU to one downslope HRU (one-to-one routing) and(or) stream segment, even if multiple cascade links are specified from an upslope HRU. The cascade link with the greatest fraction of contributing area (parameter hru_pct_up) in the upslope HRU is selected as the single outflow from the HRU.

cascade_tol—area tolerance for cascade links. This parameter can be used to ignore cascade links involving small contributing areas in the upslope HRU or ground-water reservoir that may be generated in a GIS analysis to produce the cascade parameters. Contributing areas less than cascade_tol are evenly distributed to all other cascade paths originating in the upslope HRU. Contributing areas for an HRU cascade link equals the area of the HRU (parameter hru_area) times the fraction of outflow for the path (parameter hru_pct_up). Contributing areas for a ground-water reservoir cascade link equals the area of the ground-water reservoir (variable gwres_area computed in the Cascade Module) times the fraction of outflow for the link (parameter gw_pct_up).

hru_pct_up and gw_pct_up—fraction of total outflow from an upslope HRU and ground-water reservoir, respectively, to be added as inflow to a downslope HRU (or ground-water reservoir) or stream segment. This fraction represents the ratio of the contributing area to the total area of the upslope HRU (or ground-water reservoir) to be routed to a downslope HRU (or ground-water reservoir).

hru_strmseg_down_id and gw_strmseg_down_id—identification number of a stream segment that receives outflow from an HRU and ground-water reservoir, respectively. When either parameter is greater than zero, the specified stream segment receives the outflow for the cascade path and the corresponding parameter (hru_down_id and gw_down_id) is ignored.

Table A1-7.   Input parameters required for the PRMS Cascade Module: cascade_prms.

[Equation number refers to equations listed in the main body of report— equation variable of parameter name is defined in first listed equation. HRU: hydrologic response unit; nhru, number of HRUs; ncascade, number of cascade paths associated with HRUs; ncascdgw, number of cascade paths associated with PRMS ground-water reservoirs; ngw, number of PRMS ground-water reservoirs ; one, a dimension of one]

Parameter name	Description	Dimension variable	Units	Type	Range	Default value	Equation number
hru_area	Area of HRU	nhru	acres	real	0.1 to 1.0e9	1.0	22, 37
casacade_flg[1]	Type of cascade routing (0=allow many-to-many; 1=only allow one-to-one)	one	dimensionless	integer	0 or 1	0	
cascade_tol[1]	Minimum area of upslope HRU for computing cascading flow	one	acres	real	0.0 to 99.0	5.0	
	Parameters for routing surface runoff and interflow among HRUs and to stream segments						
hru_up_id	Identifier of HRU that contributes flow for each cascade link	ncascade	dimensionless	integer	1 to nhru	1	
hru_strmseg_down_id[1]	Identifier of stream segment that receives flow for each cascade link	ncascade	dimensionless	integer	0 to nsegment + 1	0	
hru_down_id	Identifier of HRU that receives flow for each cascade link; if hru_strmseg_down_id is not 0 for a cascade link, hru_down_id is ignored	ncascade	dimensionless	integer	1 to nhru	1	
hru_type	Type of each HRU (0=inactive; 1=land; 2=lake)	nhru	dimensionless	integer	0 to 2	1	
hru_pct_up[1]	Decimal fraction of area in the upslope HRU that contributes Hortonian runoff to the downslope HRU	ncascade	dimensionless	real	0.0 to 1.0	1.0	36
	Parameters for routing flow among ground-water reservoirs and to streams, used for PRMS-only simulations						
gw_pct_up[1]	Decimal fraction of source ground-water reservoir associated with each ground-water cascade link	ncascdgw	dimensionless	real	0.0 to 1.0	1.0	
hru_gwres	Identifier of ground-water reservoir associated with an HRU	nhru	dimensionless	integer	1 to ngw	1	
gw_strmseg_down_id[1]	Identifier of stream segment that receives flow for each ground-water cascade link	ncascdgw	dimensionless	integer	0 to nsegment + 1	0	
gw_up_id	Identifier of HRU that contributes flow to each ground-water cascade link	ncascdgw	dimensionless	integer	1 to ngw	1	
gw_down_id	Identifier of ground-water reservoir that receives flow from each ground-water cascade link; gw_down_id is ignored if gw_strmseg_down_id is not 0 for a cascade link	ncascdgw	dimensionless	integer	1 to ngw	1	

[1]Additional description of parameter provided on p. 151.

Observed Data Module

The PRMS Observed Data Module (obs_prms) makes available the measured data for each time step specified in the PRMS Data File(s), and verifies that the data are within a valid range. If measured values are outside the valid range for an input variable, a warning message is printed. Data that are made available are precipitation and maximum and minimum air temperature. Optional measured data if specified in the PRMS Data File include solar radiation and streamflow. A single parameter is used by the Observed Data Module and it is only needed when the xyz_dist module is active. The input parameter for the module is defined in table A1-8.

Table A1-8.   Input parameter required for the PRMS Observed Data Module: obs_prms.

[nmonths, number of months in a year]

Parameter name	Description	Dimension variable	Units	Type	Range	Default value
rain_code	Use of measured precipitation values (1=if psta_nuse stations have precipitation; 2=if any precipitation station has precipitation; 3=if xyz regression indicates precipitation; 4=if rain_day variable is set to 1 in a PRMS Data File; 5=if psta_freq_use stations have precipitation)	nmonths	dimensionless	integer	1 to 5	2

Potential Solar-Radiation Module

The PRMS Potential Solar-Radiation Module (soltab_hru_prms) calculates tables of 366 (maximum number of days in a year) values of potential solar radiation and hours of sunlight for each HRU on the basis of representative slope, aspect, and latitude of each HRU. The module also computes a table of the potential solar radiation at the watershed centroid with a horizontal slope. Input parameters for the module are defined in table A1-9.

Table A1-9.   Input parameters required for the PRMS potential solar-radiation module: soltab_hru_prms.

[Equation number refers to equations listed in the main body of report—parameter name is defined following the equation. HRU, hydrologic response unit; nhru, number of HRUs]

Parameter name	Description	Dimension variable	Units	Type	Range	Default value	Equation number
basin_lat	Latitude of watershed centroid	one	degrees latitude	real	-90.0 to 90.0	40.0	13a
hru_aspect	Aspect of HRU	nhru	degrees azimuth	real	0.0 to 360.0	0	
hru_lat	Latitude of HRU centroid	nhru	degrees latitude	real	-90.0 to 90.0	40.0	13a
hru_slope	HRU slope, specified as change in vertical length divided by change in horizontal length	nhru	dimensionless	real	0.0 to 10.0	0.0	15b

Temperature Distribution Modules

Four PRMS Temperature Distribution Modules (temp_1sta_prms, temp_laps_prms, temp_dist2_prms, and xyz_dist) are included in GSFLOW. The modules are used to distribute maximum and minimum temperatures to each HRU. The user selects one of the four modules for a particular simulation by setting the control parameter `temp_module` in the GSFLOW Control File to the one of the temperature module names. If module xyz_dist is selected for distributing temperature then it also must be selected for distributing precipitation. Modules temp_1sta_prms, temp_laps_prms, and temp_dist2_prms are explained in this section and module xyz_dist is described in the subsequent section "Temperature and Precipitation Distribution Module" because when specified it also must distribute precipitation to each HRU.

Module temp_1sta_prms distributes temperatures to HRUs using temperature data measured at one station and a monthly parameter based on a lapse rate with elevation. Module temp_laps_prms distributes temperatures to HRUs using a lapse rate computed from daily measured data at two temperature stations. Module temp_dist2_prms distributes temperatures to HRUs using a lapse rate computed from from two or more stations weighted by the inverse of the square of the distance between the centroid of an HRU and each station location. Input parameters for modules temp_1sta_prms, temp_laps_prms, and temp_dist2_prms are defined in table A1-10. Some parameters are common to the three modules and are listed first, followed by parameters specific for module temp_1sta_prms, then for temp_laps_prms, and last for temp_dist2_prms.

Table A1-10. Input parameters required for the PRMS Temperature Distribution Modules: temp_1sta_prms, temp_laps_prms, and temp_dist2_prms.

[Equation number refers to equations listed in the main body of report—parameter name is defined following the equation. HRU, hydrologic response unit; `nhru`, number of HRUs; `ntemp`, number of measurement stations that measure air temperature; `nmonths`, number of months in a year; `elev_units`, PRMS Basin Module parameter to define units of feet (0) or meters (1)]

Parameter name	Description	Dimension variable	Units	Type	Range	Default value	Equation number
		Parameters common to three modules					
basin_tsta	Identifier of the measurement station used to compute basin air temperature	one	dimensionless	integer	1 to ntemp	1	
hru_area	Area of HRU	nhru	acres	real	0.1 to 1.0e9	1.0	
hru_elev	Mean land-surface altitude of HRU	nhru	elev_units	real	-300.0 to 30,000.0	0.0	1, 2, 5a
temp_units	Units of measured air temperature (0=degrees Fahrenheit; 1=degrees Celsius)	one	dimensionless	integer	0 or 1	0.0	
tmax_adj[1]	Maximum daily HRU temperature adjustment factor, which is estimated on the basis of slope and aspect	nhru	temp_units	real	-10.0 to 10.0	0.0	1, 2, 5a
tmin_adj[1]	Minimum daily HRU temperature adjustment factor, which is estimated on the basis of slope and aspect	nhru	temp_units	real	-10.0 to 10.0	0.0	1, 2, 5a
tsta_elev	Altitude of the air temperature measurement station	ntemp	elev_units	real	-300.0 to 30,000.0	0.0	1, 4, 5a
		Additional parameters for module temp_1sta_prms					
hru_tsta	Identifier of the measurement station used to compute HRU daily maximum and minimum air temperatures	nhru	dimensionless	integer	0 to ntemp	1	
tmax_lapse[1]	Monthly maximum air temperature lapse rate, representing the change in maximum air temperature per 1,000 feet or meters of altitude change for each month, January to December	nmonths	temp_units	real	-10.0 to 10.0	3.0	1
tmin_lapse[1]	Monthly minimum air temperature lapse rate, representing change in minimum air temperature per 1,000 feet or meters of altitude change for each month, January to December	nmonths	temp_units	real	-10.0 to 10.0	3.0	1

Table A1-10. Input parameters required for the PRMS Temperature Distribution Modules: temp_1sta_prms, temp_laps_prms, and temp_dist2_prms—Continued.

[Equation number refers to equations listed in the main body of report—parameter name is defined following the equation. HRU, hydrologic response unit; nhru, number of HRUs; ntemp, number of measurement stations that measure air temperature; nmonths, number of months in a year; elev_units, PRMS Basin Module parameter to define units of feet (0) or meters (1)]

Parameter name	Description	Dimension variable	Units	Type	Range	Default value	Equation number
	Additional parameters for module temp_laps_prms						
hru_tlaps	Identifier of lapse measurement station used for air-temperature lapse rate calculations	nhru	dimensionless	integer	1 to ntemp	1	
hru_tsta	Identifier of base measurement station used for air temperature lapse rate calculations	nhru	dimensionless	integer	1 to ntemp	1	
	Additional parameters for module temp_dist2_prms						
hru_xlong	Longitude of HRU centroid	nhru	feet	real	-1e+09 to 1e+09	0.0	5b
hru_ylat	Latitude of HRU centroid	nhru	feet	real	-1e+09 to 1e+09	0.0	5b
lapsemax_max	Monthly maximum lapse rate from historical data used to constrain highest daily maximum lapse rate	nmonths	temp_units	real	-2.0 to 4.0	3.0	
lapsemax_min	Monthly minimum lapse rate from historical data used to constrain lowest daily maximum lapse rate	nmonths	temp_units	real	-7.0 to -3.0	-6.5	
lapsemin_max	Monthly maximum lapse rate from historical data to constrain highest daily minimum lapse rate	nmonths	temp_units	real	-2.0 to 4.0	3.0	
lapsemin_min	Monthly minimum lapse rate from historical data used to constrain lowest daily minimum lapse rate	nmonths	temp_units	real	-7.0 to -3.0	-4.0	
monmax	Monthly maximum air temperature from historical data used to constrain lowest daily maximum air temperatures	nmonths	temp_units	real	45.0 to 115.0	100.0	
monmin	Monthly minimum air temperature from historical data used to constrain lowest daily minimum air temperatures	nmonths	temp_units	real	-35.0 to 45.0	-20.0	
tsta_xlong	Longitude of measurement station that measures air temperature	ntemp	feet	real	-1e+09 to 1e+09	0.0	5b
tsta_ylat	Latitude of measurement station that measures air temperature	ntemp	feet	real	-1e+09 to 1e+09	0.0	5b

[1]Additional description of parameter provided below.

Additional information for four parameters shown in table A1-10 follows:

tmax_adj and tmin_adj—maximum and minimum air-temperature adjustments, respectively, for each HRU based on slope and aspect of the HRU, in degrees Fahrenheit or degrees Celsius.

tmax_lapse and tmin_lapse—monthly maximum and minimum air temperature lapse rates, respectively, in degrees Fahrenheit or degrees Celsius. Twelve values each are specified that define the change in maximum and minimum air temperature per 1,000 feet or meters elevation change during each month from January to December.

Temperature and Precipitation Distribution Module

The Temperature and Precipitation Distribution Module (xyz_dist) distributes daily precipitation and minimum and maximum air temperatures to each HRU using a multiple linear regression of measured data from a group of measurement stations or from the results of an atmospheric model. The user selects the xyz_dist module for a particular simulation by setting the control parameters `temp_module` and `precip_module` in the GSFLOW Control File to `xyz_dist`. Important geographic factors affecting the spatial distribution of precipitation and air temperature distributions within a watershed are longitude (X), latitude (Y), and elevation (Z). The method used by module xyz_dist is described in Hay and others (2000). Input parameters for the module xyz_dist are defined in table A1-11.

Table A1-11.   Input parameters required for the PRMS Temperature and Precipitation Module: xyz_dist.

[Equation number refers to equations listed in the main body of report— equation variable of parameter name is defined in first listed equation. HRU, hydrologic response unit; nhru, number of HRUs; ntemp, number of measurement stations that measure air temperature; nrain, number of measurement stations that measure precipitation; nlapse, number of lapse rates in the X, Y, and Z directions; nmonths, number of months in a year]

Parameter name	Description	Dimension variable	Units	Type	Range	Default value	Equation number
adjmix_rain[1]	Monthly adjustment factor for a mixed precipitation event as a decimal fraction	nmonths	dimensionless	real	0.0 to 3.0	1.0	6
adjust_rain[1]	Monthly factor as a decimal fraction used to adjust rain values	nmonths	dimensionless	real	0.0 to 1.0	0.01	9a
adjust_snow[1]	Monthly factor as a decimal fraction used to adjust snow values	nmonths	dimensionless	real	0.0 to 1.0	0.01	9a
basin_tsta_hru	Identifier of HRU used to compute watershed air temperatures	one	dimensionless	integer	0 to nhru	1	
conv_flag	Conversion of altitude (0=no conversion; 1=feet to meters; 2=meters to feet)	one	dimensionless	integer	0 to 2	0	
elev_units	Units of altitude (0=feet; 1=meters)	one	dimensionless	integer	0 or 1	0	
hru_area	Area of HRU	nhru	acres	real	0.1 to 1.0e9	1.0	
precip_units	Specify units of precipitation (0=inches; 1=millimeters)	one	dimensionless	integer	0 or 1	0	
rain_code	Indicates use of XYZ distribution technique for each time step (1=if psta_nuse stations have precipitation; 2=if any precipitation station has precipitation; 3=always; 4=if rain_day variable is set to 1 in a PRMS Data File; 5=if psta_freq_use stations have precipitation)	nmonths	dimensionless	integer	1 to 5	2	
solrad_elev	Altitiude of each measurement station that measures solar radiation and used in calculating degree-day curves	one	meters	real	1000.0 to 10,000.0	0.0	
temp_units	Units of air temperature (0=degrees Fahrenheit; 1=degrees Celsius)	one	dimensionless	integer	0 or 1	0	
tmax_adj	Adjustment to maximum air temperature for HRU, estimated on basis of slope and aspect	nhru	temp_units	real	-10.0 to 10.0	0.0	1,3a
tmax_allrain	Monthly minimum air temperature at an HRU that results in all precipitation during a day being rain	nmonths	temp_units	real	0.0 to 90.0	40.0	
tmax_allsnow	Monthly maximum air temperature at which precipitation is all snow for the HRU	one	temp_units	real	-10.0 to 40.0	32.0	6
tmin_adj	Minimum daily temperature adjustment factor	nhru	temp_units	real	-100.0 to 100.0	0.0	1,3a

Table A1-11.   Input parameters required for the PRMS Temperature and Precipitation Module: xyz_dist—Continued

[Equation number refers to equations listed in the main body of report— equation variable of parameter name is defined in first listed equation. HRU, hydrologic response unit; nhru, number of HRUs; ntemp, number of measurement stations that measure air temperature; nrain, number of measurement stations that measure precipitation; nlapse, number of lapse rates in the X, Y, and Z directions; nmonths, number of months in a year]

Parameter name	Description	Dimension variable	Units	Type	Range	Default value	Equation number
		Location and altitude parameters (x, y, and z)					
hru_elev[1]	Mean land-surface altitude of HRU	nhru	elev_units	real	-300.0 to 30,000.0	0.0	
hru_x[1]	Longitude (X) for HRU in albers projection	nhru	meters	real	-1.e-7 to 1.e7	0.0	
hru_y[1]	Latitude (Y) for HRU in albers projection	nhru	meters	real	-1.e-7 to 1.e7	0.0	
psta_elev[1]	Altitude of each measurement station that measures precipitation	nrain	elev_units	real	-300.0 to 30,000.0	0.0	
psta_y[1]	Latitude (Y) for each measurement station that measures precipitation in albers projection	nrain	meters	real	-1.e-7 to 1.e7	0.0	
tsta_elev[1]	Altitude of each measurement station that measures air temperature	ntemp	elev_units	real	-300.0 to 30,000.0	0.0	
tsta_x[1]	Longitude (X) for each measurement station that measures air temperature in albers projection	ntemp	meters	real	-1.e-7 to 1.e7	0.0	
tsta_y[1]	Latitude (Y) for each measurement station that measures air temperature in albers projection	ntemp	meters	real	-1.e-7 to 1.e7	0.0	
		Multiple linear regression parameters					
max_lapse[1]	Maximum air temperature regression coefficient for longitude, latitude, and altitude, respectively by month, starting with January	nlapse by nmonths	temp_units	real	-100.0 to 100.0	0.0	3b
min_lapse[1]	Minimum air temperature regression coefficient for longitude, latitude, and altitude, respectively by month starting with January	nlapse by nmonths	temp_units	real	-100.0 to 100.0	0.0	3b
ppt_lapse[1]	Precipitation regression coefficient for longitude, latitude, and altitude, respectively by month, starting with January	nlapse by nmonths	inches	real	-10.0 to 10.0	0.0	9b, 9c
		Designated station parameters					
psta_freq_nuse[1]	Defines measurement stations used to determine if precipitation is occurring in watershed (0=no; 1=yes)	nrain	dimensionless	integer	0 or 1	1	
psta_nuse	Defines which measurement stations will be used in the distribution regression of precipitation (0=no; 1=yes)	nrain	dimensionless	integer	0 or 1	1	
tsta_nuse	Defines which measurement stations will be used in distribution regression of air temperatures (0=no; 1=yes)	ntemp	dimensionless	integer	0 or 1	1	

Table A1-11.   Input parameters required for the PRMS Temperature and Precipitation Module: xyz_dist—Continued

[Equation number refers to equations listed in the main body of report— equation variable of parameter name is defined in first listed equation. HRU, hydrologic response unit; nhru, number of HRUs; ntemp, number of measurement stations that measure air temperature; nrain, number of measurement stations that measure precipitation; nlapse, number of lapse rates in the X, Y, and Z directions; nmonths, number of months in a year]

Parameter name	Description	Dimension variable	Units	Type	Range	Default value	Equation number
	Mean monthly climate parameters
psta_month_ppt[1]	Monthly average precipitation at each measurement station	nrain by nmonths	precip_units	real	0.0 to 200.0	0.0	
tsta_month_max[1]	Monthly average maximum air temperature at measurement station	ntemp by nmonths	temp_units	real	-100.0 to 200.0	0.0	
tsta_month_min[1]	Monthly average minimum air temperature at each measurement station	ntemp by nmonths	temp_units	real	-100.0 to 200.0	0.0	
	Transformation parameters for dependent variables
ppt_add[1]	Calculated mean of precipitation for watershed	one	precip_units	real	-10.0 to 10.0	0.0	
ppt_div[1]	Calculated standard deviation of precipitation for watershed	one	precip_units	real	-10.0 to 10.0	0.0	
tmax_add[1]	Calculated mean of maximum air temperature for watershed	one	temp_units	real	-100.0 to 100.0	0.0	
tmax_div[1]	Calculated standard deviation of maximum air temperature for watershed	one	temp_units	real	-100.0 to 100.0	0.0	
tmin_add[1]	Calculated mean of minimum air temperature for watershed	one	temp_units	real	-100.0 to 100.0	0.0	
tmin_div[1]	Calculated standard deviation of minimum air temperature for watershed	one	temp_units	real	-100.0 to 100.0	0.0	
	Transformation parameters for independent variables
x_add[1]	Calculated mean of measurement station longitude (X) coordinates for watershed	one	meters	real	-1.e-7 to 1.e7	0.0	
x_div[1]	Calculated standard deviation of measurement station longitude (X) coordinates for watershed	one	meters	real	-1.e-7 to 1.e7	0.0	
y_add[1]	Calculated mean of measurement station latitude (Y) coordinates for watershed	one	meters	real	-1.e-7 to 1.e7	0.0	
y_div[1]	Calculated standard deviation of measurement station latitude (Y) coordinates for watershed	one	meters	real	-1.e-7 to 1.e7	0.0	
z_add[1]	Calculated mean of measurement station altitude (Z) coordinates for watershed	one	meters	real	-1.e-7 to 1.e7	0.0	
z_div[1]	Calculated standard deviation of measurement station altitude (Z) coordinates for watershed	one	meters	real	-1.e-7 to 1.e7	0.0	

Additional information for selected parameters defined in table A1-11 follows:

adjmix_rain—adjustment factor to the fraction of measured daily precipitation in the PRMS Data File that is estimated to be rain. The factor is used to adjust rain in relation to precipitation before computing daily volumes of rain and snow in each HRU.

adjust_rain and adjust_snow—adjustment factors to precipitation values (rain and snow, respectively) specified in the PRMS Data File. The adjustments are done before computing daily volumes of rain and snow in each HRU.

hru_elev, hru_x, hru_y, psta_elev, psta_x, psta_y, tsta_elev, tsta_x, tsta_y—location and elevation (longitude (x), latitude (y), and altitude (elev)) for the centroid of each HRU, and location and elevation of precipitation (psta) and air-temperature (tsta) measurement stations. These values are computed for the watershed and should not be adjusted in any model-calibration process.

max_lapse, min_lapse, and ppt_lapse—maximum (max) and minimum (min) air-temperature and precipitation (ppt) lapse rates, respectively. These two-dimensional parameters (dimensions: nlapse by nmonths, or 3 by 12) are used to define a multiple-linear regression (MLR) coefficient for each independent variable (longitude (x), latitude (y), or altitude (z)) by month.

psta_month_ppt—Mean monthly precipitation at a precipitation station. The values are used in place of the station mean when data are missing at a precipitation station.

tsta_month_max and tsta_month_min— Mean monthly maximum and minimum air temperature at a measurement station, respectively. The values are used in place of the station mean when data are missing at a temperature station.

ppt_add, tmax_add, and tmin_add—mean values of the precipitation (precip) and maximum (tmax) and minimum (tmin) air-temperature for the watershed. These values are computed for the watershed and should not be adjusted in any model calibration process.

ppt_div, tmax_div, and tmin_div—standard deviation values of the precipitation (precip) and maximum (tmax) and minimum (tmin) air-temperature for the watershed. These values are computed for the watershed and should not be adjusted in any model-calibration process.

x_add, y_add, and z_add—mean values of the longitude (x), latitude (y), and altitude (z) for the climate-stations used in the regression analysis. These values are computed for the watershed and should not be adjusted in any model-calibration process.

x_div, y_div, and z_div—standard deviation values of the longitude (x), latitude (y), and altitude (z) for the climate-stations used in the regression analysis. These values are computed for the watershed and should not be adjusted in any model-calibration process.

Precipitation Distribution Modules

Four PRMS Precipitation Distribution Modules (precip_prms, precip_laps_prms, precip_dist2_prms, and xyz_dist) are included in GSFLOW. The modules are used to distribute precipitation to each HRU and to determine the form of precipitation (rain, snow, or a mixture of both). The user selects only one of the modules for a simulation by setting the control parameter precip_module in the GSFLOW Control File to one of those four precipitation module names. If module xyz_dist is selected for distributing precipitation then it also must be selected for distributing temperature. Modules precip_prms, precip_laps_prms, and precip_dist2_prms are explained in this section and module xyz_dist is described in the preceding section "Temperature and Precipitation Distribution Module" because it also is used to distribute temperature to each HRU. Module precip_prms distributes precipitation to HRUs using precipitation data measured at one station and a monthly adjustment factor for the form of precipitation. Module precip_laps_prms distributes precipitation to HRUs using a lapse rate computed from daily measured data at two precipitation stations. Module precip_dist2_prms distributes precipitation to HRUs using a lapse rate computed from from two or more stations weighted by the inverse of the square of the distance between the centroid of an HRU and each station location. Input parameters for modules precip_prms, precip_laps_prms, and precip_dist2_prms are defined in table A1-12.

Table A1-12.   Input parameters required for the PRMS Precipitation Distribution Modules: precip_prms, precip_laps_prms, and precip_dist2_prms.

[Equation number refers to equations listed in the main body of report— equation variable of parameter name is defined in first listed equation. HRU, hydrologic response unit; nhru, number of HRUs; ntemp, number of measurement stations that measure air temperature; nrain, number of measurement stations that measure precipitation; nlapse, number of lapse rates in the X, Y, and Z directions; nmonths, number of months in a year; elev_units, PRMS Basin Module parameter to define units of feet (0) or meters (1)]

Parameter name	Description	Dimension variable	Units	Type	Range	Default value	Equation number
	Parameters common to all three modules
adjmix_rain	Monthly rain adjustment factor for a mixed precipitation event (usually 1.0)	nmonths	dimensionless	real	0.0 to 3.0	1.0	6
hru_area	Area of HRU	nhru	acres	real	0.1 to 1.0e9	1.0	
precip_units	Units of precipitation (0=inches; 1=millimeters)	one	dimensionless	integer	0 or 1	0	
temp_units	Units of air temperature (0=degrees Fahrenheit; 1=degrees Celsius)	one	dimensionless	integer	0 or 1	0	
tmax_allrain	Monthly minimum air temperature at an HRU that results in all precipitation during a day being rain	nmonths	temp_units	real	0.0 to 90.0	40.0	
tmax_allsnow	Monthly maximum air temperature at which precipitation is all snow for the HRU	one	temp_units	real	-10.0 to 40.0	32.0	6
	Additional parameters for module precip_prms
hru_psta	Identifier of measurement station used as base in calculating precipitation lapse rate	nhru	dimensionless	integer	1 to nrain	1	
rain_adj[1]	Monthly factor as a decimal fraction used to adjust rain at the HRU	nhru by nmonths	dimensionless	real	0.2 to 5.0	1.0	7
snow_adj[1]	Monthly factor as a decimal fraction used to adjust snow at the HRU	nhru by nmonths	dimensionless	real	0.2 to 5.0	1.0	7
	Additional parameters for module precip_laps_prms
hru_elev	Mean land-surface altitude of HRU	nhru	elev_units	real	-300.0 to 30,000.0	0.0	1, 8
hru_plaps	Identifier of lapse measurement station used in calculating precipitation lapse rate	nhru	dimensionless	integer	1 to nrain	1	
hru_psta	Identifier of measurement station used as base in calculating precipitation lapse rate	nhru	dimensionless	integer	1 to nrain	1	
padj_rn[1]	Mean monthly factor to adjust rain lapse rate computed between station_psta and station_plaps when precipitation is rain (positive factors are used as multipliers and negative factors are made positive and substituted for the computed lapse rate)	nrain by nmonths	inches per day	real	-2.0 to 10.0	1.0	8
padj_sn[1]	Mean monthly factor to adjust snow lapse rate computed between station_psta and station_plaps (positive factors are used as multipliers and negative factors are made positive and substituted for the computed lapse rate)	nrain by nmonths	inches per day	real	-2.0 to 10.0	1.0	8

Table A1-12.   Input parameters required for the PRMS Precipitation Distribution Modules: precip_prms, precip_laps_prms, and precip_dist2_prms.—Continued

[Equation number refers to equations listed in the main body of report— equation variable of parameter name is defined in first listed equation. HRU, hydrologic response unit; nhru, number of HRUs; ntemp, number of measurement stations that measure air temperature; nrain, number of measurement stations that measure precipitation; nlapse, number of lapse rates in the X, Y, and Z directions; nmonths, number of months in a year; elev_units, PRMS Basin Module parameter to define units of feet (0) or meters (1)]

Parameter name	Description	Dimension variable	Units	Type	Range	Default value	Equation number
	Additional parameters for module precip_laps_prms—Continued
pmn_mo	Mean monthly precipitation at base and lapse stations	nrain by nmonths	inches	real	0.0 to 100.0	1.0	8
psta_elev	Land surface altitude of base and lapse stations	nrain	elev_units	real	-300.0 to 30,000.0	0.0	8
	Additional parameters for module precip_dist2_prms
hru_xlong	Longitude of HRU centroid	nhru	feet	real	-1e+09 to 1e+09	0.0	10b
hru_ylat	Latitude of HRU centroid	nhru	feet	real	-1e+09 to 1e+09	0.0	10b
maxmon_prec[1]	Maximum monthly precipitation at all measurement stations	nmonths	inches	real	0.0 to 15.0	5.0	
psta_mon	Mean monthly precipitation at each measurement station	nrain by nmonths	inches	real	0.0 to 50.0	1.0	10c, d
psta_xlong	Longitude of each measurement station that measures precipitation	nrain	feet	real	-1e+09 to 1e+09	0.0	10b
psta_ylat	Latitude of each measurement station that measures precipitation	nrain	feet	real	-1e+09 to 1e+09	0.0	10b
rain_mon	Mean monthly rain on each HRU that can be obtained from National Weather Service's spatial distribution of mean annual precipitation for the 1971-2000 climate normal period	nhru by nmonths	inches	real	0.0 to 50.0	1.0	10c
snow_mon	Mean monthly snow on each HRU that can be obtained from National Weather Service's spatial distribution of mean annual precipitation for the 1971–2000 climate normal period	nhru by nmonths	inches	real	0.0 to 50.0	1.0	10d

[1] Additional description of parameter provided below.

Additional information for selected parameters defined in table A1-12 follows:

rain_adj and snow_adj—monthly factors that adjust measured rainfall and snowfall at a precipitation station, respectively, to an HRU. The factors account for differences in elevation, spatial variation, topography, and gage catch efficiency.

padj_rn and padj_sn—monthly rain and snow adjustment factors, respectively, for each precipitation station. These monthly factors are used to adjust the precipitation lapse rate (hru_plaps) computed between stations. Positive factors are multiplied by the lapse rate. Negative factors are made positive and substituted for the computed lapse rate.

maxmon_prec—maximum monthly precipitation. Precipitation is assumed to be in error if measured precipitation at any measurement station is greater than the maxmon_prec value for the month.

Solar-Radiation Distribution Modules

Two PRMS Solar-Radiation Distribution Modules (ccsolrad_hru_prms and ddsolrad_hru_prms) are included in GSFLOW. The modules are used to distribute solar radiation to each HRU. The user selects one module for a particular simulation by setting the control parameter `solrad_module` in the GSFLOW Control File to the one of the module names. Both modules include algorithms for estimating solar radiation when solar-radiation data are not available. The estimation algorithm in module ccsolrad_hru_prms uses a relation between solar radiation and cloud cover, whereas the estimation algorithm in module ddsolrad_hru_prms uses a relation between maximum air temperature and degree-day. Input parameters for the two modules are defined in table A1-13.

Table A1-13.   Input parameters required for PRMS Solar-Radiation Modules: `ccsolrad_hru_prms` and `ddsolrad_hru_prms`.

[Equation number refers to equations listed in the main body of report— equation variable of parameter name is defined in first listed equation. HRU, hydrologic response unit; `nhru`, number of HRUs; `ntemp`, number of measurement stations that measure air temperature; `nsol`, number of measurement stations that measure solar radiation; `nmonths`, number of months in a year; `temp_units`, PRMS Temperature Distribution Modules parameter to define units of degrees Fahrenheit (0) or Celsius (1)]

Parameter name	Description	Dimension variable	Units	Type	Range	Default value	Equation number
		Parameters common to both modules					
`basin_solsta`	Identifier of measurement station used in computing solar radiation	`one`	dimensionless	integer	1 to `nsol`	1	
`hru_area`	Area of HRU	`nhru`	acres	real	0.1 to 1.0e9	1.0	
`ppt_rad_adj`	Precipitation threshold used to determine if solar radiation is adjusted for cloud cover.	`nmonths`	inches	real	0.0 to 0.5	0.02	
`rad_conv`[1]	Factor to convert measured solar radiation to langleys	`one`	Converts measured solar radiation to langleys	real	0.1 to 100.0	1.0	
`radj_sppt`	Precipitation-day adjustment factor to solar radiation for a summer day with precipitation greater than `ppt_rad_adj` as a decimal fraction	`one`	dimensionless	real	0.0 to 1.0	0.44	15b
`radj_wppt`	Precipitation-day adjustment factor to solar radiation for a winter day with precipitation greater than `ppt_rad_adj` as a decimal fraction	`one`	dimensionless	real	0.0 to 1.0	0.5	15b
`radmax`	Maximum fraction of potential solar radiation that reaches land surface as a decimal fraction	`one`	dimensionless	real	0.0 to 1.0	0.8	
		Additional parameters for module ccsolrad_hru_prms					
`basin_tsta`	Identifier of measurement station used in computing air temperature	`one`	dimensionless	integer	1 to `ntemp`	1	
`ccov_intcp`[1]	Intercept in the regression equation that relates cloud cover to daily minimum and maximum air temperature by month, starting with January	`nmonths`	dimensionless	real	0.0 to 5.0	1.83	16a
`ccov_slope`[1]	Slope in the regression equation that relates cloud cover to daily minimum and maximum air temperature by month, starting with January	`nmonths`	dimensionless	real	-0.5 to -0.01	-0.13	16a
`crad_coef`[1]	Constant used in the cloud-cover to solar-radiation relation, a value can be obtained from Thompson (1976, fig. 1)	`one`	dimensionless	real	0.1 to 0.7	0.4	16b

Table A1-13. Input parameters required for PRMS Solar-Radiation Modules: ccsolrad_hru_prms and ddsolrad_hru_prms—Continued

[Equation number refers to equations listed in the main body of report— equation variable of parameter name is defined in first listed equation. HRU, hydrologic response unit; nhru, number of HRUs; ntemp, number of measurement stations that measure air temperature; nsol, number of measurement stations that measure solar radiation; nmonths, number of months in a year; temp_units, PRMS Temperature Distribution Modules parameter to define units of degrees Fahrenheit (0) or Celsius (1)]

Parameter name	Description	Dimension variable	Units	Type	Range	Default value	Equation number
	Additional parameters for module ccsolrad_hru_prms—Continued						
crad_exp[1]	Exponent used in the cloud-cover to solar-radiation relation, a value of 0.61 is suggested by Thompson (1976)	one	dimensionless	real	0.2 to 0.8	0.61	16b
	Additional parameters for module ddsolrad_hru_prms						
dday_slope[1]	Slope of monthly degree-day to temperature relation	nmonths	degree-day per temp_units	real	0.2 to 0.7	0.4	
dday_intcp[1]	Intercept of monthly degree-day to temperature relation	nmonths	degree-day	real	-60.0 to 4.0	-10.0	
radadj_intcp[1]	Intercept of solar radiation adjustment to temperature	one	degree-day	real	0.0 to 1.0	0.0	
radadj_slope[1]	Slope of solar radiation adjustment to temperature	one	degree-day per temp_units	real	0.0 to 1.0	0.0	
tmax_allrain	Monthly minimum air temperature at an HRU that results in all precipitation during a day being rain	nmonths	temp_units	real	0.0 to 90.0	40.0	
tmax_index	Maximum monthly air temperature used to adjust solar radiation for precipitation	nmonths	temp_units	real	-10.0 to 110.0	50.0	

[1] Additional description of parameter provided below.

Additional information for selected parameters defined in table A1-13 follows:

ccov_intcp and ccov_slope—intercept and slope, respectively, in the temperature/cloud-cover relation:
    cloud cover=ccov_intcp+ccov_slope(tmaxf-tminf)
    See table A1-2 for definition of tmaxf and tminf.

crad_coef and crad_exp–coefficient and exponent, respectively, in the cloud cover/solar radiation relation (Thompson, 1976):
    solar radiation=crad_coef+(1-crad_coef)(1-cloud cover)$^{crad\_exp}$

dday_intcp and dday_slope—intercept and slope, respectively, in the temperature/degree day relation:
    degree-day coefficient=(dday_slope)(tmaxf)+dday_intcp
    See table A1-2 for definition of tmaxf.

radadj_intcp and radadj_slope—intercept and slope, respectively, in the temperature range adjustment to solar radiation:
    ppt_adj=(radadj_slope(tmax-tmin))+radadj_intcp
    See table A1-2 for definition of tmaxf and tminf.

rad_conv—conversion factor to langleys for observed radiation. For example, if the units for solar radiation are watt-hours per square meter, the multiplication factor rad_conv should be set to 0.0860.

Potential Evapotranspiration Modules

Three PRMS Potential Evapotranspiration Modules (potet_hamon_hru_prms, potet_jh_prms, and potet_pan_prms) are included in GSFLOW. The modules are used to calculate the amount of potential evapotranspiration and to determine if a time step is one of active transpiration in an HRU. The user selects one of the modules for a simulation by setting the control parameter et_module in the GSFLOW Control File to one of the module names. Module potet_hamon_hru_prms uses the Hamon formulation (Hamon, 1961; Murray, 1967; and Federer and Lash, 1978) to calculate potential evapotranspiration. Module potet_jh_prms uses the Jensen-Haise formulation (Jensen and others, 1969) to calculate potential evapotranspiration. Module potet_pan_prms uses measured pan evaporation and a monthly coefficient to calculate potential evapotranspiration. Input parameters for the three modules are defined in table A1-14.

Table A1-14.   Input parameters required for the PRMS Potential-Evapotranspiration Modules: potet_hamon_hru_prms, potet_jh_prms, and potet_pan_prms.

[Equation number refers to equations listed in the main body of report— equation variable of parameter name is defined in first listed equation. HRU, hydrologic response unit; nhru, number of HRUs; nevap, number of measurement stations that measure air temperature; nmonths, number of months in a year; temp_units, PRMS Temperature Distribution Modules parameter to define units of degrees Fahrenheit (0) or Celsius (1)].

Parameter name	Description	Dimension variable	Units	Type	Range	Default value	Equation number
			Parameters common to all three modules				
hru_area	Area of HRU	nhru	acres	real	0.1 to 1.0e9	1.0	
transp_beg	Begin month for transpiration computations at HRU	nhru	month	integer	1 to 12	4	
transp_end	Last month for transpiration computations at HRU	nhru	month	integer	1 to 12	10	
			Additional parameters for module potet_hamon_hru_prms				
hamon_coef	Monthly air temperature coefficient used in the Hamon potential evapotranspiration equation	nmonths	inch-cubic meter per gram	real	0.004 to 0.008	0.0055	18
temp_units	Units of measured air temperature (0=degrees Fahrenheit; 1=degrees Celsius)	one	dimensionless	integer	0 or 1	0	
transp_tmax	Maximum temperature used to determine when transpiration begins in an HRU	nhru	degree-day	real	0.0 to 1000.0	500.0	
			Additional parameters for module potet_jh_prms				
jh_coef	Monthly air temperature coefficient used in Jensen-Haise potential evapotranspiration equation	nmonths	temp_units	real	0.005 to 0.06	0.014	20a
jh_coef_hru	Air temperature coefficient used in Jensen-Haise potential evapotranspiration equation for each HRU	nhru	temp_units	real	5.0 to 20.0	13.0	20a
temp_units	Units of measured air temperature (0=degrees Fahrenheit; 1=degrees Celsius)	one	dimensionless	integer	0 or 1	0	
transp_tmax	Maximum temperature used to determine when transpiration begins in each HRU	nhru	degree-day	real	0.0 to 1000.0	500.0	
			Additional parameters for module potet_pan_prms				
hru_pansta	Identifier of measurement station that measures pan evaporation	nhru	dimensionless	integer	1 to nevap	1	
epan_coef	Monthly pan evaporation coefficient used to convert value to potential evapotranspiration	nmonths	dimensionless	real	0.2 to 3.0	1.0	21

PRMS Canopy Interception Module: intcp_prms

The PRMS Canopy Interception Module (intcp_prms) calculates the amount of rain and snow that is intercepted by vegetation, the amount of evaporation of intercepted rain and snow, and the amount of net rain and snow throughfall that reaches the soil or snowpack. Input parameters for the module are defined in table A1-15.

Table A1-15.   Input parameters required for PRMS Canopy Interception Module: intcp_prms.

[Equation number refers to equations listed in the main body of report— equation variable of parameter name is defined in first listed equation. HRU, hydrologic response unit; nhru, number of HRUs; nmonths, number of months in a year]

Parameter name	Description	Dimension variable	Units	Type	Range	Default value	Equation number
cov_type	Plant type on HRU (0=bare soil; 1=grasses; 2=shrubs; 3=trees)	nhru	dimensionless	integer	0 to 3	3	
covden_sum	Summer plant canopy density as a decimal fraction of the HRU area	nhru	dimensionless	real	0.0 to 1.0	0.5	22, 23
covden_win	Winter plant canopy density as a decimal fraction of the HRU area	nhru	dimensionless	real	0.0 to 1.0	0.5	22, 23
epan_coef	Monthly evaporation pan coefficient	nmonths	dimensionless	real	0.2 to 3.0	1.0	
hru_area	Area of HRU	nhru	acres	real	0.1 to 1.0e9	1.0	22
hru_type	Type of HRU (0=inactive; 1=land; 2=lake)	nhru	dimensionless	integer	0 to 2	1	
potet_sublim	Fraction of potential evapotranspiration sublimated from snow surface as a decimal fraction	one	dimensionless	real	0.1 to 0.75	0.5	
snow_intcp	Maximum snow storage in the plant canopy for plant type on HRU	nhru	inches	real	0.0 to 5.0	0.1	22
srain_intcp	Maximum summer rain storage in the plant canopy for plant type on HRU	nhru	inches	real	0.0 to 5.0	1	22
wrain_intcp	Maximum winter rain storage in the plant canopy for plant type on HRU	nhru	inches	real	0.0 to 5.0	0.1	22

Snow Computation Module

The PRMS Snow-Computation Module (snowcomp_prms) initiates development of a snowpack and simulates snow accumulation and depletion processes using an energy-budget approach. Input parameters for the module are defined in table A1-16.

Table A1-16. Input parameters required for PRMS Snow-Computation Module: snowcomp_prms.

[Equation number refers to equations listed in the main body of report— equation variable of parameter name is defined in first listed equation. HRU, hydrologic response unit; nhru, number of HRUs; nmonths, number of months in a year; ndepl, number of snow depletion curves; temp_units, PRMS Temperature Distribution Modules parameter to define units of degrees Fahrenheit (0) or Celsius (1)]

Parameter name	Description	Dimension variable	Units	Type	Range	Default value	Equation number
albset_rna	Decimal fraction of rain in a mixed rain and snow event above which snow albedo is not reset (applied when snowpack is accumulating)	one	dimensionless	real	0.0 to 1.0	0.8	
albset_rnm	Decimal fraction of rain in a mixed rain and snow event above which snow albedo is not reset (applied when snowpack is melting)	one	dimensionless	real	0.0 to 1.0	0.6	
albset_sna	Minimum snow fall, in water equivalent, needed to reset snow albedo when snowpack is accumulating as a decimal fraction	one	dimensionless	real	0.001 to 1.0	0.05	
albset_snm	Minimum snow fall, in water equivalent, needed to reset snow albedo when snowpack is melting as a decimal fraction	one	dimensionless	real	0.001 to 1.0	0.2	
cecn_coef	Monthly convection-condensation energy coefficient	nmonths	calories per degree Celsius above 0	real	0.0 to 20.0	5.0	
cov_type	Plant cover type for HRU (0=bare soil; 1=grasses; 2=shrubs; 3=trees)	nhru	dimensionless	integer	0 to 3	3	
covden_sum	Summer plant cover density for plant type on HRU as a decimal fraction	nhru	dimensionless	real	0.0 to 1.0	0.5	
covden_win	Winter plant cover density for plant type on HRU as a decimal fraction	nhru	dimensionless	real	0.0 to 1.0	0.5	
den_init	Density of new-fallen snow as a decimal fraction	one	dimensionless	real	0.01 to 0.5	0.10	24
den_max	Average maximum snowpack density as a decimal fraction of the liquid water equivalent	one	dimensionless	real	0.1 to 0.8	0.6	24
emis_noppt	Emissivity of air on days without precipitation	one	dimensionless	real	0.757 to 1.0	0.757	
freeh2o_cap	Free-water holding capacity of snowpack expressed as decimal fraction of total snowpack water equivalent	one	dimensionless	real	0.01 to 0.2	0.05	
hru_area	Area of HRU	nhru	acres	real	0.1 to 1.0e9	1.0	
hru_deplcrv	Identifier of snowpack areal-depletion curve for HRU	nhru	dimensionless	integer	1 to ndepl	1	
hru_type	Type of HRU (0=inactive; 1=land; 2=lake)	nhru	dimensionless	integer	0 to 2	1	
melt_force[1]	Julian date to force snowmelt	one	Julian day	integer	1 to 366	90	
melt_look[1]	Julian date to start looking for when snowmelt begins	one	Julian day	integer	1 to 366	90	

Table A1-16. Input parameters required for PRMS Snow-Computation Module: snowcomp_prms—Continued.

[Equation number refers to equations listed in the main body of report— equation variable of parameter name is defined in first listed equation. HRU, hydrologic response unit; nhru, number of HRUs; nmonths, number of months in a year; ndepl, number of snow depletion curves; temp_units, PRMS Temperature Distribution Modules parameter to define units of degrees Fahrenheit (0) or Celsius (1)]

Parameter name	Description	Dimension variable	Units	Type	Range	Default value	Equation number
potet_sublim	Decimal fraction of potential evapotranspiration that is sublimated from snow surface	one	dimensionless	real	0.1 to 0.75	0.5	30
rad_trncf	Transmission coefficient for short-wave radiation through winter plant canopy on an HRU as a decimal fraction	nhru	dimensionless	real	0.0 to 1.0	0.5	
settle_const	Snowpack settlement-time constant	one	per day	real	0.01 to 0.5	0.10	24
snarea_curve	Snow area-depletion curve values, 11 for each curve as a decimal fraction	11 by ndepl	dimensionless	real	0.0 to 1.0	1.0	
snarea_thresh	Maximum water equivalent threshold, water equivalent in an HRU less than threshold results in use of snow-covered-area curve	nhru	inches	real	0.0 to 200.0	50.0	
tmax_allsnow	Monthly maximum air temperature at which precipitation is all snow for the HRU	one	temp_units	real	-10.0 to 40.0	32.0	
tstorm_mo	Monthly storm prevalence (0=frontal storms prevalent; 1=convective storms prevalent)	nmonths	dimensionless	integer	0 or 1	0	

[1] Additional description of parameter provided below.

Additional information for selected parameters defined in table A1-16 follows:

melt_force—Julian day to force the use of the isothermal form of the snowpack energy equation. This date should always be later than melt_look, varies by region, and is dependent on the persistence of the snowpack.

melt_look—Julian day when the snowpack simulation changes from the accumulation phase to the melt phase. This primarily affects albedo computations. April 1 (Julian day = 90) works well for most western mountain watersheds. For eastern and some western watersheds, where rain-on-snow or intermittent snowpacks occur during the snow season, the melt phase may be more appropriate for the whole season. In this case, set melt_look to the start of the snow season. At the beginning of the water year, the snowpack simulation is reset to the accumulation phase.

Surface Runoff and Infiltration Modules

Two different PRMS Surface Runoff and Infiltration Modules (srunoff_smidx_casc and srunoff_carea_casc) are included in GSFLOW. The modules are used to compute surface runoff and infiltration for each HRU. The user selects one of the two modules for a particular simulation by setting the control parameter `srunoff_module` in the GSFLOW Control File to either module name. Module srunoff_smidx_casc uses antecedent soil moisture and a non-linear variable-source-area method, whereas module srunoff_carea_casc uses antecedent soil moisture and a linear variable-source-area method. Input parameters for the two modules are defined in table A1-17.

Table A1-17. Input parameters required for the two PRMS Surface Runoff and Infiltration Modules: srunoff_carea_casc and srunoff_smidx_casc.

[Equation number refers to equations listed in the main body of report— equation variable of parameter name is defined in first listed equation. HRU, hydrologic response unit; nhru, number of HRUs]

Parameter name	Description	Dimension variable	Units	Type	Range	Default value	Equation number
	Parameters common to both modules						
carea_max	Maximum possible area contributing to surface runoff, expressed as a decimal fraction of HRU area	nhru	dimensionless	real	0.0 to 1.0	0.6	34
hru_area	Area of HRU	nhru	acres	real	0.1 to 1.0e9	1.0	22, 31,32, 33, 36, 37
hru_percent_imperv	Decimal fraction of HRU area that is impervious	nhru	dimensionless	real	0.0 to 1.0	0.0	
hru_type	Type of HRU (0=inactive; 1=land; 2=lake)	nhru	dimensionless	integer	0 to 2	1	
imperv_stor_max	Maximum retention storage for HRU impervious area	nhru	inches	real	0.0 to 10.0	0.0	31
snowinfil_max	Daily maximum snowmelt infiltration for the HRU	nhru	inches	real	0.0 to 20.0	2.0	37
soil_moist_max	Maximum available capillary water-holding capacity of soil zone in an HRU	nhru	inches	real	0.0 to 20.0	6.0	
	Additional parameters for module srunoff_carea_casc						
carea_min	Minimum possible area contributing to surface runoff, as a decimal fraction of HRU area	nhru	dimensionless	real	0.0 to 1.0	0.2	34
soil_rechr_max	Maximum quantity of water in the capillary reservoir (value must be less than or equal to soil_moist_max)	nhru	inches	real	0.0 to 10.0	2.0	34a
	Additional parameters for module srunoff_smidx_casc						
smidx_coef[1]	Coefficient in non-linear contributing area algorithm	nhru	dimensionless	real	0.0001 to 1.0	0.01	34b
smidx_exp[1]	Exponent in non-linear contributing area algorithm	nhru	per inch	real	0.2 to 0.8	0.3	34b

[1] Additional description of parameter provided below.

Additional information for selected parameters defined in table A1-17 follows:

smidx_coef and smidx_exp—coefficient and exponent, respectively, in the non-linear contributing area algorithm.

Soil Zone Module

The PRMS Soil Zone Module (soilzone_gsflow) calculates inflows to and outflows from the soil zone of each HRU and includes inflows from infiltration, ground water, and upslope HRUs, and outflows to gravity drainage, interflow, and surface runoff to downslope HRUs. Inflow from ground water is not allowed when using the PRMS-only model. The Soil Zone Module is a combination and extension of the PRMS Soil-Moisture Balance (smbal_prms) and Subsurface Flow (ssflow_prms) Modules. For more information about this module see section, "Changes to PRMS" in body of this report. Input parameters for the module are defined in table A1-18.

Table A1-18.   Input parameters required for PRMS Soil-Zone Module: soilzone_gsflow.

[Equation number refers to equations listed in the main body of report—equation variable of parameter name is defined in first listed equation. HRU, hydrologic response unit; nhru, number of HRUs; nhrucell, number of unique intersections between gravity reservoirs in PRMS soil zone and MODFLOW finite-difference cells; nssr, number of PRMS subsurface reservoirs]

Parameter name	Description	Dimension variable	Units	Type	Range	Default value	Equation number
cov_type	Plant cover type (0=bare soil; 1=grasses; 2=shrubs; 3=trees)	nhru	dimensionless	integer	0 to 3	3	
covden_sum	Summer plant cover density for plant type as a decimal fraction	nhru	dimensionless	real	0.0 to 1.0	0.5	
covden_win	Winter plant cover density for plant type as a decimal fraction	nhru	dimensionless	real	0.0 to 1.0	0.5	
fastcoef_lin	Linear flow-routing coefficient for fast interflow	nhru	per day	real	0.0 to 1.0	0.1	67a
fastcoef_sq	Non-linear flow-routing coefficient for fast interflow	nhru	per inch-day	real	0.0 to 1.0	0.8	67a
gvr_hru_id	Identifier of HRU corresponding to each gravity reservoir	nhrucell	dimensionless	integer	1 to nhru	1	
gvr_hru_pct	Decimal fraction of HRU area associated with gravity reservoir	nhrucell	dimensionless	real	0.0 to 1.0	0.0	
hru_area	Area of HRU	nhru	acres	real	0.1 to 1.0e9	1.0	22, 39, 45, 46, 58
hru_type	Type of HRU (0=inactive; 1=land; 2=lake)	nhru	dimensionless	integer	0 to 2	1	
pref_flow_den	Decimal fraction of the soil zone available for preferential flow	nhru	dimensionless	real	0.0 to 1.0	0.2	38, 47
sat_threshold	Maximum volume of water per unit area in the soil zone (set to 999.0 for infinite volume)	nhru	inches	real	1.0 to 999.0	999.0	47, 65
slowcoef_lin	Linear flow-routing coefficient for slow interflow	nhru	per day	real	0.0 to 1.0	0.015	50, 56
slowcoef_sq	Non-linear flow-routing coefficient for slow interflow	nhru	per inch-day	real	0.0 to 1.0	0.1	50, 56
soil_moist_init	Initial value of available water in the capillary reservoir	nhru	inches	real	0.0 to 20.0	3.0	
soil_moist_max	Maximum volume of water per unit area in the capillary reservoir	nhru	inches	real	0.0 to 20.0	6.0	44, 46, 63a
soil_rechr_init	Initial value in capillary reservoir where evaporation and transpiration can occur simultaneously (value must be less than or equal to soil_moist_max)	nhru	inches	real	0.0 to 10.0	1.0	
soil_rechr_max	Maximum value in capillary reservoir where evaporation and transpiration can occur simultaneously (value must be less than or equal to soil_moist_max)	nhru	inches	real	0.0 to 10.0	2.0	
soil_type	Soil type in HRU (1=sand; 2=loam; 3=clay)	nhru	dimensionless	integer	1 to 3	2	

Table A1-18.   Input parameters required for PRMS Soil-Zone Module: soilzone_gsflow—Continued.

[Equation number refers to equations listed in the main body of report—equation variable of parameter name is defined in first listed equation. HRU, hydrologic response unit; nhru, number of HRUs; nhrucell, number of unique intersections between gravity reservoirs in PRMS soil zone and MODFLOW finite-difference cells; nssr, number of PRMS subsurface reservoirs]

Parameter name	Description	Dimension variable	Units	Type	Range	Default value	Equation number
soil2gw_max	Maximum value of soil-water excess routed directly to PRMS ground-water reservoir	nhru	inches	real	0.0 to 5.0	0.0	
ssr2gw_exp	Exponent in the equation used to compute gravity drainage to PRMS ground-water reservoir or MODFLOW finite-difference cell	nssr or nhrucell	dimensionless	real	0.0 to 3.0	1.0	59
ssr2gw_rate	Linear coefficient in the equation used to compute gravity drainage to PRMS ground-water reservoir or MODFLOW finite-difference cell	nssr or nhrucell	inches per day	real	0.0 to 1.0	0.1	59
ssrmax_coef	Maximum amount of gravity drainage to PRMS ground-water reservoir or MODFLOW finite-difference cell	nssr or nhrucell	inches	real	1.0 to 20.0	1.0	59
ssstor_init	Initial storage in PRMS subsurface reservoir or gravity reservoir	nssr or nhrucell	inches	real	0.0 to 20.0	0.0	

Ground-Water Flow Module

The PRMS Ground-Water Flow Module (gwflow_casc_prms) sums inflow to and computes storage and outflow from each PRMS ground-water reservoir. Outflows can be routed to downslope ground-water reservoirs or to a stream. This module is used for PRMS-only simulations. Input parameters for module gwflow_casc_prms are defined in table A1-19.

Table A1-19.   Input parameters required for PRMS Ground-Water Flow Module: gwflow_casc_prms, included with PRMS-only simulations.

[HRU, hydrologic response unit; nhru, number of HRUs; nssr, number of PRMS subsurface reservoirs; ngw, number of PRMS ground-water reservoirs]

Parameter name	Description	Dimension variable	Units	Type	Range	Default value
hru_area	Area of HRU	nhru	acres	real	0.1 to 1.0e9	1.0
hru_gwres	Identifier of PRMS ground-water reservoir associated with HRU	nhru	dimensionless	integer	1 to ngw	1
gwstor_init	Initial storage in ground-water reservoir	ngw	inches	real	0.0 to 20.0	0.1
gwflow_coef[1]	Linear coefficient to route water in ground-water reservoir to streams	ngw	per day	real	0.0 to 1.0	.015
gwsink_coef[1]	Linear coefficient to route water in ground-water reservoir to ground-water sink	ngw	per day	real	0.0 to 1.0	0.0

[1]Additional description of parameter provided below.

Additional information for selected parameters defined in table A1-19 follows:

gwflow_coef— ground-water routing coefficient. The coefficient is multiplied by the storage in the ground-water reservoir to calculate ground-water flow computation to downslope HRUs and(or) to streamflow.

gwsink_coef—ground-water sink coefficient. The coefficient is multiplied by the storage in the ground-water reservoir to calculate seepage from each ground-water reservoir to the ground-water sink. Water in the ground-water sink is no longer available for flow within the watershed.

Streamflow Module

The PRMS Streamflow Module (strmflow_prms) calculates daily streamflow as the sum of surface runoff, interflow, flow from detainment reservoirs, and ground-water flow. The module is used for PRMS-only simulations. Input parameters for the module are defined in table A1-20.

Table A1-20.   Input parameters required for PRMS Streamflow Module: strmflow_prms, included with PRMS-only simulations.

[HRU, hydrologic response unit; nhru, number of HRUs; nssr, number of PRMS subsurface reservoirs; ngw, number of PRMS ground-water reservoirs; nsfres, number of on-stream detainment reservoirs; cfs, cubic feet per second]

Parameter name	Description	Dimension variable	Units	Type	Range	Default value
basin_cfs_init	Initial streamflow at watershed outlet, required if first time step is a storm period	one	cfs	real	0.0 to 1e+09	0.0
hru_area	Area of HRU	nhru	acres	real	0.1 to 1.0e9	1.0
hru_gwres	Identifier of PRMS ground-water reservoir associated with HRU	nhru	dimensionless	integer	1 to ngw	1
Parameters only required when detainment reservoirs are included in a PRMS-only simulation						
hru_sfres	Identifier of detainment reservoir associated with HRU	nhru	dimensionless	integer	0 to nsfres	0
mxnsos	Maximum number of values in table for Puls routing	nsfres	dimensionless	integer	0 to 10	0
o2	Outflow from detainment reservoir in table for Puls routing	mxnsos by nsfres	cfs	real	0.0 to 100,000.0	0.0
s2	Storage in detainment reservoir in table for Puls routing	mxnsos by nsfres	cfs-days	real	0.0 to 100,000.0	0.0
sfres_coef	Linear coefficient to route detainment reservoir storage to streams	nsfres	per day	real	0.0 to 1.0	0.1
sfres_din1	Inflow to detainment reservoir from previous time step	nsfres	cfs	real	0.0 to 1.0	0.1
sfres_init	Initial storage in detainment reservoir	nsfres	cfs-days	real	0.0 to 2,000,000.0	0.0
sfres_qro	Initial daily mean outflow from detainment reservoir	nsfres	cfs	real	0.0 to 1.0	0.1
sfres_type	Type of detainment reservoir (8=Puls; 9=linear)	nsfres	dimensionless	integer	8 or 9	8
upst_res1	Identifier of first upstream detainment reservoir	nsfres	dimensionless	integer	0 to nsfres	0
upst_res2	Identifier of second upstream detainment reservoir	nsfres	dimensionless	integer	0 to nsfres	0
upst_res3	Identifier of third upstream detainment reservoir	nsfres	dimensionless	integer	0 to nsfres	0

Hydrologic-Response-Unit Summary Module

The PRMS Hydrologic-Response-Unit Summary Module (hru_sum_prms) calculates daily, monthly, yearly, and total summaries of volumes and flows for each HRU. Input parameters are defined in table A1-21.

Table A1-21.   Input parameters required for PRMS Hydrologic-Response-Unit Summary Module: hru_sum_prms.

[HRU, hydrologic response unit; nhru, number of HRUs]

Parameter name	Description	Dimension variable	Units	Type	Range	Default value
covden_sum	Summer plant cover density for plant type on HRU as a decimal fraction	nhru	dimensionless	real	0.0 to 1.0	0.5
covden_win	Winter plant cover density for plant type on HRU as a decimal fraction	nhru	dimensionless	real	0.0 to 1.0	0.5
hru_percent_imperv	Decimal fraction of HRU area that is impervious	nhru	dimensionless	real	0.0 to 1.0	0.0
hru_type	Type of HRU (0=inactive; 1=land; 2=lake)	nhru	dimensionless	integer	0 to 2	1
moyrsum	Monthly and yearly summaries for each HRU (0 for no summaries; 1 for summaries)	one	dimensionless	integer	0 or 1	0
pmo	Month for which monthly summary is written to PRMS Water-Budget (0=no monthly summary; 1=January, 2=February, and so forth)	one	dimensionless	integer	0 to 12	0

Basin Summary Module

The PRMS Basin Summary Module (basin_sum_prms) sums daily, monthly, yearly, and total summary values of volumes and flows as area-weighted watershed averages. Input parameters for the module are defined in table A1-22.

Table A1-22.   Input parameters required for PRMS Basin Summary Module: basin_sum_prms.

[HRU, hydrologic response unit; nhru, number of HRUs; ntemp, number of measurement stations that measure air temperature; nobs, number of streamflow-gaging stations]

Parameter name	Description	Dimension variable	Units	Type	Range	Default value
basin_tsta	Identifier of measurement station used in computing air temperature	one	dimensionless	integer	0 to ntemp	1
basin_tsta_hru	Identifier of HRU used in computing watershed temperatures	one	dimensionless	integer	0 to nhru	1
objfunc_q	Streamflow-gaging station used in objective function calculations	one	dimensionless	integer	0 to nobs	1
print_freq	Frequency of output written in PRMS Water-Budget File (0=none; 1=simulation totals; 2=yearly; 4=monthly; 8=daily; or additive combinations—for example, use 3 for output of yearly and simulation totals)	one	dimensionless	integer	0 to 15	1
print_objfunc	Objective function output in PRMS Water-Budget File (0=no; 1=yes)	one	dimensionless	integer	0 to 1	0
print_type	Type of output written in PRMS Water-Budget File (0=measured and predicted flow only; 1=water balance table; 2=detailed output)	one	dimensionless	integer	0 to 2	1
runoff_units	Units of measured streamflows written in PRMS Water-Budget File (0=cubic feet per second; 1=cubic meters per second)	one	dimensionless	integer	0 or 1	0

GSFLOW Modules

Input instructions for GSFLOW modules are described next. Several parameters are used by more than one module. Because the modules are defined independently of one another, descriptions of parameters used by more than one module are repeated in each module. For example, the parameter gvr_cell_id is used by GSFLOW modules gsflow_setconv, gsflow_prms2mf, gsflow_mf2prms, and gsflow_budget and a description of the parameter is repeated in each module.

GSFLOW Computation-Control Modules

Two GSFLOW modules are used to control the sequence of computations within GSFLOW. The module gsflow_prms is used to call PRMS and GSFLOW modules for each daily time step in the proper sequence, which varies for each simulation mode (integrated, PRMS-only, and MODFLOW-only). This module replaces the call_modules.c subroutine in the Modular Modeling System. Module gsflow_modflow is used to call the MODFLOW-2005 packages and run procedures of the PRMS Soil-Zone Module (soilzone_gsflow) as well as other GSFLOW modules in the proper sequence. It replaces the MODFLOW-2005 main program. Module gsflow_prms requires one parameter, model_mode, which is specified in the GSFLOW Control File. Module gsflow_modflow requires two parameters: (1) mxsziter, which is specified in the PRMS Parameter File; and (2) modflow_name, which is specified in the GSFLOW Control File. Table A1-23 defines the input parameters for the two GSFLOW Computation-Control Modules (gsflow_prms and gsflow_modflow).

Table A1-23. Input parameters required for GSFLOW Computation-Control Modules: gsflow_prms and gsflow_modflow.

Parameter name	Description	Dimension variable	Units	Type	Range	Default value
		Parameter for gsflow_prms				
model_mode[1]	Model to run	one	dimensionless	integer	GSFLOW, PRMS, or MODFLOW	GSFLOW
		Parameters for gsflow_modflow				
model_mode[1]	Model to run	one	dimensionless	integer	GSFLOW, PRMS, or MODFLOW	GSFLOW
modflow_name[1]	Pathname of MODFLOW Name File	one	dimensionless	character	1 to 256	modflow.nam
mxsziter[2]	Maximum number of iterations soil-zone flow to MODFLOW finite-difference cells are computed each time step	one	dimensionless	integer	2 to 200	15

[1]Parameter specified in GSFLOW Control File.

[2]Parameter is not required in MODFLOW-only simulations.

GSFLOW Conversion Factors Module

A set of variables are determined in GSFLOW Conversion Factors Module (gsflow_setconv) that are used in other GSFLOW modules to convert units between PRMS and MODFLOW-2005 during a simulation. The variables are summarized in table 7 in the main body of this report. Input parameters for the module are defined in table A1-24.

Table A1-24. Input parameters required for GSFLOW Conversion Factors Module: gsflow_setconv.

[HRU, hydrologic response unit; nhru, number of HRUs; nhrucell, number of unique intersections between gravity reservoirs in PRMS soil zone and MODFLOW finite-difference cells]

Parameter name	Description	Dimension variable	Units	Type	Range	Default value
gvr_cell_id	Finite-difference cell associated with a gravity reservoir	nhrucell	dimensionless	integer	0 to ngwcell	0
gvr_cell_pct	Decimal fraction of HRU area associated with a finite-difference cell	nhrucell	dimensionless	real	0.0 to 1.0	0.0

GSFLOW Integration Modules

Two GSFLOW modules (gsflow_prms2mf and gsflow_mf2prms) are used to integrate the spatial units and transfer dependent variables (model states and fluxes) and volumetric flow rates between PRMS and MODFLOW-2005. Module gsflow_prms2mf distributes gravity drainage from gravity reservoirs in the soil zone of PRMS to finite-difference cells in MODFLOW-2005. The module also distributes surface runoff and interflow from HRUs to stream segments and lakes in MODFLOW-2005. Module gsflow_mf2prms distributes ground-water discharge from finite-difference cells in MODFLOW-2005 to gravity reservoirs in the soil zone of PRMS. Input parameters for the two modules are defined in table A1-25.

Table A1-25.   Input parameters required GSFLOW Integration Modules: gsflow_prms2mf and gsflow_mf2prms.

[HRU, hydrologic response unit; nhru, number of HRUs; nhrucell, number of unique intersections between gravity reservoirs in PRMS soil zone and MODFLOW finite-difference cells; ngwcell, number of MODFLOW finite-difference cells in a layer (includes active and inactive cells; nreach, number of MODFLOW stream reaches; nsegment, number of MODFLOW stream segments]

Parameter name	Description	Dimension variable	Units	Type	Range	Default value
	Parameters common to both modules					
gvr_cell_id	Finite-difference cell associated with a gravity reservoir	nhrucell	dimensionless	integer	0 to ngwcell	0
gvr_hru_id	HRU associated with a gravity reservoir	nhrucell	dimensionless	integer	1 to nhru	1
gvr_hru_pct	Decimal fraction of HRU area associated with a gravity reservoir	nhrucell	dimensionless	real	0.0 to 1.0	0.0
hru_area	Area of HRU	nhru	acres	real	0.1 to 1.0e9	1.0
hru_type	Type of HRU (0=inactive; 1=land; 2=lake)	nhru	dimensionless	integer	0 to 2	1
lake_hru_id	MODFLOW lake number associated with an HRU	nhru	dimensionless	integer	0 to nhru	0
	Additional parameters for gsflow_prms2mf					
gvr_cell_pct	Decimal fraction of HRU area associated with a finite-difference cell	nhrucell	dimensionless	real	0.0 to 1.0	0.0
hru_segment	HRU associated with a stream segment	nhru	dimensionless	integer	0 to nsegment	
local_reachid	Stream reach within a stream segment	nreach	dimensionless	integer	0 to nreach	0
mxsziter	Maximum iterations for computing soil-zone flow to finite-difference cells during a time step	one	dimensionless	integer	2 to 200	15
numreach_segment	Number of stream reaches in a stream segment	nsegment	dimensionless	integer	0 to nreach	1
reach_segment	Stream segment associated with a stream reach	nreach	dimensionless	integer	0 to nsegment	0
segment_pct_area	Decimal fraction of HRU area that contributes flow to a stream reach	nreach	dimensionless	real	0.0 to 1.0	0.0
szconverge[1]	Convergence criterion for checking soil-zone flows	one	inches	real	1.e-15 to 1.e-1	1.e-8

[1]Additional description of parameter provided below.

Additional information for one of the parameters defined in table A1-25 follows:

szconverge—convergence criterion for checking soil-zone states that is specified as the smallest change in gravity drainage from the PRMS gravity-flow reservoir between iterations.

GSFLOW Budget Module

The GSFLOW Budget Module (gsflow_budget) calculates a watershed budget for GSFLOW and adjusts the final storage in gravity reservoirs using flows to and from finite-difference cells at the end of each time step. Input parameters for gsflow_budget are defined in table A1-26.

Table A1-26. Input parameters required for GSFLOW Budget Module: gsflow_budget.

[HRU, hydrologic response unit; nhru, number of HRUs; nhrucell, number of unique intersections between gravity reservoirs in PRMS soil zone and MODFLOW finite-difference cells; ngwcell, number of MODFLOW finite-difference cells in a layer (includes active and inactive cells); cfs, cubic foot per second]

Parameter name	Description	Dimension variable	Units	Type	Range	Default value
basin_cfs_init	Initial streamflow at outlet	one	cfs	real	0.0 to 1e+09	0.0
gvr_cell_id	Identifier of finite-difference cell associated with a gravity reservoir	nhrucell	dimensionless	integer	0 to ngwcell	0
gvr_hru_id	Identifier of HRU associated with a gravity reservoir	nhrucell	dimensionless	integer	1 to nhru	1
gvr_hru_pct	Decimal fraction of HRU area associated with a gravity reservoir	nhrucell	dimensionless	real	0.0 to 1.0	.0
hru_type	Type of HRU (0=inactive; 1=land; 2=lake)	nhru	dimensionless	integer	0 to 2	1
lake_hru_id	MODFLOW lake number associated with an HRU	nhru	dimensionless	integer	0 to nhru	0
hru_area	Area of HRU	nhru	acres	real	0.1 to 1.0e9	1.0

GSFLOW Summary Module

The GSFLOW Summary Module (gsflow_sum) calculates summary tables of the water balance at the end of each time step. Input parameters for gsflow_sum are defined in table A1-27. Parameters id_obsrunoff, and runoff_units are specified in the PRMS Parameter File. Parameters csv_output_file, gsf.rpt, gsflow_output_file, model_output_file, and rpt_days are specified in the GSFLOW Control File.

Table A1-27. Input parameters required for GSFLOW Summary Module: gsflow_sum.

[HRU, hydrologic response unit; nhru, number of HRUs; ntemp, number of measurement stations that measure air temperature; nobs, number of streamflow gaging stations]

Parameter name	Description	Dimension variable	Units	Type	Range	Default value
csv_output_file[1]	Pathname for GSFLOW Comma-Separated-Values (CSV) File	one	dimensionless	character	1 to 256	gsflow.csv
gsf_rpt[1]	Switch to specify whether or not the GSFLOW Comma-Separated-Values (CSV) File is generated (0=no, 1=yes)	one	dimensionless	integer	0or1	1
gsflow_output_file[1]	Pathname for GSFLOW Water Budget File	one	dimensionless	character	1 to 256	gsflow.out
id_obsrunoff	Identifier for streamflow-gaging station at outlet	one	dimensionless	integer	0 to nobs	0
model_output_file[1]	Pathname for PRMS Water Budget File	one	dimensionless	character	1 to 256	prms.out
rpt_days[1]	Frequency that summary tables are written to GSFLOW Water-Budget File (0=none, >0 frequency in days, e.g., 1=daily, 7=every 7th day)	one	days	integer	1 to 365	7
runoff_units	Units of measured streamflow (0=cubic feet per second; 1=cubic meters per second)	one	dimensionless	integer	0 or 1	0

[1]Parameter specified in GSFLOW Control File.

MODFLOW Input Files

Input files for most of the MODFLOW packages were not changed for use with GSFLOW and the following input instructions were extracted from existing MODFLOW documentation. However, GSFLOW does restrict the values of some MODFLOW input variables. Information regarding these input data restrictions is differentiated from the original input instructions in the following sections by placing braces ({ }) around the added text. Additionally, input instructions related to parameters were removed for simplicity, and no description of the removed parameter items is included in this appendix. MODFLOW packages within GSFLOW that previously read data using parameters can still be used in GSFLOW. Input instructions for using parameters in MODFLOW are provided by Harbaugh (2005). Some parameter-related variables are described in this appendix and must be assigned a zero value when parameters are not used in a simulation.

Input instructions for the Ground-Water Flow Process of MODFLOW-2005 are given separately for each package used by GSFLOW. The Basic Package reads several files. Each file is described in a separate subsection. The final sections in this chapter describe the Array Reading Utility Subroutines and the List Utility Subroutine.

Data for cells in the model grid can be categorized in two ways according to the density or sparseness of the data. Layer data refers to any type of data for which a value is required for every cell in one or more layers of the grid. Examples of layer data include hydraulic conductivity and specific storage. List data refers to any type of data for which data values are required for selected cells in the grid. Examples include the well withdrawal rate as simulated with the Well (WEL) Package and head-dependent flow as simulated using a specified head and hydraulic conductance with the General-Head Boundary (GHB) Package.

Layer data are described by the number of columns and rows in parentheses after the variable. Although the program contains many three-dimensional variables that are defined through user input, these are treated as multiple-layer variables for input purposes. For example, IBOUND(NCOL, NROW) indicates that IBOUND is a layer variable. An IBOUND variable is read for each layer of the grid. Most multiple-layer variables are read by utility subroutines. which are named along with the variable. The utility subroutine U1DREL ("Utility-1D-Real") is used for reading a one-dimensional array of variables that are real numbers, U2DREL is used for reading an array of two-dimensional variables that are real numbers, and U2DINT is used for reading an array of two-dimensional variables that are integers.

List data of variables include the layer, row, and column indices that are used to identify a particular cell and multiple lines of data can be read. List data are usually read by the list utility subroutine, ULSTRD (Utility-List-Read). The use of ULSTRD is mentioned as part of the note that describes the repeated item.

The input data for each item must start on a new line. All variables for an item are assumed to be contained in a single line unless the item consists of a multi-valued variable. That is, a multi-valued variable may occupy multiple lines. When reading a layer variable, the data for each row must start on a new line.

Each input variable has a data type, which can be Real, Integer, or Character. Integers are whole numbers and must not include a decimal point or exponent. Real numbers can include a decimal point and an exponent. If no decimal point is included in the entered value, then the decimal point is assumed to be at the right side of the value. Any printable character is allowed for character variables. All variables starting with the letters I-N are integers. Variables starting with the letters A-H and O-Z are either real numbers or character data, and most of these variables are real numbers; that is, there are few character variables in MODFLOW. A variable starting with A-H or O-Z is assumed to be a real number unless otherwise stated in the variable definitions. This data-type convention is used for all input variables regardless of whether a variable is an actual program variable.

Both free and fixed format is used as specified throughout the input instructions. With fixed format, a data value has a specified column location, called the field, within a line. If a numeric value does not require the entire field width, the value should be right justified within the field, and blanks should fill the remaining width of the field. The justification of character data within a field does not matter unless specifically noted.

Free format is similar to Fortan list directed input. With free format, values are not required to occupy a fixed number of columns in a line. Each value can occupy one or more columns as required to represent the value; however, the values must be included in the prescribed order. One or more spaces, or a single comma optionally combined with spaces, must separate adjacent values. Also, a numeric value of zero must be explicitly represented with 0 and not by one or more spaces when free format is used. Free-format input implemented in MODFLOW-2005 does not allow for null values in which input values are left unchanged from their previous values. A forward slash (/) cannot be used to terminate an input item without including values for all the variables; required data values for all input variables must be explicitly specified on an input item. MODFLOW does not require apostrophes to delineate apostrophes unless a blank or a comma is part of a character variable.

Name File

Although the Name File is actually used as part of the Basic Package, it is described separately because it is fundamental for identifying which input files will be used in the MODFLOW-components of GSFLOW. The Name File contains the names of the input and output files used for MODFLOW in a GSFLOW simulation and controls whether or not parts of the model program are active. (Temporary OPEN/CLOSE files, described in section "Array Reading Utility Subroutines," are not included in the Name File.) The Name File is read on FORTRAN unit 99. The Name File is constructed as follows.

FOR EACH SIMULATION

1. Ftype   Nunit   Fname   [Fstatus]

The Name File contains one of the above lines (item 1) for each file. All variables are free format. The length of each line must be 299 characters or less. The lines can be in any order except for the line where Ftype (file type) is LIST as described in the subsequent "Explanation of Variables."

Comment lines are indicated by the # character in column one and can be located anywhere in the file. Any text characters can follow the # character. Comment lines have no effect on the simulation; their purpose is to allow users to provide documentation about a particular simulation. All comment lines after the first item-1 line are written in the listing file.

The Name File must include two files required for each simulation, the BAS Package File and the Discretization File.

Explanation of Variables in the Name File:

Ftype—is the file type, which must be one of the following character values. Ftype may be entered in any combination of uppercase and lowercase. {Note that the file types listed below only include those that can be used in the MODFLOW-only or GSFLOW mode of a GSFLOW simulation; see table 1 in body of report.}

LIST for the Listing File—this type must be present and must be the first file in the Name File.

DIS for the Discretization File

BAS6 for the Basic Package

OC for the Output Control Option

CHD for the Time-Variant Specified-Head Option

BCF6 for the Block-Centered Flow Package

LPF for the Layer Property Flow package

HFB6 for the Horizontal Flow Barrier Package

WEL for the Well Package

MNW1 for the Multi-Node Well Package

GHB for the General-Head Boundary Package

FHB for the and Flow and Head Boundary Package

SFR for the Streamflow Routing (SFR2) Package

LAK for Lake Package

GAGE for the Gage Package

UZF for the Unsaturated-Zone Flow Package

SIP for the Strongly Implicit Procedure Package

PCG for the Preconditioned Conjugate-Gradient Package

DE4 for the Direct Solution Package

DATA(BINARY) for binary (unformatted) files, such as those used to save cell-by-cell budget data and binary (unformatted) head and drawdown data.

DATA for formatted (text) files, such as those used to save formatted head and drawdown and for input of data from files that are separate from the primary package input files.

Nunit—is the Fortran unit to be used when reading from or writing to the file. Any legal unit number on the computer being used can be specified except unit 99. Unit 99 is used for the Name File and for reading multi-valued variables using the OPEN/CLOSE option of the utility subroutines (see section, "Array Reading Utility Subroutines"). The unit number for each file must be unique.

Fname—is the name of the file, which is a character value. Pathnames may be specified as part of Fname.

Fstatus—is the optional file status, which applies only to file types Data and Data(Binary). Two values are allowed: OLD and REPLACE. OLD indicates that the file should already exist. REPLACE indicates that if the file already exists, then it should be deleted before opening a new file. The default is to open the existing file if the file exists or create a new file if the file does not exist.

Basic Package Input Instructions

The Basic Package reads several files: Name File (described above), Discretization File, Basic Package file, Multiplier Array File, Zone Array File, Output Control Option File, and the Constant-Head Boundary (that is, the Time-Variant Specified-Head) Option File. The Name File, Discretization File, and Basic Package File are required for all simulations. The other files are optional. Input instructions for the Multiplier Array and Zone Array Files are part of MODFLOW parameters and are described in Harbaugh (2005, p. 8-15 and 8-16) but not herein.

Discretization File

Discretization information is read from the file that is specified by DIS as the file type.

FOR EACH SIMULATION

0. [#Text]
    Item 0 is optional—# must be in column 1. Item 0 can be repeated multiple times.
1. NLAY   NROW   NCOL   NPER   ITMUNI   LENUNI
2. LAYCBD(NLAY)
3. DELR(NCOL) - U1DREL
    Figure A1-3 illustrates the orientation of DELR and DELC.
4. DELC(NROW) - U1DREL
5. Top(NCOL,NROW) - U2DREL
6. BOTM(NCOL,NROW) - U2DREL

Item 6 is repeated for each model layer and Quasi-3D confining bed in the grid. Thus, the number of BOTM variables must be NLAY plus the number of Quasi-3D confining beds. The BOTM variables are read in sequence going down from the top of the system. For example, in a 3-layer model with a Quasi-3D confining bed below layer 2, there would be four BOTM arrays. The arrays would be the bottom of layer 1, the bottom of layer 2, the bottom of the Quasi-3D confining bed below layer 2, and the bottom of layer 3.

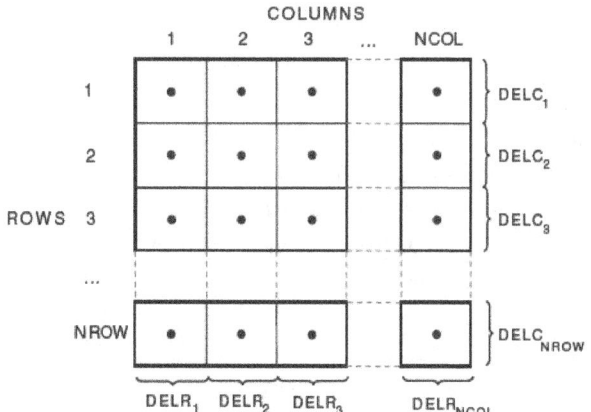

Figure A1-3.   Plan view of grid showing DELR and DELC values.

FOR EACH STRESS PERIOD

7. PERLEN   NSTP   TSMULT   Ss/Tr

Explanation of Variables Read from the Discretization File:

Text—is a character variable (199 characters) that starts in column 2. Any characters can be included in Text. The # character must be in column 1. Lines beginning with # are restricted to the first lines of the file. Text is written to the Listing File.

NLAY—is the number of layers in the model grid.

NROW—is the number of rows in the model grid.

NCOL—is the number of columns in the model grid.

NPER—is the number of stress periods in the simulation.

ITMUNI—indicates the time unit of model data, which must be consistent for all data values that involve time. For example, if the chosen time unit is years (ITMUNI=5), then stress-period length, time-step length, transmissivity, and so on, must all be expressed using years for their time units. {Unlike a normal MODFLOW run, ITMUNI cannot be undefined when using GSFLOW}. When the time unit is defined, MODFLOW uses it to print a table of elapsed simulation time:

1 – seconds   4 - days

2 – minutes   5 - years

3 – hours

LENUNI—indicates the length unit of model data, which must be consistent for all data values that involve length. For example, if feet is the chosen length unit, grid spacing, head, hydraulic conductivity, water volumes, and so forth, must all be expressed using feet for their length units. {Unlike a normal MODFLOW run, ITMUNI cannot be undefined when using GSFLOW}. Be sure to use consistent units for all input data even when LENUNI indicates an undefined length unit:

1 – feet

2 – meters

3 – centimeters

LAYCBD—is a flag, with one value for each model layer, which indicates whether or not a layer has an implicit or Quasi-three-dimensional confining bed below it. 0 indicates no confining bed, and not zero indicates a confining bed. LAYCBD for the bottom layer must be 0.

DELR—is the cell width along rows. Read one value for each of the NCOL columns. This is a multi-value one-dimensional variable with one value for each column.

DELC—is the cell width along columns. Read one value for each of the NROW rows. This is a multi-value one-dimensional variable with one value for each row.

Top—{in a normal MODFLOW simulation, Top is the top elevation of layer 1. However, for a GSFLOW simulation, Top represents the elevation of the soil-zone base}

BOTM—is the bottom elevation of a model layer or a Quasi-3d confining bed.

PERLEN—is the length of a stress period.

NSTP—is the number of time steps in a stress period.

TSMULT—{in a normal MODFLOW simulation, TSMULT is the multiplier for the length of successive time steps. However, for a GSFLOW simulation, TSMULT must be set to a value of 1.0. NSTP and PERLEN must be specified such that the time step length is equal to 1 day and the period length is equal to multiples of 1 day.}

Ss/Tr—is a character variable that indicates whether the stress period is transient or steady state. The only allowed options are Ss and Tr, but these are case insensitive. {In a GSFLOW simulation, Ss/Tr can be set to Ss for the first stress period only.}

Basic Package File

The Basic (BAS) Package file is specified with BAS6 as the file type.

FOR EACH SIMULATION

0.   [#Text]
     Item 0 is optional—# must be in column 1. Item 0 can be repeated multiple times.
1.   Options (199 text characters)
     If there are no options to specify, then a blank line must be included for item 1.
2.   IBOUND(NCOL,NROW) or (NCOL,NLAY) — U2DINT
3.   HNOFLO (10-space field unless item 1 contains 'FREE'.)
4.   STRT(NCOL,NROW) or (NCOL,NLAY) — U2DREL
     A layer variable is read for each layer in the grid.

Explanation of Variables Read from the BAS Package File:

Text—is a character variable that starts in column 2. The first two comment lines will become variable HEADNG, which is used as a printout title throughout the program. (If there are no comment lines, then HEADNG will be blank.) HEADNG is limited to 80 columns, but subsequent Text lines can be up to 199 columns. Any characters can be included in Text. The # character must be in column 1. Lines beginning with # are restricted to the first lines of the file. Text is written to the Listing File.

Options—is a character variable that is scanned for words (separated by one or more spaces) that specify program options. Two options are currently (2007) recognized. Unrecognized words are ignored, and a word may be specified in either uppercase or lowercase. A blank line is acceptable and indicates no options. The option XSECTION(courier) is not supported in GSFLOW.

CHTOCH indicates that flow between adjacent constant-head cells should be calculated.

FREE indicates that free format is used for input variables throughout the Basic Package and other packages as indicated in their input instructions. Be sure that all variables read using free format have a non-blank value and that a comma or at least one blank separates all adjacent values.

IBOUND—is the boundary variable. One value is read for every finite-difference cell. These values are read one layer at a time. Note that although IBOUND is read as one or more two-dimensional variables, IBOUND is stored internally as a three-dimensional variable.

If IBOUND(J,I,K) < 0, cell J,I,K has a constant head.

If IBOUND(J,I,K) = 0, cell J,I,K is no flow.

If IBOUND(J,I,K) > 0, cell J,I,K is variable head.

HNOFLO—is the value of head to be assigned to all no-flow cells (IBOUND = 0). Because head at no-flow cells is unused in model calculations, this does not affect model results but serves to identify no-flow cells when head is printed. This value is used also as drawdown at no-flow cells if the drawdown option is used. Even if the user does not anticipate having no-flow cells, a value for HNOFLO must be entered.

STRT—is initial (starting) head—that is, head at the beginning of the simulation. STRT must be specified for all simulations, including steady-state simulations. One value is read for every finite-difference cell. For simulations in which the first stress period is steady state, the values used for STRT generally do not affect the simulation (exceptions may occur if cells go dry and (or) rewet). The execution time, however, will be less if STRT includes hydraulic heads that are close to the steady-state solution.

Output Control Option File

Input to the Output Control Option File of the Ground-Water Flow Process is read from the file that is specified as type OC in the Name File. If no OC file is specified, default output control is used. Under the default, head and overall budget are written to the Listing File at the end of every MODFLOW stress period. The default printout format for head and drawdown is 10G11.4.

Output Control data may be specified as words or numeric codes. One of these methods must be used throughout any simulation.

Output control using words

Recognized words are shown in ***BOLD ITALICS***; these words must be entered exactly as shown except that they may be entered in either uppercase or lowercase. Optional parts of lines are shown in brackets. One or more spaces must separate each word or variable, and the total line length must not exceed 199 characters.

FOR EACH SIMULATION

0. [#Text]
    Item 0 is optional—# must be in column 1. Item 0 can be repeated multiple times.
1. Any combination of the following lines:
    HEAD   PRINT   FORMAT   IHEDFM
    Specifies the format for writing head to the Listing File.

    HEAD   SAVE   FORMAT   CHEDFM ***[LABEL]***
    Specifies the format for writing head to a file other than the Listing File. Omit this line to obtain a binary (unformatted) file. Binary files usually are smaller than text files, but they are not generally transportable among different computer operating systems or different Fortran compilers.

    HEAD   SAVE   UNIT   IHEDUN
    Specifies the file unit for writing head to a file other than the Listing File.

    DRAWDOWN   PRINT   FORMAT   IDDNFM
    Specifies the format for writing drawdown to the Listing File.

    DRAWDOWN   SAVE   FORMAT   CDDNFM ***[LABEL]***
    Specifies the format for writing drawdown to a file other than the Listing File. Omit this line to obtain an unformatted (binary) file. Binary files usually are smaller than text files, but they are not generally transportable among different computer operating systems or different Fortran compilers.

    DRAWDOWN   SAVE   UNIT   IDDNUN
    Specifies the file unit for writing drawdown to a file other than the Listing File.

    IBOUND   SAVE   FORMAT   CBOUFM ***[LABEL]***
    Specifies the format for writing IBOUND to a file.

    IBOUND   SAVE   UNIT   IBOUUN
    Specifies the file unit for writing IBOUND to a file.

    COMPACT BUDGET [**AUX** or ***AUXILIARY***]
    COMPACT BUDGET indicates that the cell-by-cell budget file(s) will be written in a more compact form than is used in the 1988 version of MODFLOW (referred to as MODFLOW-88) (McDonald and Harbaugh, 1988); however, programs that read these data in the form written by MODFLOW-88 will be unable to read the new compact file. If this option is not used, MODFLOW-2005 will write the files using the MODFLOW-88 form. The optional word ***AUX*** (or ***AUXILIARY***) indicates that auxiliary data that are defined in packages (see input data for the RIV, WEL, DRN, and GHB Packages) should be saved in the budget file along with budget data.

FOR EACH TIME STEP FOR WHICH OUTPUT IS DESIRED

2. ***PERIOD***   IPEROC   ***STEP***   ITSOC
3. Any combination of the following lines:
    PRINT HEAD [list layers if all layers not desired]
    Head is written to the Listing File.

PRINT DRAWDOWN [list layers if all layers not desired]
   Drawdown is written to the Listing File.

PRINT BUDGET
   Overall volumetric budget is written to the Listing File.

SAVE HEAD [list layers if all layers not desired]
   Head is written to a file other than the Listing File.

SAVE DRAWDOWN [list layers if all layers not desired]
   Drawdown is written to a file other than the Listing File.

SAVE IBOUND [list layers if all layers not desired]
   IBOUND is written to a file other than the Listing File. This option is provided to allow changes in IBOUND to be recorded in simulations where IBOUND changes during a simulation.

SAVE BUDGET
   Cell-by-cell budget data are written to the files that are designated in the packages that compute budget terms.

Item 2 and one or more item-3 lines are specified each time for which output is desired. These lines must be in the order of increasing simulation time.

   Explanation of Variables Read by Output Control Using Words:

Text—is a character variable (199 characters) that starts in column 2. Any characters can be included in Text. The # character must be in column 1. Lines beginning with # are restricted to the first lines of the file. Text is written to the Listing File.

IHEDFM—is a code for the format in which heads will be printed. (Positive values indicate wrap format; negative values indicate strip format.)

0	- 10G11.4	11	- 20F5.4
1	- 11G10.3	12	- 10G11.4
2	- 9G13.6	13	- 10F6.0
3	- 15F7.1	14	- 10F6.1
4	- 15F7.2	15	- 10F6.2
5	- 15F7.3	16	- 10F6.3
6	- 15F7.4	17	- 10F6.4
7	- 20F5.0	18	- 10F6.5
8	- 20F5.1	19	- 5G12.5
9	- 20F5.2	20	- 6G11.4
10	- 20F5.3	21	- 7G9.2

CHEDFM—is a character value that specifies the format for saving heads, and can only be specified if the word method of output control is used. The format must contain 20 characters or less and must be a valid Fortran format that is enclosed in parentheses. The format must be enclosed in apostrophes if it contains one or more blanks or commas. The optional word ***LABEL*** after the format is used to indicate that each layer of output should be preceded with a line that defines the output (simulation time, the layer being output, and so forth). If there is no line specifying CHEDFM, then heads are written to a binary (unformatted) file. Binary files are usually more compact than text files, but they are not generally transportable among different computer operating systems or different Fortran compilers.

IHEDUN—is the unit number on which heads will be saved.

IDDNFM—is a code for the format in which drawdowns will be printed. The codes are the same as for IHEDFM.

CDDNFM—is a character value that specifies the format for saving drawdown, and can only be specified if the word method of output control is used. The format must contain 20 characters or less and must be a valid Fortran format that is enclosed in parentheses. The format must be enclosed in apostrophes if it contains one or more blanks or commas. The optional word ***LABEL*** after the format is used to indicate that each layer of output should be preceded with a line that defines the output (simulation time, the layer being output, and so forth). If there is no line specifying CDDNFM, then drawdown is written to a binary (unformatted) file. Binary files are usually more compact than text files, but they are not generally transportable among different computer operating systems or different Fortran compilers.

IDDNUN—is the unit number on which drawdowns will be saved.

CBOUFM—is a character value that specifies the format for saving IBOUND, and can only be specified if the word method of output control is used. The format must contain 20 characters or less and must be a valid Fortran format that is enclosed in parentheses. The format must be enclosed in apostrophes if it contains one or more blanks or commas. The optional word **LABEL** is used to indicate that each layer of output should be preceded with a line that defines the output (simulation time, the layer being output, and so forth). If there is no line specifying CBOUFM, then IBOUND is written using format (20I4).

IBOUND is never written as a binary (unformatted) file.

IBOUUN—is the unit number on which IBOUND will be saved.

IPEROC—is the stress period number at which output is desired.

ITSOC—is the time step number (within a stress period) at which output is desired.

Figure A1-4 shows an example of output control using words. Note that the first line cannot be blank, but after the first line, blank lines are ignored when the word method is used to specify Output Control data. Indented lines are allowed because of the use of free format input.

Output Control Using Numeric Codes

All variables are free format if the word FREE is specified in item 1 of the Basic Package input file; otherwise, the variables all have 10-character fields.

FOR EACH SIMULATION
0. [#Text]
    Item 0 is optional—# must be in column 1. Item 0 can be repeated multiple times.
1. IHEDFM   IDDNFM   IHEDUN   IDDNUN

FOR EACH TIME STEP

2. INCODE   IHDDFL   IBUDFL   ICBCFL
3. Hdpr   Ddpr   Hdsv   Ddsv
    (Item 3 is read 0, 1, or NLAY times, depending on the value of INCODE.)

Explanation of Variables Read by Output Control Using Numeric Codes:

Text—is a character variable (199 characters) that starts in column 2. Any characters can be included in Text. The # character must be in column 1. Lines beginning with # are restricted to the first lines of the file. Text is written to the Listing File.

IHEDFM—is a code for the format in which heads will be printed. See the description above in the explanation of variables read by output control using words.

IDDNFM—is a code for the format in which drawdowns will be printed. The codes are the same as for IHEDFM.

```
HEAD PRINT FORMAT 15
HEAD SAVE FORMAT (20F10.3) LABEL
HEAD SAVE UNIT 30
COMPACT BUDGET
DRAWDOWN PRINT FORMAT 14
PERIOD 1 STEP 1
    PRINT HEAD 2 6
    PRINT DRAWDOWN
    PRINT BUDGET
    SAVE BUDGET
    SAVE HEAD
PERIOD 1 STEP 7
    SAVE HEAD 1 3 5
    PRINT DRAWDOWN
    SAVE BUDGET
PERIOD 2 STEP 5
    PRINT HEAD
    PRINT BUDGET
    SAVE BUDGET
    SAVE HEAD
```

Figure A1-4. Example Output Control input using words.

IHEDUN—is the unit number on which heads will be saved.

IDDNUN—is the unit number on which drawdowns will be saved.

INCODE—is the code for reading item 3.

> If INCODE < 0, item 3 flags are used from the last time step. Item 3 is not read.
>
> If INCODE = 0, all layers are treated the same way. Item 3 will consist of one line.
>
> If INCODE > 0, item 3 will consist of one line for each layer.

IHDDFL—is a head and drawdown output flag. This flag allows item 3 flags to be specified in an early time step and then used or not used in subsequent time steps. Thus, IHDDFL can be used to avoid resetting item 3 flags every time step.

> If IHDDFL = 0, no heads or drawdowns will be printed or saved regardless of which item 3 flags are specified.
>
> If IHDDFL ≠ 0, heads and drawdowns will be printed or saved according to the item 3 flags.

IBUDFL—is a budget print flag.

> If IBUDFL = 0, overall volumetric budget will not be printed.
>
> If IBUDFL ≠ 0, overall volumetric budget will be printed.

ICBCFL—is a flag for writing cell-by-cell flow data.

> If ICBCFL = 0, cell-by-cell flow terms are not written to any file.
>
> If ICBCFL ≠ 0, cell-by-cell flow terms are written to the Listing File or a Budget File depending on flags set in the component of flow packages, that is, IWELCB, IRCHCB, and so forth.

Hdpr—is the output flag for head printout.

> If Hdpr = 0, head is not printed for the corresponding layer.
>
> If Hdpr ≠ 0, head is printed for the corresponding layer.

Ddpr—is the output flag for drawdown printout.

> If Ddpr = 0, drawdown is not printed for the corresponding layer.
>
> If Ddpr ≠ 0, drawdown is printed for the corresponding layer.

Hdsv—is the output flag for head save.

> If Hdsv = 0, head is not saved for the corresponding layer.
>
> If Hdsv ≠ 0, head is saved for the corresponding layer.

Ddsv—is the output flag for drawdown save.

> If Ddsv = 0, drawdown is not saved for the corresponding layer.
>
> If Ddsv ≠ 0, drawdown is saved for the corresponding layer.

Constant-Head (Time-Variant Specified-Head) Option File

Input to the Constant-Head (CHD; Time-Variant Specified-Head) Option File is read from the file that has file type CHD in the Name File. Optional variables are shown in brackets. All variables are free format if the option FREE is specified in the Basic Package input file; otherwise, the non-optional variables have 10-character fields and the optional variables are free format.

Once a cell is made constant head, the cell stays constant head throughout the remainder of the simulation. For example, if a cell is listed in the CHD file as constant head in stress period 1 and not listed in stress period 2, then the cell continues to be constant head in stress period 2 and throughout the remainder of the stress periods. The head is adjusted only in the stress periods in which a cell is listed. For the stress periods in which a constant-head cell is not listed, the head stays at the value that it had at the end of the previous stress period.

FOR EACH SIMULATION

0. [#Text]
    Item 0 is optional—# must be in column 1. Item 0 can be repeated multiple times.
1. MXACTC   [Option]
    FOR EACH STRESS PERIOD

2. ITMP   NP
3. Layer   Row   Column   Shead   Ehead   [xyz]
    ITMP repetitions of item 3 are read by subroutine ULSTRD if ITMP > 0. **SFAC** of the ULSTRD utility subroutine
    applies to Shead and Ehead. Item 3 is not read if ITMP is negative or 0.

    Explanation of Variables Read by the CHD Option:

Text—is a character variable (199 characters) that starts in column 2. Any characters can be included in Text. The # character
    must be in column 1. Lines beginning with # are restricted to the first lines of the file. Text is written to the Listing File.

MXACTC—is the maximum number of constant-head boundary cells in use during any stress period.

Option—is an optional list of character values.
    AUXILIARY **abc** or **AUX abc**—defines an auxiliary variable, named abc, which will be read for each constant-head
    boundary as part of item 3. Up to five variables can be specified, each of which must be preceded by AUXILIARY or
    AUX. These variables will not be used by the Ground-Water Flow Process, but they will be available for use by other
    processes. The auxiliary variable values will be read after the Ehead variable.

ITMP—is a flag and a counter.
    If ITMP < 0, data from the preceding stress period will be reused.
    If ITMP ≥ 0, ITMP is the number of constant-head boundaries read for the current stress period.

NP—{this value should be set to zero when parameters are not used.}

Layer—is the layer number of the constant-head boundary.

Row—is the row number of the constant-head boundary.

Column—is the column number of the constant-head boundary.

Shead—is the head at the boundary at the start of the stress period.

Ehead—is the head at the boundary at the end of the stress period.

[xyz]—represents the values of the auxiliary variables for a constant-head boundary that have been defined in item 1. The
    values of auxiliary variables must be present in each repetition of items 2 and 3 if they are defined in item 1. The values
    must be specified in the order used to define the variables in item 1.

Block-Centered Flow Package

Input for the Block-Centered Flow (BCF) Package is read from the file that is type BCF6 in the Name File. The BCF
Package is an alternative to the LPF and HUF Packages. Only one of these packages should be used during a simulation.

FOR EACH SIMULATION

1. IBCFCB   HDRY   IWDFLG   WETFCT   IWETIT   IHDWET
    These six variables are free format if the option FREE is specified in the Basic Package input file; otherwise, the
    variables all have 10-character fields.

2. `Ltype(NLAY)`

    Read one value for each layer. These values are free format if the word `FREE` is specified in item 1 of the Basic Package input file; otherwise, the values are read using fixed format fields that are each 2 characters wide with 40 values per line. Use only as many lines as required for the number of model layers.

3. `TRPY(NLAY)` — U1DREL

    A subset of the following two-dimensional variables is used to describe each layer. The variables needed for each layer depend on the layer-type code (LAYCON, which is defined as part of `Ltype`), whether the simulation has any transient stress periods (at least one stress period defined in the Discretization File specifies `Ss/Tr` as TR), and if the wetting capability is active (`IWDFLG` not 0). Unneeded variables must be omitted. In no situation will all variables be required. The required variables (items 4-9) for layer 1 are read first; then the variables for layer 2 and so forth.

4. `[Sf1(NCOL,NROW)]` — U2DREL If there is at least one transient stress period.

    If `LAYCON` is 0 or 2 (see `Ltype`), then read item 5.

5. `[Tran(NCOL,NROW)]` — U2DREL.

    Otherwise, if `LAYCON` is 1 or 3 (see `Ltype`), read item 6.

6. `[HY(NCOL,NROW)]` — U2DREL

7. `[Vcont(NCOL,NROW)]` — U2DREL. If not the bottom layer.

8. `[Sf2(NCOL,NROW)]` — U2DREL. If there is at least one transient stress period and `LAYCON` (see `Ltype`) is 2 or 3.

9. `[WETDRY(NCOL,NROW)]` — U2DREL. If `IWDFLG` is not 1 and `LAYCON` is 1 or 3 (see `Ltype`).

Explanation of Variables Read by the BCF Package:

`IBCFCB`—is a flag and a unit number.

> If `IBCFCB` > 0, cell-by-cell flow terms will be written to this unit number when `SAVE BUDGET` or a non-zero value for `ICBCFL` is specified in Output Control. The terms that are saved are storage, constant-head flow, and flow between adjacent cells.

> If `IBCFCB` = 0, cell-by-cell flow terms will not be written.

> If `IBCFCB` < 0, cell-by-cell flow for constant-head cells will be written in the listing file when `SAVE BUDGET` or a non-zero value for `ICBCFL` is specified in Output Control. Cell-by-cell flow to storage and between adjacent cells will not be written to any file.

`HDRY`—is the head that is assigned to cells that are converted to dry during a simulation. Although this value plays no role in the model calculations, `HDRY` values are useful as indicators when looking at the resulting heads that are output from the model. `HDRY` is thus similar to `HNOFLO` in the Basic Package, which is the value assigned to cells that are no-flow cells at the start of a model simulation.

`IWDFLG`—is a flag that determines if the wetting capability is active.

> If `IWDFLG` = 0, the wetting capability is inactive.

> If `IWDFLG` is not 0, the wetting capability is active.

`WETFCT`—is a factor that is included in the calculation of the head that is initially established at a cell when that cell is converted from dry to wet. (See `IHDWET`.)

`IWETIT`—is the iteration interval for attempting to wet cells. Wetting is attempted every `IWETIT` iteration. This applies to outer iterations and not inner iterations. If `IWETIT` is 0, the value is changed to 1.

`IHDWET`—is a flag that determines which equation is used to define the initial head at cells that become wet:

> If `IHDWET` = 0, equation 5-32A in Harbaugh (2005) is used: $h = BOT + WETFCT (h_n - BOT)$

> If `IHDWET` is not 0, equation 5-32B in Harbaugh (2005) is used: $h = BOT + WETFCT (THRESH)$

`Ltype`—contains a combined code for each layer that specifies both the layer type (LAYCON) and the method of computing interblock conductance. Use as many lines as needed to enter a value for each layer. Values are two-digit numbers.

The left digit defines the method of calculating interblock transmissivity. The methods are described by Goode and Appel (1992).

0 or blank—harmonic mean (the method used in MODFLOW-88).

1—arithmetic mean

2—logarithmic mean

3—arithmetic mean of saturated thickness and logarithmic-mean hydraulic conductivity.

The right digit defines the layer type (LAYCON), which is the same as in MODFLOW-88:

0—confined—Transmissivity and storage coefficient of the layer are constant for the entire simulation.

1—unconfined—Transmissivity of the layer varies and is calculated from the saturated thickness and hydraulic conductivity. The storage coefficient is constant. This type code is valid only for layer 1.

2—confined/unconfined—Transmissivity of the layer is constant. The storage coefficient may alternate between confined and unconfined values. Vertical flow from above is limited if the layer desaturates.

3—confined/unconfined—Transmissivity of the layer varies and is calculated from the saturated thickness and hydraulic conductivity. The storage coefficient may alternate between confined and unconfined values. Vertical flow from above is limited if the aquifer desaturates.

TRPY—is a one-dimensional variable containing a horizontal anisotropic factor for each layer and is the ratio of transmissivity or hydraulic conductivity (whichever is being used) along a column to transmissivity or hydraulic conductivity along a row. Set to 1.0 for isotropic conditions. This is a single variable with one value per layer. Do not read a variable for each layer—that is, include only one array control line for the entire variable.

Sf1—is the primary storage coefficient. Read only if one or more transient stress periods are specified in the Discretization File. For LAYCON equal to 1, Sf1 will always be specific yield, whereas for LAYCON equal to 2 or 3, Sf1 will always be confined storage coefficient. For LAYCON equal to 0, Sf1 would normally be confined storage coefficient; however, a LAYCON value of 0 also can be used to simulate water-table conditions where drawdowns everywhere are expected to remain a small fraction of the saturated thickness, and where there is no layer above, or flow from above is negligible. In this case, specific yield values would be entered for Sf1.

Tran—is the transmissivity along rows. Tran is multiplied by TRPY to obtain transmissivity along columns. Read only for layers where LAYCON is 0 or 2.

HY—is the hydraulic conductivity along rows. HY is multiplied by TRPY to obtain hydraulic conductivity along columns. Read only for layers where LAYCON is 1 or 3.

Vcont—is the vertical hydraulic conductivity divided by the thickness from a layer to the layer below (also called leakance). The value for a cell is the hydraulic conductivity divided by thickness for the material between the node in that cell and the node in the cell below. Because there is not a layer beneath the bottom layer, Vcont cannot be specified for the bottom layer.

Sf2—is the secondary storage coefficient. Read only for layers where LAYCON is 2 or 3 and only if there are one or more transient stress periods specified in the Discretization File. The secondary storage coefficient is always specific yield.

WETDRY—is a combination of the wetting threshold (THRESH) and a flag to indicate which neighboring cells can cause a cell to become wet. If WETDRY < 0, only the cell below a dry cell can cause the cell to become wet. If WETDRY > 0, the cell below a dry cell and the four horizontally adjacent cells can cause a cell to become wet. If WETDRY is 0, the cell cannot be wetted. The absolute value of WETDRY is the wetting threshold. When the sum of BOT and the absolute value of WETDRY at a dry cell is equaled or exceeded by the head at an adjacent cell, the cell is wetted. Read only if LAYCON is 1 or 3 and IWDFLG is not 0.

Layer-Property Flow Package

Input for the Layer-Property Flow (LPF) Package is read from the file that is type LPF in the Name File. Free format is used for reading all values. The LPF Package is an alternative to the BCF and HUF Packages. Only one of these packages can be used during a simulation.

FOR EACH SIMULATION

0. [#Text]
      Item 0 is optional—# must be in column 1. Item 0 can be repeated multiple times.
1. ILPFCB   HDRY   NPLPF [Options]
2. LAYTYP(NLAY)
3. LAYAVG(NLAY)
4. CHANI(NLAY)
5. LAYVKA(NLAY)
6. LAYWET(NLAY)
7. [WETFCT   IWETIT   IHDWET]
      (Include item 7 only if LAYWET indicates at least one wettable layer.)
8. and 9. These items are not included in these instructions because they relate to parameters.

A subset of the following two-dimensional variables is used to describe each layer. All the variables that apply to layer 1 are read first, followed by layer 2, followed by layer 3, and so forth. A variable not required due to simulation options (for example, Ss and Sy for a completely steady-state simulation) must be omitted from the input file.
These variables are read by the array-reading utility subroutine, U2DREL.

10. HK(NCOL,NROW)
11. [HANI(NCOL,NROW)]
      Include item 11 only if CHANI is less than or equal to 0.
12. VKA(NCOL,NROW)
13. [Ss(NCOL,NROW)]
      Include item 13 only if at least one stress period is transient.
14. [Sy(NCOL,NROW)]
      Include item 14 only if at least one stress period is transient and LAYTYP is not 0.
15. [VKCB(NCOL,NROW)]
      Include item 15 only if LAYCBD (in the Discretization File) is not 0.
16. [WETDRY(NCOL,NROW)]
      Include item 16 only if LAYWET is not 0 and LAYTYP is not 0.

Explanation of Variables Read by the LPF Package:

Text—is a character variable (199 characters) that starts in column 2. Any characters can be included in Text. The # character must be in column 1. Lines beginning with # are restricted to the first lines of the file. Text is written to the Listing File.

ILPFCB—is a flag and a unit number.

If ILPFCB > 0, cell-by-cell flow terms will be written to this unit number when SAVE BUDGET or a non-zero value for ICBCFL is specified in Output Control. The terms that are saved are storage, constant-head flow, and flow between adjacent cells.

If ILPFCB = 0, cell-by-cell flow terms will not be written.

If ILPFCB < 0, cell-by-cell flow for constant-head cells will be written in the listing file when SAVE BUDGET or a non-zero value for ICBCFL is specified in Output Control. Cell-by-cell flow to storage and between adjacent cells will not be written to any file.

HDRY—is the head that is assigned to cells that are converted to dry during a simulation. Although this value plays no role in the model calculations, HDRY values are useful as indicators when looking at the resulting heads that are output from the model. HDRY is thus similar to HNOFLO in the Basic Package, which is the value assigned to cells that are no-flow cells at the start of a model simulation.

NPLPF—{this value should be set to zero when parameters are not used.}

Options—are optional key words that activate options:

STORAGECOEFFICIENT indicates that variable Ss is read as storage coefficient rather than specific storage.

CONSTANTCV indicates that vertical conductance for an unconfined cell is computed from the cell thickness rather than the saturated thickness.

THICKSTRT indicates that layers having a negative LAYTYP are confined, and their cell thickness for conductance calculations will be computed as STRT-BOT rather than TOP-BOT, where STRT is the ground-water head at the beginning of the simulation.

NOCVCORRECTION indicates that vertical conductance is not corrected when the vertical flow correction is applied.

LAYTYP—contains a flag for each layer that specifies the layer type.

0 – confined

>0 – convertible

<0 – convertible unless the THICKSTRT option is in effect. When THICKSTRT is in effect, a negative value of LAYTYP indicates that the layer is confined, and its saturated thickness will be computed as STRT-BOT, where STRT is the ground-water head at the beginning of the simulation.

LAYAVG—contains a flag for each layer that defines the method of calculating interblock transmissivity.

0—harmonic mean

1—logarithmic mean

2—arithmetic mean of saturated thickness and logarithmic-mean hydraulic conductivity.

CHANI—contains a value for each layer that is a flag or the horizontal anisotropy. If CHANI ≤ 0, then variable HANI defines horizontal anisotropy. If CHANI is >0, then CHANI is the horizontal anisotropy for the entire layer, and HANI is not read.

LAYVKA—contains a flag for each layer that indicates whether variable VKA is vertical hydraulic conductivity or the ratio of horizontal to vertical hydraulic conductivity.

0—indicates VKA is vertical hydraulic conductivity

not 0—indicates VKA is the ratio of horizontal to vertical hydraulic conductivity, where the horizontal hydraulic conductivity is specified as HK in item 10.

LAYWET—contains a flag for each layer that indicates whether wetting is active.

0—indicates wetting is inactive

not 0—indicates wetting is active

WETFCT—is a factor that is included in the calculation of the head that is initially established at a cell when the cell is converted from dry to wet. (See IHDWET.)

IWETIT—is the iteration interval for attempting to wet cells. Wetting is attempted every IWETIT iteration. If using the PCG solver (Hill, 1990), this applies to outer iterations, not inner iterations. If IWETIT ≤ 0, the value is changed to 1.

IHDWET—is a flag that determines which equation is used to define the initial head at cells that become wet:

If IHDWET = 0, equation 5-32A in Harbaugh (2005) is used: h = BOT + WETFCT (h_n - BOT)

If IHDWET is not 0, equation 5-32B in Harbaugh (2005) is used: h = BOT + WETFCT(THRESH)

HK—is the hydraulic conductivity along rows. HK is multiplied by horizontal anisotropy (see CHANI and HANI) to obtain hydraulic conductivity along columns.

HANI—is the ratio of hydraulic conductivity along columns to hydraulic conductivity along rows, where HK of item 10 specifies the hydraulic conductivity along rows. Thus, the hydraulic conductivity along columns is the product of the values in HK and HANI. Read only if CHANI ≤ 0.

VKA—is either vertical hydraulic conductivity or the ratio of horizontal to vertical hydraulic conductivity depending on the value of LAYVKA. If LAYVKA is 0, VKA is vertical hydraulic conductivity. If LAYVKA is not 0, VKA is the ratio of horizontal to vertical hydraulic conductivity. In this case, HK is divided by VKA to obtain vertical hydraulic conductivity, and values of VKA typically are ≥ 1.0.

Ss—is specific storage unless the STORAGECOEFFICIENT option is used. When STORAGECOEFFICIENT is used, Ss is confined storage coefficient. Read only for a transient simulation (at least one transient stress period).

Sy—is specific yield. Read only for a transient simulation (at least one transient stress period) and if the layer is convertible (LAYTYP is not 0).

VKCB—is the vertical hydraulic conductivity of a quasi-three-dimensional confining bed below a layer. Read only if there is a confining bed. Because the bottom layer cannot have a confining bed, VKCB cannot be specified for the bottom layer.

WETDRY—is a combination of the wetting threshold and a flag to indicate which neighboring cells can cause a cell to become wet. If WETDRY < 0, only the cell below a dry cell can cause the cell to become wet. If WETDRY > 0, the cell below a dry cell and the four horizontally adjacent cells can cause a cell to become wet. If WETDRY is 0, the cell cannot be wetted. The absolute value of WETDRY is the wetting threshold. When the sum of BOT and the absolute value of WETDRY at a dry cell is equaled or exceeded by the head at an adjacent cell, the cell is wetted. Read only if LAYTYP is not 0 and LAYWET is not 0.

Hydrogeologic-Unit Flow Package

Input for the Hydrologic-Unit Flow (HUF) Package is read from the file that is type HUF2 in the Name File. Free format is used for reading all values. The HUF Package is an alternative to the BCF and LPF Packages. The three packages cannot be used simultaneously. The HUF Package is documented in Anderman and Hill (2000). Additional capbilities of the HUF Package are described in Anderman and others (2002) and Anderman and Hill (2003). Users of GSFLOW are encouraged to review these documents when using the HUF Package. Input instructions are from Anderman and Hill (2003, appendix A).

FOR EACH SIMULATION

0. [#Text]

    Item 0 is optional—# must be in column 1. Item 0 can be repeated multiple times.

1. IHUFCB   HDRY   NHUF   NPHUF   IOHUF   IOHUFHEADS   IOHUFFLOWS

2. LTHUF(NLAY)

3. LAYWT(NLAY)

4. [WETFCT IWETIT IHDWET]

    Include item 4 only if LAYWT indicates at least one wettable layer.

5. WETDRY(NCOL,NROW)

    Repeat Item 5 for each layer for which LAYWET is not 0. Arrays are read by the array-reading utility module, U2DREL.

6. HGUNAM

7. TOP(NCOL,NROW)

8. THCK(NCOL,NROW)

    Items 6-8 are repeated for each hydrogeologic unit (NHUF times). Items 7 and 8 are read by the array-reading utility subroutine, U2DREL.

9. HGUNAM   HGUHANI   HGUVANI

    Repeat Item 9 for each hydrogeologic unit. If HGUNAM is set to "ALL", HGUHANI and HGUVANI are set for all hydrogeologic units and only one Item 9 is necessary. Otherwise, HGUNAM must correspond to one of the names defined in Item 6, and there must be NHUF repetitions of Item 9. The repetitions can be in any order.

10. PARNAM   PARTYP   Parval   NCLU

11. HGUNAM   Mltarr   Zonarr   IZ

    Each Item 11 record is called a parameter cluster. Repeat Item 11 NCLU times.

    Repeat Items 10-11 for each parameter to be defined (NPHUF times).

12. [**PRINT** HGUNAM PRINTCODE PRINTFLAGS]

Item 12 is optional and is included only for hydrogeologic units for which printing is desired. Item 12 must start with the word PRINT. If HGUNAM is set to ALL, PRINTCODE and PRINTFLAGS are set for all hydrogeologic units, and only one Item 12 is necessary. Otherwise, HGUNAM must correspond to one of the names defined in Item 6.

Explanation of Variables Read by the HUF Package:

Text—is a character variable (199 characters) that starts in column 2. Any characters can be included in Text. The # character must be in column 1. Lines beginning with # are restricted to the first lines of the file. Text is written to the Listing File.

IHUFCB—is a flag and a unit number.

If IHUFCB > 0, cell-by-cell flow terms will be written to this unit number when SAVE BUDGET or a non-zero value for ICBCFL is specified in Output Control. The terms that are saved are storage, constant-head flow, and flow between adjacent cells.

If IHUFCB = 0, cell-by-cell flow terms will not be written.

If IHUFCB < 0, cell-by-cell flow for constant-head cells will be written in the listing file when SAVE BUDGET or a non-zero value for ICBCFL is specified in Output Control. Cell-by-cell flow to storage and between adjacent cells will not be written to any file.

HDRY—is the head that is assigned to cells that are converted to dry during a simulation. Although this value plays no role in the model calculations, HDRY values are useful as indicators when looking at the resulting heads that are output from the model. HDRY is thus similar to HNOFLO in the Basic Package, which is the value assigned to cells that are no-flow cells at the start of a model simulation.

NHUF— is the number of hydrogeologic units defined using the HUF package.

NPHUF—{this value should be set to zero when parameters are not used.

IOHUF—is a flag and a unit number.

0 – interpolated heads will not be written.

>0 – calculated heads will be interpolated and written on unit IOHUF for each hydrogeologic unit using the format defined in the output-control file.

IOHUFHEADS—is a flag and unit number.

0 – interpolated heads will not be written.

>0 – calculated heads will be interpolated and written on unit IOHUFHEADS for each hydrogeologic unit using the format defined in the output-control file.

IOHUFFLOWS—is a flag and unit number.

0 – interpolated flows will not be written.

>0 – calculated cell-by-cell flows will be interpolated and written on unit IOHUFFLOWS for each hydrogeologic unit using the format defined in the output-control file.

LTHUF—is a flag specifying the layer type. Read one value for each layer; each element holds the code for the respective layer. There is a limit of 200 layers. Use as many records as needed to enter a value for each layer.

0 – indicates a confined layer

not 0 – indicates a convertible layer

LAYWT—is a flag that indicates if wetting is active. Read one value per layer.

0 – indicates wetting is inactive

1 – indicates wetting is active

WETFCT—is a factor that is included in the calculation of the head that is initially established at a cell when the cell is converted from dry to wet. (See IHDWET.)

IWETIT—is the iteration interval for attempting to wet cells. Wetting is attempted every IWETIT iteration. If using the PCG solver (Hill, 1990), this applies to outer iterations, not inner iterations. If IWETIT is 0, the value is changed to 1.

IHDWET—is a flag that determines which equation is used to define the initial head at cells that become wet:

If IHDWET = 0, equation 5-32A in Harbaugh (2005) is used: h = BOT + WETFCT (h$_n$ - BOT)

If IHDWET is not 0, equation 5-32B in Harbaugh (2005) is used: h = BOT + WETFCT(THRESH).

WETDRY—is a combination of the wetting threshold and a flag to indicate which neighboring cells can cause a cell to become wet. If WETDRY < 0, only the cell below a dry cell can cause the cell to become wet. If WETDRY > 0, the cell below a dry cell and the four horizontally adjacent cells can cause a cell to become wet. If WETDRY is 0, the cell cannot be wetted. The absolute value of WETDRY is the wetting threshold. When the sum of BOT and the absolute value of WETDRY at a dry cell is equaled or exceeded by the head at an adjacent cell, the cell is wetted. Read only if LTHUF is not 0 and LAYWT is 1.

HGUNAM— is the name of the hydrogeologic unit. This name can consist of up to 10 characters and is not case sensitive.

TOP—is the elevation of the top of the hydrogeologic unit.

THCK— is the thickness of the hydrogeologic unit.

HGUHANI— is a flag and a horizontal anisotropy value for a hydrogeologic unit. Horizontal anisotropy is the ratio of hydraulic conductivity along columns to hydraulic conductivity along rows. Read one value for each hydrogeologic unit unless HGUNAM is set to ALL.

    0 – indicates that horizontal anisotropy will be defined using a HANI parameter.

    > 0 – HGUHANI is the horizontal anisotropy of the entire hydrogeologic unit.

HGUVANI— is a flag that indicates whether array VK is vertical hydraulic conductivity or the ratio of horizontal to vertical hydraulic conductivity. Read only one value for each hydrogeologic unit unless HGUNAM is set to ALL.

    0 – indicates VK is hydraulic conductivity (VK parameter must be used).

    > 0 –indicates VK is the ratio of horizontal to vertical hydraulic conductivity and HGUVANI is the vertical anisotropy of the entire hydrogeologic unit. Value is ignored if a VANI parameter is defined for the corresponding hydrogeologic unit.

PARNAM— is the name of a parameter to be defined. This name can consist of up to 10 characters and is not case sensitive.

PARTYP— is the type of parameter to be defined. For the HUF Package, the allowed parameter types are:

    HK—defines variable HK, horizontal hydraulic conductivity.

    HANI—defines variable HANI, horizontal anisotropy.

    VK—defines variable VK, vertical hydraulic conductivity, for units for which HGUVANI is set to zero.

    VANI—defines variable VANI, vertical anisotropy, for units for which HGUVANI is set to greater than zero.

    SS— defines variable Ss, the specific storage.

    SY— defines variable Sy, the specific yield.

    SYTP— when all model layers are confined, defines the storage coefficient for the top active cell at each row, column location. The value specified is not multiplied by model layer thickness. When SYTP is specified as the parameter type, HGUNAM of item 11 needs to be set to SYTP.

Parval— is the initial value of the parameter; however, this value can be replaced by a value specified in the Sensitivity Process input file.

NCLU— is the number of clusters required to define the parameter. Each Item-11 record is a cluster (variables HGUNAM, Mltarr, Zonarr, and IZ).

HGUNAM— is the hydrogeologic unit to which the parameter applies. When PARTYP = SYTP, HGUNAM must be set to SYTP.

Mltarr— is the name of the multiplier array to be used to define array values that are associated with a parameter. The name "NONE" means that there is no multiplier array, and the array values will be set equal to Parval.

Zonarr— is the name of the zone array to be used to define array elements that are associated with a parameter. The name "ALL" means that there is no zone array and that all elements in the hydrogeologic unit are part of the parameter.

IZ— is up to 10 zone numbers (separated by spaces) that define the array elements that are associated with a parameter. The first zero or non-numeric value terminates the list. These values are not used if Zonarr is specified as "ALL".

PRINTCODE— determines the format for printing the values of the hydraulic-property arrays for the hydrogeologic unit as defined by parameters. The print codes are the same as those used in an array control record (Harbaugh and others, 2000, p. 87).

PRINTFLAGS— determines the hydraulic-property arrays to be printed and must be set to "ALL" or any of the following: "HK", "HANI", "VK", "SS", or "SY". Arrays will be printed only for those properties that are listed. When VK is specified, the property printed depends on the setting of HGUVANI.

Model-Layer Variable-Direction Horizontal Anisotropy (LVDA) Capability

Input for the Model-Layer Variable-Direction Horizontal Anisotropy (LVDA) Capability within the HUF Package is read from the file that has type LVDA in the Name File and must be used with the HUF Package. Free format is used for reading all values. This file defines the horizontal anisotropy direction for each cell of the finite-difference grid. This additional capability of the HUF Package is described in Anderman and others (2002). Users of GSFLOW are encouraged to review this document when using the HUF Package with the LVDA Capability. Although the HUF input file is not changed when using the LVDA capability, the definition of two parameter types (HK and HANI) does change. The new definitions of the two parameters are described in "Explanation of Variables Read by the LVDA Capability."

FOR EACH SIMULATION

0. [#Text]
    Item 0 is optional—# must be in column 1. Item 0 can be repeated multiple times.
1. NPLVDA
2. PARNAM   PARTYP   Parval   NCLU
3. Layer   Mltarr   Zonarr   IZWT
    Each Item 3 record is called a parameter cluster. Repeat Item 3 NCLU times. Repeat Items 2 and 3 for each parameter to be defined (that is, NPLVDA times).

Explanation of Variables Read by the LVDA Capability

Text—is a character variable (199 characters) that starts in column 2. Any characters can be included in Text. The # character must be in column 1. Lines beginning with # are restricted to the first lines of the file. Text is written to the Listing File.

NPLVDA— is the number of LVDA parameters.

PARNAM— is the name of a parameter to be defined. This name can consist of up to 10 characters and is not case sensitive.

PARTYP— is the type of parameter to be defined. For the LVDA Capability, the only allowed parameter type is:

LVDA— defines the angle between the grid axis and the principal direction of horizontal hydraulic conductivity. Angle is positive in a clockwise direction with the positive x direction being zero. The angle can vary between –90 and 90 degrees. Hydraulic conductivity along the principal axis (Kmax of figure 2 in Anderman and others, 2002) is defined in the HUF Package input file by an HK parameter; the hydraulic conductivity along the minor axis (Kmin of figure 2 in Anderman and others, 2002) axis is defined by a HANI parameter in conjunction with the HK parameter. Using the LVDA Capability, HANI values need to be ≤1.

Parval— is the initial value of the parameter; however, this value can be replaced by a value specified in the Sensitivity Process input file.

NCLU— is the number of clusters required to define the parameter. Each Item-3 record is a cluster (variables Layer, Mltarr, Zonarr, and IZ).

Layer— is the layer to which the direction applies.

Mltarr— is the name of the multiplier array to be used to define array values that are associated with a parameter. The name "NONE" means that there is no multiplier array, and the array values will be set equal to Parval.

Zonarr— is the name of the zone array to be used to define array elements that are associated with a parameter. The name "ALL" means that there is no zone array and that all elements in the hydrogeologic unit are part of the parameter.

IZWT— is up to 10 zone numbers (separated by spaces) that define the array elements that are associated with a parameter. The first zero or non-numeric value terminates the list. These values are not used if Zonarr is specified as "ALL".

Horizontal Flow Barrier Package

Input for the Horizontal Flow Barrier (HFB) Package is read from the file that has file type HFB6 in the Name File. All variables are read in free format.

FOR EACH SIMULATION

0. [#Text]
     Item 0 is optional—# must be in column 1. Item 0 can be repeated multiple times.
1. NPHFB   MXFB   NHFBNP

2. and 3. These items are not included in these instructions because they relate to parameters.

4. Layer   IROW1   ICOL1   IROW2   ICOL2   Hydchr
     NHFBNP repetitions of item 2 are read.
5. NACTHFB

Text—is a character variable (199 characters) that starts in column 2. Any characters can be included in Text. The # character must be in column 1. Lines beginning with # are restricted to the first lines of the file. Text is written to the Listing File.

NPHFB— {this value should be set to zero when parameters are not used.}

MXFB— {this value should be set to zero when parameters are not used.}

NHFBNP—is the number of HFB barriers.

Layer—is the number of the model layer in which the horizontal flow barrier is located.

IROW1—is the row number of the cell on one side of the horizontal flow barrier.

ICOL1—is the column number of the cell on one side of the horizontal flow barrier.

IROW2—is the row number of the cell on the other side of the horizontal flow barrier.

ICOL2—is the column number of the cell on the other side of the horizontal flow barrier.

Hydchr—is the hydraulic characteristic of the horizontal flow barrier. The hydraulic characteristic is the barrier hydraulic conductivity divided by the width of the horizontal flow barrier.

NACTHFB—{this value must be zero because parameters are not supported in GSFLOW.}

Well Package

Input to the Well (WEL) Package is read from the file that has type `WEL` in the Name File. Optional variables are shown in brackets. All variables are free format if the option `FREE` is specified in the Basic Package input file; otherwise, the non-optional variables have 10-character fields and the optional variables are free format.

FOR EACH SIMULATION

0. `[#Text]`
    Item 0 is optional—# must be in column 1. Item 0 can be repeated multiple times.
1. `MXACTW    IWELCB    [Option]`

    FOR EACH STRESS PERIOD

2. `ITMP    NP`
3. `Layer    Row    Column    Q    [xyz]`
    `ITMP` repetitions of item 3 are read by subroutine ULSTRD if `ITMP` > 0. (`SFAC` of the ULSTRD utility subroutine applies to `Q`.) Item 3 is not read if `ITMP` is negative or zero.

Explanation of Variables Read by the WEL Package:

`Text`—is a character variable (199 characters) that starts in column 2. Any characters can be included in `Text`. The # character must be in column 1. Lines beginning with # are restricted to the first lines of the file. `Text` is written to the Listing File.

`MXACTW`—is the maximum number of wells in use during any stress period.

`IWELCB`—is a flag and a unit number.

   If `IWELCB` > 0, cell-by-cell flow terms will be written to this unit number when `SAVE BUDGET` or a non-zero value for `ICBCFL` is specified in Output Control.

   If `IWELCB` = 0, cell-by-cell flow terms will not be written.

   If `IWELCB` <0, well recharge for each well will be written to the listing file when `SAVE BUDGET` or a non-zero value for `ICBCFL` is specified in Output Control.

`Option`—is an optional list of character values.

   `AUXILIARY abc` or `AUX abc`—defines an auxiliary variable, named `abc`, which will be read for each well as part of item 3. Up to five variables can be specified, each of which must be preceded by `AUXILIARY` or `AUX`. These variables will not be used by the Ground-Water Flow Process, but they will be available for use by other processes. The auxiliary variable values will be read after the `Q` variable.

   `CBCALLOCATE` or `CBC`—indicates that memory should be allocated to store cell-by-cell flow for each well in order to make these flows available for use in other packages.

`ITMP`—is a flag and a counter.

   If `ITMP` < 0, well data from the last stress period will be reused.

   If `ITMP` ≥ 0, `ITMP` will be the number of wells read for the current stress period.

`NP`—{ this value must be zero because parameters are not supported in GSFLOW.}

`Layer`—is the layer number of the finite-difference cell that contains the well.

`Row`—is the row number of the finite-difference cell that contains the well.

`Column`—is the column number of the finite-difference cell that contains the well.

`Q`—is the volumetric recharge rate. A positive value indicates recharge and a negative value indicates discharge (pumping).

`[xyz]`—represents the values of the auxiliary variables for a well that have been defined in item 1. The values of auxiliary variables must be present in each repetition of item 3 if they are defined in item 1. The values must be specified in the order used to define the variables in item 1.

Multi-Node Wells Package

Input for the Multi-Node, Drawdown-Limited Well (MNW) Package is initiated by specifying MNW1 in the Name File. Data are read from MNW Package input files as 256-character-wide, alphanumeric lines to facilitate the addition of comments within the model input files and the use of keys to identify input variables. All integer, real, and character variables are read from the alphanumeric lines. The lines are initially read by the subroutine NCREAD. Lines that begin with a '#' sign in the first column are treated as comments, are not passed to any other routines, and are discarded. Once NCREAD has acquired a valid data line, the line is checked for a '!!' sign that designates the beginning of any in-line comments. If a '!!' sign is detected, the '!!' sign and all text to the right of the '!!' sign are removed from the line before passing it to any other routines.

Alphanumeric strings are used in the MNW Package to identify variables (keys) and make logical decisions (flags). Specification of these keys and flags is case insensitive because all letters are capitalized before performing any logical tests. Keys precede the variable to be read, which is acquired by identifying the key and reading the first value that follows the key. Logical decisions are based on the presence (true) or absence (false) of a flag. In this report, **bold**, uppercase letters are used to denote the part of the key that is tested. Key:data pairs that are not delimited by parentheses are mandatory and must be included, and Key:data pairs that are delimited within parentheses are optional because default values are used if they are not specified by the user.

The MNW Package is documented in Halford and Hanson (2002) and users of GSFLOW are encouraged to review that document when using the MNW Package.

FOR EACH SIMULATION

1. MXMNW   IWL2CB   IWELPT   **REF**erence   SP: kspref
    Item 1 is required.
2. LOSSTYPE   (PLossMNW)
    Item 2 is required.
3a. **KEY**:DATA   **FILE**:filename   **WEL1**:iunw1
3b. **KEY**:DATA   **FILE**:filename   **BYNODE**:iunby   **ALLTIME**
3c. **KEY**:DATA   **FILE**:filename   **QSUM**:iunqs   **ALLTIME**
    Items 3a, 3b, and 3c are optional.

FOR EACH STRESS PERIOD

4. ITMP   **ADD**
5. Layer   Row   Column   Qdes   (MN or MULTI)   QWval   Rw   Skin   Hlim   Href   (**DD**)
    Iqwgrp   **Cp**: C   (**QCUT** or **Q-%CUT**: Qfrcmn, Qfrcmx)   **DEFAULT**   **SITE**: MNWsite
    The first four values in item 5 for the variables Layer, Row, Column, and Qdes are read initially as a free format. If this fails, the four values are read as fixed format entries from the first 40 columns using the format I10, I10, I10, and F10.0. In all instances, these values must be specified. The following eight values for the remaining variables are optional; space-delimited or comma-delimited entries but must be entered in the sequence specified for item 5. The alphanumeric flags MN and DD can appear anywhere between columns 41 and 256, inclusive. Input item 5 normally consists of one line for each well cell defined or modified. If ITMP is 0 or less, item 5 is not read and should not be specified.

Explanation of Variables Read by the MNW Package:

MXMNW—is an integer variable equal to the maximum number of well cells to be defined.

IWL2CB—is a flag and an integer unit number.

    If IWL2CB > 0, it is the unit number on which cell-by-cell flow terms will be recorded whenever ICBCFL is set.

    If IWL2CB = 0, cell-by-cell flow terms will not be printed or recorded.

    If IWL2CB < 0, well recharge, water levels in the well and cell, drawdown in the well, and the flow-rate-weighted water-quality value of the Iqwgrp will be printed whenever ICBCFL is set.

IWELPT—is a flag. If IWELPT is not equal to 0, no well information will be printed.

REFerence SP: kspref—is the set of water levels in the HNEW matrix at the beginning of the stress period kspref that will be used as default reference values for calculating drawdown. kspref defaults to 1 if it is not specified by the user.

LOSSTYPE—is a flag to determine the user-specified model for well loss.

> If LOSSTYPE is set to SKIN, head loss is defined with a coefficient representing skin effects (Halford and Hanson, 2002, p. 8). Model is linear.

> If LOSSTYPE is set to NONLINEAR, head loss is defined with a coefficient representing skin effects and a non-linear well loss coefficient (Halford and Hanson, 2002, p. 8).

PLossMNW—is variable P in equation 2 of Halford and Hanson (2002), and is the power of the nonlinear discharge component of well loss, which usually varies between 1.5 and 3.5 (Rorabaugh, 1953).

FILE: filename—is the name of an auxiliary output file.

WEL1: iunw1—is a unit number. The unit number will be associated with the auxiliary output file defined by the WEL1 filename. Output is a WEL1 input file with the flow rates specified at the end of each stress period.

BYNODE: iunby—is a unit number. The unit number will be associated with the auxiliary output file defined by the BYNODE filename. Output is flow rate at each well node.

QSUM: iunqs—is a unit number. The unit number will be associated with the auxiliary output file defined by the QSUM filename. Output is total flow rate from each multi-node well.

ALLTIME—is a flag that indicates flow rates should be written to BYNODE or QSUM files at every time step regardless of the settings in the output control (OC) file.

ITMP—is an integer flag and a counter.

> If ITMP < 0, wells from the previous stress period will be reused and input from item 4 will not be read.

> If ITMP = 0, no wells will be simulated and input from item 4 will not be read.

> If ITMP > 0, is the number of items of drawdown-limited well data that will be read for the current stress period. If the key **ADD** is <u>not</u> detected on item 4, the maximum number of drawdown-limited wells for the current stress period will be ITMP. If the key **ADD** is detected on item 4, ITMP wells will be added to the existing list of drawdown-limited wells.

ADD—is a flag that indicates whether or not the well cells read for the current stress period will augment or replace the well cells that were previously defined.

Layer—is the layer number of the finite-difference cell that contains the well.

Row—is the row number of the finite-difference cell that contains the well.

Column—is the column number of the finite-difference cell that contains the well.

Qdes—is the desired volumetric pumping or recharge rate (variable type is real). A positive value indicates recharge and a negative value indicates discharge. The actual volumetric recharge rate will range from 0 to Qdes and is not allowed to switch directions between discharge and recharge conditions during any stress period.

(MN)—is a flag that indicates this entry is part of a multi-node well. The flag **MN** is not included on the first entry of a multi-node well and is exclusive of the flag **MULTI**.

(**MULTI**)—is a flag that indicates this entry is the end of a multi-node well and all intervening nodes between this entry and the previous **MULTI** flag are part of a multi-node well. Intervening nodes will be assigned the same cell-to-well conductance that was specified in this entry. The flag **MULTI** is not included on the first entry of a multi-node well and is exclusive of the flag **MN**.

Qwval—is a real variable equal to the water-quality value that is to be flow-rate averaged amongst wells in the same Iqwgrp. Negative water-quality values and positive flow terms are not averaged. Water-quality values can be re-specified for each stress period.

Rw—is a real-valued flag and a variable used to define the cell-to-well conductance.

If Rw > 0, the variable represents the radius of the well and the cell-to-well conductance is calculated with equation 5 in Halford and Hanson (2002).

If Rw = 0, the head in the cell is assumed to be equivalent to the head in the well and the cell-to-well conductance is set to 1,000 times the transmissivity of the cell. The cell is not allowed to be part of a multi-node well.

If Rw < 0, the absolute value of the variable is the cell-to-well conductance.

Skin—is a real variable that defines the friction losses to the well owing to the screen and to formation damage. The variable is either a skin or the coefficient **B** depending on the LOSSTYPE, and is used in equation 5 in Halford and Hanson (2002) when Rw > 0.

Hlim—is a real variable equal to the limiting water level, which is a minimum for discharging wells and a maximum for recharging wells. If the flag **DD** is set, the value of Hlim read is a drawdown from the reference elevation. For Qdes < 0, Hlim = Href – Hlim and for Qdes > 0, Hlim = Href + Hlim.

Href—is a real variable equal to the reference elevation. If the value of Href read is greater than the maximum water level from the HNEW matrix at the beginning of the stress period kspref, Href is set to the simulated water level at the location of the drawdown-limited well.

DD—is a flag that indicates the value of Hlim read is a drawdown or build-up from the reference elevation.

Iqwgrp—is an integer variable equal to the water-quality identifier. Flow-rate averaged water-quality values are reported for each group of wells with the same Iqwgrp and Qwval entries that are not negative.

Cp: C —is a real variable that is a coefficient for nonlinear head losses (see equation 2 in Halford and Hanson, 2002). The variable is used only when the LOSSTYPE is NONLINEAR. Default value is 0 if not specified.

QCUT—is a flag that indicates pumping limits will be specified as a rate (units of length cubed per time).

Q-%CUT—is a flag that indicates pumping limits will be specified as a percentage of the specified rate.

Qfrcmn—is a real variable equal to the minimum pumping rate that a well must exceed to remain active.

Qfrcmx—is a real variable equal to the minimum potential pumping rate that must be exceeded to reactivate a well.

DEFAULT—is a flag that sets this entry of Qfrcmn and Qfrcmx as the new default values.

SITE: MNWsite—is an optional label for identifying wells. An individual file of time, discharge, water level in well, concentration, net-inflow, net-outflow, and node-by-node flows will be written for each well with a unique MNW site label. Individual well files are tab delimited. Only one label should be applied to a multi-node well.

General-Head Boundary Package

Input to the General-Head Boundary (GHB) Package is read from the file that has file type GHB in the Name File. Optional variables are shown in brackets. All variables are free format if the option FREE is specified in the Basic Package input file; otherwise, the non-optional variables have 10-character fields and the optional variables are free format.

FOR EACH SIMULATION

0. [#Text]
    Item 0 is optional—# must be in column 1. Item 0 can be repeated multiple times.
1. MXACTB   IGHBCB   [Option]

FOR EACH STRESS PERIOD

2. ITMP   NP
3. Layer   Row   Column   Bhead   Cond   [xyz]
    ITMP repetitions of item 3 are read by subroutine ULSTRD if ITMP > 0. (SFAC of the ULSTRD utility subroutine applies to Cond.) Item 3 is not read if ITMP is negative or 0.

Explanation of Variables Read by the GHB Package:

Text—is a character variable (199 characters) that starts in column 2. Any characters can be included in Text. The # character must be in column 1. Lines beginning with # are restricted to the first lines of the file. Text is written to the Listing File.

MXACTB—is the maximum number of general-head boundary cells in use during any stress period.

IGHBCB—is a flag and a unit number.

   If IGHBCB > 0, cell-by-cell flow terms will be written to the unit number when SAVE BUDGET or a non-zero value for ICBCFL is specified in Output Control.

   If IGHBCB = 0, cell-by-cell flow terms will not be written.

   If IGHBCB < 0, boundary leakage for each GHB cell will be written to the listing file when SAVE BUDGET or a non-zero value for ICBCFL is specified in Output Control.

Option—is an optional list of character values.

   AUXILIARY abc or AUX abc defines an auxiliary variable, named abc, which will be read for each general-head boundary as part of item 3. Up to five variables can be specified, each of which must be preceded by AUXILIARY or AUX. These variables will not be used by the Ground-Water Flow Process, but they will be available for use by other processes. The auxiliary variable values will be read after the Cond variable.

   CBCALLOCATE or CBC indicates that memory should be allocated to store cell-by-cell flow for each general-head boundary in order to make these flows available for use in other packages.

ITMP—is a flag and a counter.

   If ITMP < 0, general-head boundary data from the preceding stress period will be reused.

   If ITMP ≥ 0, ITMP is the number of general-head boundaries read for the current stress period.

NP—{this value must be zero because parameters are not supported in GSFLOW.}

Layer—is the layer number of the cell affected by the head-dependent boundary.

Row—is the row number of the cell affected by the head-dependent boundary.

Column—is the column number of the cell affected by the head-dependent boundary.

Bhead—is the boundary head.

Cond—is the hydraulic conductance of the interface between the aquifer cell and the boundary.

[xyz]—represents the values of the auxiliary variables for a boundary that have been defined in item 1. The values of auxiliary variables must be present in each repetition of item 3 if they are defined in item 1. The values must be specified in the order used to define the variables in item 1.

Flow and Head Boundary Package

Input for the Flow and Head Boundary (FHB) Package is initiated by specifying FHB in the Name File. All input is free format, which requires each of the numbered data groups to start on a new input item. More than one item can be used for any data group and numbers within data groups must be separated by at least one space or a comma. Integer data types cannot include a decimal point. Blank spaces are not treated as zeros.

The FHB Package is documented in Leake and Lilly (1997) and users of GSFLOW are encouraged to review that document when using the FHB Package

FOR EACH SIMULATION

1. NBDTIM   NFLW   NHED   IFHBSS   IFHBCB   NFHBX1   NFHBX2
    Omit item 2 if NFHBX1 = 0. Input item 2 consists of one line for each of NFHBX1 auxiliary variables.
2. VarName   Weight
    Omit item 3 if NFHBX2 = 0. Input item 3 consists of one line for each of NFHBX2 auxiliary variables.
3. VarName   Weight
    Items 4a and 4b are required for all simulations. Include NBDTIM times in item 4b.
4a. IFHBUN   CNSTM   IFHBPT
4b. BDTIM(NBDTIM)
    Omit items 5a and 5b if NFLW = 0. Input item 5b consists of one set of numbers for each of NFLW cells. Each set of
       numbers includes layer, row, and column indices, an integer auxiliary variable, and NBDTIM values of specified flow.
5a. IFHBUN   CNSTM   IFHBPT
5b. Layer   Row   Column   IAUX   FLWRAT(NBDTIM)
    Omit items 6a and 6b if NFHBX1 = 0 or if NFLW = 0. Include one set of items 6a and 6b for each of NFHBX1 auxiliary
       variables. Input item 6b consists of one set of numbers for each of NFLW cells. Each set includes NBDTIM values of
       the variable.
6a. IFHBUN   CNSTM   IFHBPT
6b. AuxVar(NBDTIM)
    Omit items 7a and 7b if NHED = 0. Input item 7b consists of one set of numbers for each of NFLW cells. Each set of
       numbers includes layer, row, and column indices, an integer auxiliary variable, and NBDTIM values of specified head.
7a. IFHBUN   CNSTM   IFHBPT
7b. Layer   Row   Column   IAUX   SBHED(NBDTIM)
    Omit items 8a and 8b if NFHBX2 = 0 or if NHED = 0. Include one set of items 8a and 8b for each of NFHBX2 auxiliary
       variables. Input item 8b consists of one set of numbers for each of NHED cells. Each set includes NBDTIM values of
       the variable.
8a. IFHBUN   CNSTM   IFHBPT
8b. AuxVar(NBDTIM)

Explanation of Variables Read by the FHB Package:

NBDTIM—is an integer variable equal to the number of times at which flow and head will be specified for all selected cells.

   If NBDTIM=1, specified flow and head values will remain constant for the entire simulation.

   If NBDTIM>1, specified flow and head values will be computed for each time step using linear interpolation.

NFLW—is an integer variable equal to the number of cells at which flows will be specified.

NHED—is an integer variable equal to the number of cells at which head will be specified.

IFHBSS—is an integer variable equal to the FHB steady-state option flag. If the simulation is transient, the flag is read but not used. For steady-state simulations, the flag controls how specified-flow, specified-head, and auxiliary-variable values will be computed for each steady-state solution.

If IFHBSS = 0, values of flow, head, and auxiliary variables will be taken at the starting time of the simulation. This results in use of the first value in arrays FLWRAT, SBHED, and AuxVar for each respective boundary cell.

If IFHBSS ≠ 0, values of flow, head and auxiliary variables will be interpolated in the same way that values are computed for transient simulations.

IFHBCB—is a flag and an integer unit number.

If IFHBCB > 0, it is the unit number on which cell-by-cell flow terms will be recorded whenever ICBCFL is set (see Harbaugh, 2005, p. 8-15).

If IFHBCB ≤ 0, cell-by-cell flow terms will not be recorded.

NFHBX1—is an integer variable equal to the number of auxiliary variables whose values will be computed for each time step for each specified-flow cell.

NFHBX2—is an integer variable equal to the number of auxiliary variables whose values will be computed for each time step for each specified-head cell.

VarName—is the name of an auxiliary variable (a character variable). Name can include up to 16 characters with no embedded blank characters.

Weight—is a real variable equal to the time-weighting factor for an auxiliary variable specifying the fraction of each time step at which the value of the variable will be interpolated. Value must be in the range from 0.0 to 1.0.

IFHBUN—is the unit number on which data lists will be read. The same or different unit numbers can be used to read lists in items 4b, 5b, 6b, 7b, and 8b.

CNSTM—is a constant multiplier (a real variable) for data list BDTIM (item 4b), FLWRAT (part of item 5b), SBHED (part of item 7b), and auxiliary variables in items 6b and 8b.

IFHBPT—is an integer flag for printing values of data lists in items 4b, 5b, 6b, 7b, and 8b.

If IFHBPT >0, data list read at the beginning of the simulation will be printed

If IFHBCB ≤0, data list read at the beginning of the simulation will not be printed.

BDTIM—is simulation time (a real variable) at which values of specified flow and (or) values of specified head will be read. NBDTIM values are required.

Layer—is the layer index of specified-flow cell (item 5b) or specified-head cell (item 7b).

Row—is the row index of specified-flow cell (item 5b) or specified-head cell (item 7b).

Column—is the column index of specified-flow cell (item 5b) or specified-head cell (item 7b).

IAUX—is an integer auxiliary variable associated with each specified-flow and specified-head boundary cell.

FLWRAT—is a volumetric rate of flow (a real variable) at specified-flow cells. A list of NBDTIM values must be specified for each of NFLW specified-flow cells.

AuxVar—is the value of real auxiliary variable at specified-flow and specified-head cells. A list of NBDTIM values must be specified for each of NFLW specified-flow cells and for each of NHED specified-head cells.

SBHED—is an array of real values containing NBDTIM values of the head for each specified-head cell.

Streamflow-Routing Package

Input to the Streamflow-Routing (SFR7) Package is read from the file that has file type SFR in the Name File. The user can optionally specify that stream gages and monitoring stations along a stream channel are to be written using the Gage Package by including GAGE in the Name File (see "Gage Package" section). The modifications in SFR2 do not require any changes to the data input for SFR1.

The modification of SFR2 to simulate unsaturated flow relies on the specific yield values as specified in the Layer Property Flow (LPF) Package, the Hydrogeologic-Unit Flow (HUF) Package, or the Block-Centered Flow (BCF) Package. When the option to use vertical hydraulic conductivity in the LPF Package is specified, the layer(s) that contain cells where unsaturated flow will be simulated must be specified as convertible. That is, the variable LAYTYP specified in LPF must not be equal to zero, otherwise the model will print an error and stop execution.

Additional variables that must be specified to define hydraulic properties of the unsaturated zone are all included within the SFR2 input file. All values are entered in free format. Data input for SFR1 works without modification if unsaturated flow is not simulated.

The Streamflow Routing Package is documented in Prudic and others (2004) and Niswonger and Prudic (2005) and users of GSFLOW are encouraged to review those documents when using the SFR Package. However, unlike previous versions of the Streamflow Routing Package (Prudic and others, 2004; Niswonger and Prudic, 2005) input data for SFR2 cannot be specified using parameters in GSFLOW.

FOR EACH SIMULATION

0. [#Text]
    Item 0 is optional—# must be in column 1. Item 0 can be repeated multiple times.
1. NSTRM   NSS   NSFRPAR   NPARSEG   CONST   DLEAK   ISTCB1   ISTCB2   [ISFROPT]
    [NSTRAIL]   [ISUZN]   [NSFRSETS]   [IRTFLG]
    The first two variables (NSTRM and NSS) are used for dimensioning arrays, and must be equal to the actual number of stream reaches defined in Item 2 and the number of segments that define the complete stream network, respectively. If NSTRM is negative, then unsaturated flow is simulated beneath the stream segments; the absolute value of NSTRM is used to define the number of stream reaches and for dimensioning of arrays.

    If BCF is used and unsaturated flow is active, then ISFROPT must equal 3 or 5.

    SFR2 differs from the Stream (STR1) Package (Prudic, 1989) because the new package solves for stream depth at the midpoint of each reach instead of at the beginning of the reach. To solve for depth at the midpoint of each reach, like SFR1, SFR2 uses Newton's iterative method and consequently, a tolerance (DLEAK) is used for stopping the iterative process.

ONE LINE FOR EACH STREAM REACH

2. KRCH   IRCH   JRCH   ISEG   IREACH   RCHLEN   [STRTOP]   [SLOPE]   [STRTHICK]
    [STRHC1]   [THTS]   [THTI]   [EPS]   [UHC]
    Reach information is read in sequential order from upstream to downstream, first by segments and then by reaches. Segments should be numbered in downstream order, although this is not necessary. However, if the segments are not numbered in downstream order, the inflows and outflows from each segment will still be computed, but the computed inflows into a segment from upstream tributary streams having a higher segment number will be from the previous iteration. Reaches must be listed and read in sequentially because the order determines the connections of inflows and outflows within a stream segment.

    The stream network is assumed to remain fixed geometrically over the duration of a simulation. However, the active part of the stream network can be made to vary over time by making selected stream segments inactive for selected stress periods. This would be implemented by setting the streambed hydraulic conductivity, segment inflow, overland runoff, and direct precipitation to zero for the inactive segments in item 4 for the specific stress periods when they are known to be inactive or dry.

    If the finite-difference cell corresponding to a stream reach is inactive, the program will search for the uppermost active cell in the vertical column to apply the leakage. If there are no active cells or if the cell is a constant head, no interaction is allowed and flow in the reach is passed to the next reach.

    When STRTOP, SLOPE, STRTHICK, and STRHC1 are specified for each reach, then HCOND1, THICKM1, ELEVUP, HCOND2, THICKM2, and ELEVDN are not read using items 4b or 4c.

The residual water content for each cell is not specified by the user because it is calculated based on the specified saturated water content minus the specific yield of the active finite-difference cell corresponding to the stream reach. The calculation is performed internally to assure continuity between unsaturated and saturated zone storage.

Although unsaturated flow variables will not be used for reaches that are designated as ICALC = 0, 3, and 4 within the segment information, values must be included for all reaches when ISFROPT = 2 or 3. Dummy values may be used for reaches that are designated as ICALC = 0, 3, and 4.

FOR EACH STRESS PERIOD

3.   ITMP   IRDFLG   IPTFLG

If ITEMP > 0:

4a.   NSEG   ICALC   OUTSEG   IUPSEG   [IPRIOR]   [NSTRPTS]   FLOW RUNOFF   ETSW
      PPTSW   [ROUGHCH]   [ROUBHBK]   [CDPTH]   [FDPTH]   [AWDTH]   [BWDTH]
4b.   [HCOND1]   [THICKM1]   [ELEVUP]   [WIDTH1]   [DEPTH1]   [THTS1]   [THTI1]
      [EPS1]   [UHC1]

These variables are read for each stress period when NSTRM is positive. THICKM1 and ELEVUP are read only for the first stress period when ICALC is 1 or 2 and ISFROPT is 4 or 5. WIDTH1 is read for all stress periods when ICALC is 0 or 1 and ISFROPT is 1, but is only read for the first stress period when ISFROPT is >1. WIDTH1 and DEPTH1 are read for all stress periods when ICALC is 0 and are not dependent on the value of ISFROPT.

4c.   [HCOND2]   [THICKM2]   [ELEVDN]   [WIDTH2]   [DEPTH2]   [THTS2]   [THTI2]   [EPS2]
      [UHC2]

The same options apply to these variables as explained for the upstream variables in item 4b.

If **ICALC** = 2:

4d.   XCPT1   XCPT2   ...   XCPT8
      ZCPT1   ZCPT2   ...   ZCPT8

If ICALC = 4:

4e.   FLOWTAB(1)   FLOWTAB(2)   ...   FLOWTAB(NSTRPTS)
      DPTHTAB(1)   DPTHTAB(2)   ...   DPTHTAB(NSTRPTS)
      WDTHTAB(1)   WDTHTAB(2)   ...   WDTHTAB(NSTRPTS)

Item 4 must be repeated ITMP times. The data need not be defined in sequential order by stream segment number. If ITMP ≤ 0, then stream segment data for the previous stress period will be reused.

Item 4a will contain 8 to 13 variables; depending on the values of ICALC and IUPSEG. ICALC determines how stream depth is to be calculated; when ICALC is 1 or 2, depth is calculated using Manning's equation, which, in turn, requires a channel roughness coefficient (ICALC = 1) or a channel and bank roughness coefficient (ICALC = 2).

Items 4b and 4c may include no input when all are defined by stream reaches in item 2 or they may include as many as 9 variables depending on the value of ICALC in item 4a and(or) ISFROPT in item 1.

A stream segment that receives inflow from upstream segments is allowed to have as many as 10 upstream segments feeding it, as defined by the respective values of OUTSEG in item 4a.

Stream properties and stresses are assumed constant and uniform within a single stream segment. Additionally, hydraulic conductivity, streambed thickness, elevation of top of streambed, stream width, and stream depth may vary smoothly and linearly within a single stream segment. For these variables, data values at the upstream end of the segment are described in item 4b and data values at the downstream end of the segment are described in item 4c. Values of these variables for individual reaches of a segment are estimated using linear interpolation. To make any variable the same throughout the segment, simply specify equal values in items 4b and 4c. The two elevations in items 4b and 4c are used in conjunction with the total length of the stream segment (calculated from RCHLEN given for each reach in item 2) to compute the slope of the stream and the elevations for any intermediate reaches. The streambed thickness is subtracted from the top of streambed elevations to calculate the elevations of the bottom of the streambed (used in calculations of leakage).

If item 4d is included (for ICALC = 2), it is assumed that the cross-sectional geometry defined by these data is the same over the entire length of the segment. Similarly, if item 4e is included (for ICALC = 4), it is assumed the tabulated relation between streamflow and stream depth and width is the same over the entire length of the segment.

If the Lake (LAK3) Package (Merritt and Konikow, 2000) also is implemented, then flow out of the lake into a stream segment is dependent on the option used to compute stream depth (ICALC = 1, 2, 3, or 4). Constant discharge from a lake can be simulated no matter what value of ICALC is assigned to the stream segment emanating from the lake by assigning a positive value to FLOW in item 4a.

If a diversionary flow is large enough to warrant representation in the model, but is discharged into a pipeline, lined canal, or other structure or system that does not interact with the aquifer and the flow might exceed the available streamflow, then there is an alternative means to represent it. Instead of specifying a negative value of FLOW, we suggest representing the withdrawal by a single-reach diversionary stream segment, which would be located in the same finite-difference cell as the reach from the upstream segment (IUPSEG) from which the diversion is made; specifying the segment's streambed hydraulic conductivity equal to 0 will preclude interaction with the aquifer and setting OUTSEG = 0 will remove the flow from the system. The diversion will then be subject to the constraints associated with the value of IPRIOR.

Explanation of Variables Read by the SFR Package

Text—is a character variable (up to 199 characters) that starts in column 2. Any characters can be included in Text. The # character must be in column 1. Lines beginning with # are restricted to the first lines of the file. Text is written to the Listing File.

ITEM 1 VARIABLES:

NSTRM—an integer value equal to the number of stream reaches (finite-difference cells) that are active during the simulation. The value of NSTRM also represents the number of lines of data to be included in item 2. When NSTRM is specified as a negative integer, it also is used as a flag to read the additional variables ISFROPT, NSTRAIL, ISUZN, NSFRSETS, and IRTFLG, or changing data input simulating unsaturated flow beneath streams, and for simulating transient routing of streamflow. A negative value is automatically reset to a positive integer.

NSS—an integer value equal to the number of stream segments (consisting of one or more reaches) that are used to define the complete stream network.

NSFRPAR—{this variable must be zero because parameters are not supported in GSFLOW.}

NPARSEG— {this variable must be zero because parameters are not supported in GSFLOW.}

CONST—a real value (or conversion factor) used in calculating stream depth for a stream reach. If stream depth is not calculated using Manning's equation for any stream segment (that is, ICALC does not equal 1 or 2), then a value of zero can be entered. If Manning's equation is used, a constant of 1.486 is used for flow units of cubic feet per second, and a constant of 1.0 is used for units of cubic meters per second. The constant must be multiplied by 86,400 when using time units of days in the simulation. An explanation of time units used in MODFLOW is given by Harbaugh and others (2000, p. 10.)

DLEAK—a real value equal to the tolerance level of stream depth used in computing leakage between each stream reach and active finite-difference cell. Value is in units of length. Usually a value of 0.0001 is sufficient when units of feet or meters are used in model.

ISTCB1—an integer value used as a flag for writing stream-aquifer leakage values. If ISTCB1 > 0, it is the unit number to which unformatted leakage between each stream reach and corresponding finite-difference cell will be saved to a file whenever the cell-by-cell budget has been specified in Output Control (see Harbaugh and others, 2000, p. 52-55). If ISTCB1 = 0, leakage values will not be printed or saved. If ISTCB1 < 0, all information on inflows and outflows from each reach; on stream depth, width, and streambed conductance; and on head difference and gradient across the streambed will be printed in the main listing file whenever a cell-by-cell budget has been specified in Output Control.

ISTCB2—an integer value used as a flag for writing to a separate formatted file all information on inflows and outflows from each reach; on stream depth, width, and streambed conductance; and on head difference and gradient across the streambed. If ISTCB2 > 0, then ISTCB2 also represents the unit number to which all information for each stream reach will be saved to a separate file when a cell-by-cell budget has been specified in Output Control. If ISTCB2 < 0, it is the unit number to which unformatted streamflow out of each reach will be saved to a file whenever the cell-by-cell budget has been specified in Output Control.

If NSTRM < 0:

ISFROPT—is an integer value that defines the input structure:

ISFROPT = 1, no vertical unsaturated flow beneath streams. Streambed elevation, stream slope, streambed thickness, and streambed hydraulic conductivity are read for each reach only once at the beginning of the simulation.

ISFROPT = 2, streambed and unsaturated-zone properties are read for each reach only once at the beginning of the simulation except saturated vertical hydraulic conductivity for the unsaturated zone is the same as the vertical hydraulic conductivity of the corresponding layer in LPF and is not read in separately.

ISFROPT = 3, same as 2 except saturated vertical hydraulic conductivity for the unsaturated zone is read for each reach.

ISFROPT = 4, streambed and unsaturated-zone properties are read for the beginning and end of each stream segment. Streambed properties can vary each stress period. Saturated vertical hydraulic conductivity for the unsaturated zone is the same as the vertical hydraulic conductivity of the corresponding layer in LPF and is not read in separately.

ISFROPT = 5, same as 4 except saturated vertical hydraulic conductivity for the unsaturated zone is read for each segment at the beginning of the first stress period only.

When ISFROPT > 1, read the following variables:

NSTRAIL—an integer value that is the number of trailing-wave increments used to represent a trailing wave. Trailing waves are used to represent a decrease in the surface infiltration rate. The value can be increased to improve mass balance in the unsaturated zone. Values between 20 and 10 work well and result in unsaturated-zone mass balance errors beneath streams ranging between 0.001 and 0.01 percent. See Smith (1983) for further details.

ISUZN—an integer value that is the maximum number of vertical cells used to define the unsaturated zone beneath a stream reach. If ICALC is 1 for all segments, then ISUZN should be set to 1. ISUZN may affect model run time.

NSFRSETS—an integer value that is the maximum number of different sets of trailing waves used to allocate arrays. Arrays are allocated by multiplying NSTRAIL by NSFRSETS. A value of 30 is sufficient for problems where the stream depth varies often. NSFRSETS does not affect model run time.

If NSTRM < 0:

IRTFLG—an integer value that flags whether transient streamflow routing is active. If IRTFLG > 0 then streamflow will be routed using the kinematic-wave equation.

ITEM 2 VARIABLES:

KRCH—an integer value equal to the layer number of the cell containing the stream reach.

IRCH—an integer value equal to the row number of the cell containing the stream reach.

JRCH—an integer value equal to the column number of the cell containing the stream reach.

ISEG—an integer value equal to the number of the stream segment in which this reach is located. Stream segments contain one or more reaches and are assumed to have uniform or linearly varying characteristics.

IREACH—an integer value equal to the sequential number in a stream segment of this reach (where a reach corresponds to a single cell in the model). Numbering of reaches in a segment begins with 1 for the farthest upstream reach and continues in downstream order to the last reach of the segment.

RCHLEN—a real number equal to the length of channel of the stream reach within this finite-difference cell. The length of a stream reach can exceed the finite-difference cell dimensions because of the meandering nature of many streams. The length is used to calculate the streambed conductance for this reach. Also, the sum of the lengths of all stream reaches within a segment is used to calculate the average slope of the channel for the segment and subsequently other values, such as the elevation of the streambed and stream stage.

STRTOP—a real number equal to the top elevation of the streambed. This variable is read when ISFROPT is 1, 2, or 3.

SLOPE—a real number equal to the stream slope across the reach. This variable is read when ISFROPT is 1, 2, or 3.

STRTHICK—a real number equal to the thickness of the streambed. This variable is read when ISFROPT is 1, 2, or 3.

STRCH1—a real number equal to the hydraulic conductivity of the streambed. This variable is read when ISFROPT is 1, 2, or 3.

THTS—a real number equal to the saturated volumetric water content in the unsaturated zone. This variable is read when ISFROPT is 2 or 3.

THTI—a real number equal to the initial volumetric water content. THTI must be less than or equal to THTS and greater than or equal to THTS minus the specific yield defined in either LPF or BCF. This variable is read when ISFROPT is 2 or 3.

EPS—a real number equal to the Brooks-Corey exponent used in the relation between water content and hydraulic conductivity within the unsaturated zone (Brooks and Corey, 1966). This variable is read when ISFROPT is 2 or 3.

UHC—a real number equal to the vertical saturated hydraulic conductivity of the unsaturated zone. This variable is necessary when using BCF, whereas it is optional when using LPF. This variable is read when ISFROPT is 3.

ITEM 3 VARIABLES:

ITMP—an integer value for reusing or reading stream segment data that can change each stress period. ITMP must be equal to the number of stream segments (NSEG) for the first stress period of a simulation. For subsequent stress periods, If ITMP < NSEG then ITMP segments will be defined for the stress period and the remaining segments will be defined based on data from the previous stress period.

IRDFLG—an integer value for printing input data specified for this stress period. If IRDFLG = 0, input data for this stress period will be printed. If IRDFLG > 0, then input data for this stress period will not be printed.

IPTFLG—an integer value for printing streamflow-routing results during this stress period. If IPTFLG = 0, or whenever the variable ICBCFL is specified, the results for specified time steps during this stress period will be printed. If IPTFLG > 0, then the results during this stress period will not be printed.

ITEM 4a VARIABLES:

NSEG—an integer value of the stream segment for which information is given to identify inflow, outflow, and computation of stream depth.

ICALC—an integer value used to indicate method used to calculate stream depth in this segment:

 ICALC ≤ 0, stream depth in each reach is specified at the beginning of a stress period and remains unchanged unless flow at the midpoint of a reach is zero, then depth is set to zero in that reach. No unsaturated flow is allowed.

 ICALC > 0, stream depth is calculated and updated each iteration of the MODFLOW solver within a time step.

 ICALC = 1, stream depth is calculated using Manning's equation and assuming a wide rectangular channel. Unsaturated flow is simulated when ISFROPT > 1.

 ICALC = 2, stream depth is calculated using Manning's equation and assuming an eight-point channel cross section for each segment (which allows for the computation of a wetted perimeter and for changing hydraulic conductance of the streambed in relation to changes in flow). Unsaturated flow is simulated when ISFROPT > 1.

ICALC = 3, stream depth and width are calculated using a power function relating each to streamflow (Q) using equations 8 and 9 in Prudic and others (2004), where DEPTH(y) = CDPTH $\times Q^{\text{FDPTH}}$ and WIDTH(w) = AWDTH $\times Q^{\text{BWDTH}}$. No unsaturated flow is allowed.

ICALC = 4, stream depth and width are calculated using a table relating streamflow to depth and width. No unsaturated flow is allowed.

OUTSEG—an integer value of the downstream stream segment that receives tributary inflow from the last downstream reach of this segment. If this segment (identified by NSEG) does not feed (or discharge into) another downstream (tributary) segment, then enter a value of 0 for this variable. If the segment ends within the modeled grid and OUTSEG = 0, outflow from the segment is not routed anywhere and is no longer part of the stream network. One may wish to use this if all flow in the stream gets diverted into a lined canal or into a pipe. If the flow out of this segment discharges into a lake, set OUTSEG equal to the negative value of the lake identification number (where the minus sign is used as a flag to tell the model that flow enters a lake rather than a tributary stream segment).

IUPSEG—an integer value of the upstream segment from which water is diverted (or withdrawn) to supply inflow to this stream segment if this segment originates as a diversion from an upstream segment. If the source of a stream segment is discharge from a lake, set IUPSEG equal to the negative value of the lake identification number (where the minus sign is used as a flag to tell the model that streamflow into this segment is derived from lake outflow rather than a stream segment). If this stream segment (identified by NSEG) does not receive inflow as a diversion from an upstream segment, then set IUPSEG = 0.

IPRIOR—an integer value that only is specified if IUPSEG > 0 (do not specify a value in this field if IUPSEG = 0 or IUPSEG < 0). IPRIOR defines the prioritization system for diversion, such as when insufficient water is available to meet all diversion stipulations, and is used in conjunction with the value of FLOW (specified below).

IPRIOR = 0, then if the specified diversion flow (FLOW) is greater than the flow available in the stream segment from which the diversion is made, the diversion is reduced to the amount available, which will leave no flow available for tributary flow into a downstream tributary of segment IUPSEG.

IPRIOR = -1, then if the specified diversion flow (FLOW) is greater than the flow available in the stream segment from which the diversion is made, no water is diverted from the stream. This approach assumes that once flow in the stream is sufficiently low, diversions from the stream cease, and is the "priority" algorithm that originally was programmed into the STR1 Package (Prudic, 1989).

IPRIOR = -2, then the amount of the diversion is computed as a fraction of the available flow in segment IUPSEG; in this case, $0.0 \leq \text{FLOW} \leq 1.0$.

IPRIOR = -3, then a diversion is made only if the streamflow leaving segment IUPSEG exceeds the value of FLOW. If this occurs, then the quantity of water diverted is the excess flow and the quantity that flows from the last reach of segment IUPSEG into its downstream tributary (OUTSEG) is equal to FLOW. This represents a flood-control type of diversion, as described by Danskin and Hanson (2002).

NSTRPTS—an integer value specified only when ICALC = 4. It is used to dimension a table relating streamflow with stream depth and width as specified in item 4e. NSTRPTS must be at least 2 but not more than 50. If the table exceeds 3×50 (for streamflow, stream depth, and width) values, then MAXPTS in the allocation subroutine GWF1SFR1ALP will need to be increased from 3×50 to $3 \times$ (the desired maximum value).

FLOW—a real number that is the streamflow (in units of volume per time) entering or leaving the upstream end of a stream segment (that is, into the first reach).

If the stream is a headwater stream, FLOW defines the total inflow to the first reach of the segment. The value can be any number ≥ 0.

If the stream is a tributary stream, FLOW defines additional specified inflow to or withdrawal from the first reach of the segment (that is, in addition to the discharge from the upstream segment of which this is a tributary). This additional flow does not interact with the ground-water system. For example, a positive number might be used to represent direct outflow into a stream from a sewage treatment plant, whereas a negative number might be used to represent pumpage directly from a stream into an intake pipe for a municipal water treatment plant.

If the stream is a diversionary stream, and the diversion is from another stream segment, FLOW defines the streamflow diverted from the last reach of stream segment IUPSEG into the first reach of this segment. The diversion is computed or adjusted according to the value of IPRIOR.

If the stream is a diversionary stream, and the diversion is from a lake, FLOW defines a fixed rate of discharge diverted from the lake into the first reach of this stream segment (unless the lake goes dry) and flow from the lake is not dependent on the value of ICALC. However, if FLOW = 0, then the lake outflow into the first reach of this segment will be calculated on the basis of lake stage relative to the top of the streambed for the first reach using one of the methods defined by ICALC.

RUNOFF—a real number that is the volumetric rate of the diffuse overland runoff that enters the stream segment (in units of volume per time). The specified rate is apportioned to each reach of the segment in direct relation to the fraction of the total length of the stream channel in the segment that is present in each reach.

ETSW—a real number that is the volumetric rate per unit area of water removed by evapotranspiration directly from the stream channel (in units of length per time). ETSW is defined as a positive value.

PPTSW—a real number that is the volumetric rate per unit area of water added by precipitation directly on the stream channel (in units of length per time).

ROUGHCH—a real number that is Manning's roughness coefficient for the channel in all reaches in this segment. This variable is only specified if ICALC = 1 or 2.

ROUGHBK—a real number that is Manning's roughness coefficient for the overbank areas in all reaches in this segment. This variable is only specified if ICALC = 2.

CDPTH—a real number that is the coefficient used in the equation: ($DEPTH = CDPTH \times Q^{FDPTH}$) that relates stream depth in all reaches in this segment to streamflow. This variable is only specified if ICALC = 3.

FDPTH—a real number that is the coefficient used in the equation: ($DEPTH = CDPTH \times Q^{FDPTH}$) that relates stream depth in all reaches in this segment to streamflow. This variable is only specified if ICALC = 3.

AWDTH—a real number that is the coefficient used in the equation: ($WIDTH = AWDTH \times Q^{BWDTH}$) that relates stream width in all reaches in this segment to streamflow. This variable is only specified if ICALC = 3.

BWDTH—a real number that is the coefficient used in the equation: ($WIDTH = AWDTH \times Q^{BWDTH}$) that relates stream width in all reaches in this segment to streamflow. This variable is only specified if ICALC = 3.

ITEM 4b VARIABLES:

HCOND1—a real number that is a factor used to calculate hydraulic conductivity of the streambed at the upstream end of this segment from the parameter value (in units of length per time).

THICKM1—a real number that is the thickness of streambed material at the upstream end of this segment (in units of length).

ELEVUP—a real number that is the elevation of the top of the streambed at the upstream end of this segment (in units of length).

WIDTH1—a real number that is the average width of the stream channel at the upstream end of this segment (in units of length). This variable is only specified if ICALC < 1.

DEPTH1—a real number that is the average depth of water in the channel at the upstream end of this segment (units of length). This variable is only specified if ICALC = 0, in which case the stream stage in a reach is assumed to equal the elevation of the top of the streambed plus the depth of water.

THTS1—Saturated volumetric water content in the unsaturated zone beneath the upstream end of this segment. This variable is read for the first stress period when ICALC is 1 or 2 and ISFROPT is 4 or 5.

THTI1—Initial volumetric water content beneath the upstream end of this segment. THTI1 must be less than or equal to THTS and greater than or equal to the THTS minus the specific yield defined in either LPF or BCF. This variable is read for the first stress period when ICALC is 1 or 2 and ISFROPT is 4 or 5.

EPS1—Brooks-Corey exponent used in the relation between water content and hydraulic conductivity within the unsaturated zone beneath the upstream end of this segment. This variable is read for the first stress period when ICALC is 1 or 2 and ISFROPT is 4 or 5.

UHC1—Vertical saturated hydraulic conductivity of the unsaturated zone beneath the upstream end of this segment. This variable is necessary when using BCF, whereas it is optional when using LPF. This variable is read only for the first stress period when ICALC is 1 or 2 and ISFROPT is 5.

ITEM 4c VARIABLES:

HCOND2—a real number that is a factor used to calculate hydraulic conductivity of the streambed at the downstream end of this segment from the parameter value (in units of length per time).

THICKM2—a real number that is the thickness of streambed material at the downstream end of this segment (in units of length).

ELEVDN—a real number that is the elevation of the top of the streambed at the downstream end of this segment (in units of length).

WIDTH2—a real number that is the average width of the stream channel at the downstream end of this segment (in units of length). This variable is only specified if ICALC < 1.

DEPTH2—a real number that is the average depth of water in the channel at the downstream end of this segment (units of length). This variable is only specified if ICALC = 0, in which case the stream stage in a reach is assumed to equal the elevation of the top of the streambed plus the depth of water.

THTS2—Saturated volumetric water content in the unsaturated zone beneath the downstream end of this segment. This variable is read for the first stress period when ICALC is 1 or 2 and ISFROPT is 4 or 5.

THTI2—Initial volumetric water content beneath the downstream end of this segment. THTI1 must be less than or equal to THTS and greater than or equal to the THTS minus the specific yield defined in either LPF or BCF. This variable is read for the first stress period when ICALC is 1 or 2 and ISFROPT is 4 or 5.

EPS2—Brooks-Corey exponent used in the relation between water content and hydraulic conductivity within the unsaturated zone beneath the downstream end of this segment. This variable is read for the first stress period when ICALC is 1 or 2 and ISFROPT is 4 or 5.

UHC2—Vertical saturated hydraulic conductivity of the unsaturated zone beneath the downstream end of this segment. This variable is necessary when using BCF, whereas it is optional when using LPF. This variable is read only for the first stress period when ICALC is 1 or 2 and ISFROPT is 5.

ITEM 4d VARIABLES:

XCPTi—a real number that is the distance relative to the left bank of the stream channel (when looking downstream) for the eight points (XCPT1 through XCPT8) used to describe the geometry of this segment of the stream channel. By definition, location XCPT1 represents the left edge of the channel cross section, and its value should be set equal to 0.0; values XCPT2 through XCPT8 should be equal to or greater than the previous distance.

ZCPTi—a real number that is the height relative to the top of the lowest elevation of the streambed (thalweg). One value (ZCPT1 through ZCPT8) is needed for each of the eight horizontal distances defined by XCPTi. The location of the thalweg (set equal to 0.0) can be any location from XCPT2 through XCPT7.

ITEM 4e VARIABLES:

FLOWTAB—a real number that is the streamflow (units of volume per time) related to a given depth and width. One value is needed for each streamflow that has a corresponding value of depth and width up to the total number of values used to define the table—FLOWTAB(1) through FLOWTAB(NSTRPTS). NSTRPTS is defined in item 4a.

DPTHTAB—a real number that is the average depth (units of length) corresponding to a given flow. The number and order of values, DPTHTAB(1) through DPTHTAB(NSTRPTS) must coincide with FLOWTAB.

WDTHTAB—a real number that is the stream width (units of length) corresponding to a given flow. The number and order of values, WDTHTAB(1) through WDTHTAB(NSTRPTS), must coincide with FLOWTAB.

Lake Package

Input to the Lake Package is read from the file that has file type LAK in the Name File. The user can optionally specify that gages on a lake are to be written using the Gage Package by including GAGE in the Name File (see "Gage Package" section). Input for the Lake Package consists of nine separate items, each consisting of one or more lines. These data are used to specify information about the physical geometry of the lakes, hydraulic properties of the lakebeds, and the degree of hydraulic stress originating from atmospheric and anthropogenic sources, as well as specifying certain output control variables. Spatial and temporal units of input data specifications should be consistent with other data input for the MODFLOW run.

All input variables are read using free formats, unless specifically indicated otherwise. In free format, variables are separated by one or more spaces, or by a comma and, optionally, one or more spaces. It is important to note that, in free format, blank spaces are not read as zeroes and a blank field cannot be used to set a parameter value to zero. The Lake Package is documented in Merritt and Konikow (2000) and users of GSFLOW are encouraged to review that document when using the Lake Package.

FOR EACH SIMULATION

1. NLAKES   ILKCB
    Sublakes of multiple-lake systems are considered separate lakes for input purposes. The variable NLAKES is used, with certain internal assumptions and approximations, to dimension arrays for the simulation.

    If data are being read using the fixed format mode, then each field should be entered using I10 format.

2. THETA   [NSSITR]   [SSCNCR]
    In the original USGS Lake Package (Merritt and Konikow, 2000), NSSITR and SSCNCR were not used for transient solutions and were omitted for simulations that did not include a steady-state stress period. Although NSSITR and SSCNCR are not required input for transient stress periods, changes were made to the Lake Package such that NSSITR and SSCNCR are used for both transient and steady-state stress periods. If values for NSSITR and SSCNCR are not input, then default values of 100 and 0.0001 are assumed, respectively. For mixed steady-state/transient runs, NSSITR and SSCNCR must be included, even if the steady-state stress period is not the first one. If more than one steady-state stress period is included in the total simulation period, then the initial values of NSSITR and SSCNCR will apply to all subsequent steady-state stress periods. If there are steady-state stress periods following the first stress period of a simulations then values of NSSITR and SSCNCR are included in item 9.

    If data are being read using the fixed format mode, then the data should be entered using format (F10.4, I10, F10.4).

FOR THE FIRST STRESS PERIOD ONLY

3. `STAGES` `[SSMN]` `[SSMX]`

> This data set should consist of one line for each lake, where line 1 includes data for lake 1, and line n includes data for lake n. There must be exactly `NLAKES` lines of data.

> In the original USGS Lake Package (Merritt and Konikow, 2000), `SSMN` and `SSMX` were not needed for a transient run and were omitted when a simulation did not include a steady-state stress period. However, the Lake Package was modified such that `SSMN` and `SSMX` are used for all stress periods. If a simulation does not include a steady-state stress period, then `SSMN` and `SSMX` are omitted and they are set equal to the lake bottom and top altitudes.

> If data are being read using the fixed format mode, then each field should be entered using F10.4 format.

FOR EACH REMAINING STRESS PERIOD

4. `ITMP` `ITMP1` `LWRT`

> `ICBCFL` $\leq$ 0 also suppresses printout from the lake package.

> If data are being read using the fixed format mode, then each field should be entered using I10 format.

> If `ITMP` > 0:

5. `LKARR(NCOL, NROW)`

> An `NCOL` by `NROW` array is read for each layer in the grid by module U2DINT.

6. `BDLKNC(NCOL, NROW)`

> An `NCOL` by `NROW` array is read for each layer in the grid by module U2DREL.

7. `NSLMS`

> If data are being read using the fixed format mode, then `NSLMS` should be entered using format I5.

> If `ITMP` > 0 and `NSLMS` > 0:

8a. `IC` `ISUB(1)` `ISUB(2)` ... `ISUB(IC)`
8b. `SILLVT(1)` ... `SILLVT(IC)`

> A pair of items (items 8a and 8b) is read for each multiple-lake system, i.e., `NSLMS` pairs of items. However, `IC` = 0 will terminate the input.

> If data are being read using the fixed format mode, then each field of item 8a should be entered using I5 format and each field of item 8b should be entered using F10.4 format.

> If `ITMP1` $\geq$ 0:

9. `PRCPLK` `EVAPLK` `RNF` `WTHDRW` `[SSMN]` `[SSMX]`

> The variables `SSMN` and `SSMX` are optional. The capability for using mixed steady-state/transient runs with the Lake Package does not depend on the sequence of steady-state and transient stress periods. If the second or a subsequent stress period is steady-state, then `SSMN` and `SSMX` for those stress periods are defined in item 9a.

Explanation of Variables Read by the Lake Package:

`NLAKES`—is the number of separate lakes.

`ILKCB`—is a flag indicating whether or not to write cell-by-cell flows. If `ILKCB` > 0, flows will be written; otherwise, flows will not be written. If `ILKCB` < 0 and `ICBCFL` is not equal to 0, the cell-by-cell flows will be printed in the standard output file.

THETA—is a real variable equal to the semi-implicit ($0.5 \leq$ THETA < 1.0) or implicit (THETA $= 1.0$) solution approach for lake stage. If a negative value of THETA (abs(-0.5) $\leq$ THETA $\geq$ abs(-1.0)) is specified, then NSSITR and SSCNCR are required input regardless of whether or not the simulation includes a steady-state stress period. If THETA is negative, then an absolute value of THETA will be used for calculating lake stage. If the absolute value of THETA is less than 0.5 or greater than 1.0, then THETA is set equal to 0.5.

NSSITR—is the maximum number of iterations for Newton's solution method for lake stages during each MODFLOW iteration for steady-state aquifer head solution.

SSCNCR—is the convergence criterion for equilibrium lake stage solution by Newton's method (L).

STAGES—is the initial stage of each lake at the beginning of the run (L).

SSMN—is the minimum allowed lake stage for each lake (L).

SSMX—is the maximum allowed lake stage for each lake (L).

ITMP—options are:

> ITMP > 0, read lake definition data (items 5 and 6, and, optionally, items 7 through 9).
>
> ITMP $= 0$, no lake calculations this stress period.
>
> ITMP < 0, use lake definition data from last stress period.

ITMP1—options are:

> ITMP1 ≥ 0, read new recharge, evaporation, runoff, and withdrawal data for each lake.
>
> ITMP1 < 0, use recharge, evaporation, runoff, and withdrawal data from last stress period.

LWRT—is a flag; if LWRT > 0, suppresses printout from the Lake Package.

LKARR—a value is read for every grid cell. If LKARR(I,J,K) $= 0$, the grid cell is not a lake volume cell. If LKARR(I,J,K) > 0, its value is the identification number of the lake occupying the grid cell. LKARR(I,J,K) must not exceed the value NLAKES. If it does, or if LKARR(I,J,K) < 0, LKARR(I,J,K) is set to zero.

BDLKNC—a value is read for every grid cell. The value is the lakebed leakance (units of per time or 1/T) that will be assigned to lake/aquifer interfaces that occur in the corresponding grid cell.

NSLMS—is the number of sublake systems if coalescing/dividing lakes are to be simulated (only in transient runs). Enter 0 if no sublake systems are to be simulated.

IC—is the number of sublakes, including the center lake, in the sublake system being described in this item.

ISUB—is the identification number of the sublakes in the sublake system being described in this item. The center lake number is listed first.

SILLVT—is the sill elevation (L) that determines whether the center lake is connected with a given sublake. One value is entered in this item for each sublake in the order the sublakes are listed in the previous item.

PRCPLK—is the rate of precipitation at the surface of a lake (L/T).

EVAPLK—is the rate of evaporation from the surface of a lake (L/T).

RNF—is the volumetric rate of overland runoff from an adjacent watershed entering the lake (L^3/T). If RNF > 0, it is specified directly as a volume. If RNF < 0, it is used as a multiple applied to the product of the lake evaporation rate and the volume of the lake at its full stage (occupying all layer 1 lake cells). This is to account for the percentage of rainfall becoming runoff and the size of the watershed contributing the runoff.

WTHDRW—is the rate of water removal from a lake by means other than rainfall, evaporation, surface outflow, or ground-water seepage (L^3/T). A negative value indicates augmentation. Normally, this would be used to specify the rate of artificial withdrawal from a lake for human water use, or, if negative, artificial augmentation of a lake volume for aesthetic or recreational purposes.

Unsaturated-Zone Flow Package

Input to the Unsaturated-Zone Flow (UZF) Package is read from the file that has file type UZF in the Name File. The user can optionally specify unsaturated-zone water budgets and water content profiles for selected finite-difference cells by specifying the file type DATA in the MODFLOW Name File. The row and column cell indices for each selected finite-difference cell are included in the UZF output file. Three types of information may be printed to the specified file for each finite-difference cell depending on the value of IUZOPT specified in the UZF input file. The three output options are: option 1 prints volumes of water entering, leaving, and stored in the unsaturated zone; option 2 prints volumes and rates for water entering, leaving, and stored within the unsaturated zone; and option 3 prints the water content profile between land surface and the water table. Additionally, a time series of infiltration, unsaturated zone evapotranspiration, recharge, and ground-water discharge summed over the model domain may be printed to a specified file. The UZF Package is documented in Niswonger and others (2006a) and users of GSFLOW are encouraged to review that document when using the UZF Package.

The UZF Package input file consists of items numbered from 0 through 16, each consisting of one or more items. These data are used to specify information about the hydraulic properties of the unsaturated zone, the infiltration rate, evapotranspiration, and certain output control options. All input variables are read using free format unless indicated otherwise. In free format, variables are separated by one or more spaces or by a comma and optionally one or more spaces. A blank field cannot be used to set a variable value to zero. Units of length (L) and time (T) are used to define the dimensions of each variable.

The UZF Package relies on the specific yield values as specified in the Layer-Property Flow (LPF) Package, the Block-Centered Flow (BCF) Package, or the Hydrogeologic Unit Flow (HUF) Package. When the option to use vertical hydraulic conductivity in the LPF Package is specified, the layer(s) that contain cells where unsaturated flow will be simulated must be specified as convertible. That is, the variable LAYTYP specified in the LPF Package must not be equal to zero, otherwise the model will print an error and stop execution. The variable VKA in LPF may be specified as the ratio of horizontal to vertical hydraulic conductivity (LAYVKA is not zero) or as the vertical hydraulic conductivity (LAYVKA=0). When using the BCF Package, the right digit of LTYPE (LAYCON) must be greater or equal to one or the model will print an error and stop execution. Neither parameter estimation nor solute transport can be simulated with the UZF Package.

FOR EACH SIMULATION

0. [#Text]
    Item 0 is optional—# must be in column 1. Item 0 can be repeated multiple times.
1. NUZTOP   IUZFOPT   IRUNFLG   IETFLG   IUZFCB1   IUZFCB2   NTRAIL2   NSETS2   NUZGAG
    SURFDEP
    The variables NTRAIL2 and NSETS2 are used for dimensioning arrays.
2. [IUZFBND(NCOL,NROW)] – U2DINT

    If IRUNFLG > 0, read item 3:
3. [IRUNBND(NCOL,NROW)] – U2DINT

    If IUZFOPT = 1, read item 4:

4. [VKS(NCOL,NROW)] – U2DREL
    Vertical hydraulic conductivity is optional when either the LPF or BCF Package is used to define vertical hydraulic
        conductivity of cells (IUZFOPT=2).
5. [EPS(NCOL,NROW)] – U2DREL
6. [THTS(NCOL,NROW)] – U2DREL
7. [THTI(NCOL,NROW)] – U2DREL

    If NUZGAG > 0, item 8 is repeated NUZGAG times:

8. [IUZROW] [IUZCOL] [IFTUNIT] [IUZOPT]
    Item 8 must include exactly NUZGAG lines of data. If NUZGAG > 1, it is permissible to repeat the same cell if a summary
        of volumes and rates and water contents are desired for a given cell. The lines within item 8 can be listed in any order.

A unique unit number for IFTUNIT must be specified for each line in item 8 and each number must match that used in the MODFLOW Name File as Ftype DATA file types and file names (see Harbaugh, 2005).

FOR EACH STRESS PERIOD

9. NUZF1

    If NUZF1 > 0:

10. [FINF(NCOL,NROW)] – U2DREL

    If FINF is specified as being greater than the vertical hydraulic conductivity, then FINF is set equal to the vertical unsaturated hydraulic conductivity. Excess water is routed to streams or lakes when IRUNFLG is not zero, and if SFR2 or LAK3 is active.

    If IETFLG > 0, read items 11 through 16:

11. NUZF2

    If NUZF2 > 0:

12. [PET(NCOL,NROW)] – U2DREL
13. NUZF3

    If NUZF3 > 0:

14. [EXTDP(NCOL,NROW)] – U2DREL

    The quantity of ET removed from a cell is limited by the volume of water stored in the unsaturated zone above the extinction depth. If ground water is within the ET extinction depth, then the rate removed is based on a linear decrease in the maximum rate at land surface and zero at the ET extinction depth. The linear decrease is the same method used in the Evapotranspiration Package (McDonald and Harbaugh, 1988, chap. 10).

15. NUZF4

    If NUZF4 > 0:

16. [EXTWC(NCOL, NROW)] – U2DREL

    EXTWC must have a value between (THTS-Sy) and THTS, where Sy is the specific yield specified in either the LPF or BCF Package.

Explanation of Variables Read by the UZF Package

Text—is a character variable (up to 199 characters) that starts in column 2. Any characters can be included in Text. The # character must be in column 1. Lines beginning with # are restricted to the first lines of the file. Text is written to the Listing File.

ITEM 1 VARIABLES:

NUZTOP—is an integer value used to define which cell in a vertical column for which recharge and discharge is simulated.

If NUZTOP = 1, recharge to and discharge from only the top model layer. This option assumes land surface is defined as top of layer 1.

If NUZTOP = 2, recharge to and discharge from the specified layer in variable IUZFBND. This option assumes land surface is defined as top of layer specified in IUZFBND.

If NUZTOP = 3, recharge to and discharge from the highest active cell in each vertical column. Land surface is determined as top of layer specified in IUZFBND. A constant head node intercepts any recharge and prevents deeper percolation.

IUZFOPT—an integer value equal to 1 or 2. A value of 1 indicates that the vertical hydraulic conductivity will be specified within the UZF Package input file using array VKS. A value of 2 indicates that the vertical hydraulic conductivity will be specified within the LPF Package input file. IUZFOPT must be 1 when using the BCF Package.

IRUNFLG—an integer value that specifies whether ground water that discharges to land surface will be routed to stream segments or lakes as specified in the IRUNBND array (IRUNFLG not equal to zero) or if ground-water discharge is removed from the model simulation and accounted for in the ground-water budget as a loss of water (IRUNFLG=0). The Streamflow-Routing (SFR2) and(or) the Lake (LAK3) Packages must be active if IRUNFLG is not zero. This option is available only for steady-state stress periods in GSFLOW.

IETFLG—an integer value that specifies whether or not evapotranspiration (ET) will be simulated. ET will not be simulated if IETFLG is zero, otherwise it will be simulated. This option is for simulating ET in addition to ET simulated in PRMS. If not using this option then ET is simulated only by PRMS above the soil-zone base. If using this option then any ET demand not met by storage in the soil zone will be applied as ET demand below the base of the soil zone.

IUZFCB1—an integer value used as a flag for writing rates of ground-water recharge, ET, and ground-water discharge to land surface to a separate unformatted file using subroutine UBUDSV. If IUZFCB1 is > 0, it is the unit number to which the cell-by-cell rates will be written when SAVE BUDGET or a non-zero value for ICBCFL is specified in Output Control. If IUZFCB1 is ≤ 0, cell-by-cell rates will not be written to a file.

IUZFCB2—an integer value used as a flag for writing rates of ground-water recharge, ET, and ground-water discharge to land surface to a separate unformatted file using module UBDSV3. If IUZFCB2 is > 0, it is the unit number to which cell-by-cell rates will be written when SAVE BUDGET or a non-zero value for ICBCFL is specified in Output Control. If IUZFCB2 is ≤ 0, cell-by-cell rates will not be written to file.

NTRAIL2—an integer value that is the number of trailing-wave increments used to represent a trailing wave. Trailing waves are used to represent a decrease in the surface infiltration rate. The value can be increased to improve mass balance in the unsaturated zone. Values between 20 and 10 work well and result in unsaturated-zone mass balance errors ranging between 0.001 and 0.01 percent, although for large problems fewer trailing waves between 10 and 5 can be used and result in mass balance errors ranging between 0.01 and 0.1 percent. See Smith (1983) for further details.

NSETS2—an integer value equal to the number of wave sets used to simulate multiple infiltration periods. The number of wave sets should be set to 20 for most problems involving time varying infiltration. The total number of waves allowed within an unsaturated zone cell is equal to the product of NTRAIL2 and NSETS2. A warning message will be printed to the MODFLOW LIST file if the number of waves in a cell exceeds this value. NSETS2 affects memory usage but not computational requirements.

NUZGAG—an integer value equal to the number of locations (one cell per vertical column on layers) that will be specified for printing detailed information on the unsaturated zone water budget and water content. This output option also may be used to print time series of infiltration, evapotranspiration, unsaturated-zone storage, recharge, and ground-water discharge summed over all finite-difference cells. See definition of the UZF variable IFTUNIT for more explanation.

SURFDEP—a real value equal to the average undulation depth of the soil-zone base altitude. This value is used for calculating recharge and ground-water discharge to the soil zone when the ground-water altitude is less than a distance of SURFDEP above or below the soil-zone base.

ITEMS 2-7 VARIABLES:

IUZFBND—is an array of integer values used to define the areal extent of the active model in which recharge and discharge will be simulated.

IRUNBND—is an array of integer values used to define the stream segments within the Streamflow-Routing (SFR2) Package or lake numbers in the Lake (LAK3) Package that overland runoff from excess infiltration and ground-water discharge to land surface will be added. A positive integer value identifies the stream segment and a negative integer value identifies the lake number. {This option is available only for steady-state stress periods in GSFLOW.}

VKS—is an array of positive real values used to define the saturated vertical hydraulic conductivity of the unsaturated zone (L/T).

EPS—is an array of positive real values for each finite-difference cell used to define the Brooks-Corey epsilon of the unsaturated zone. Epsilon is used in the relation of water content to hydraulic conductivity (Brooks and Corey, 1966).

THTS—is an array of positive real values used to define the saturated water content of the unsaturated zone in units of volume of water to total volume (L^3/L^3).

THTI—is an array of positive real values used to define the initial water content for each vertical column of cells in units of volume of water at start of simulation to total volume (L^3/L^3). THTI is not specified for steady-state simulations.

ITEM 8 VARIABLES:

IUZROW—is an integer value equal to the row number of the cell that unsaturated-zone information will be printed for each time step. The variable is not used when IFTUNIT is negative.

IUZCOL—is an integer value equal to the column number of the cell that unsaturated-zone information will be printed for each time step. The variable is not used when IFTUNIT is negative.

IFTUNIT—is an integer value equal to the unit number of the output file. A positive value is for output of individual cells, whereas a negative value is for output that is summed over all finite-difference cells. The summed output includes applied infiltration, runoff, actual infiltration, ground-water discharge to land surface, ET from the unsaturated zone, ET from ground water, recharge, and change in unsaturated-zone storage.

IUZOPT—is an integer value that is a flag for the type of expanded listing desired in the output file. The variable is not used when IFTUNIT is negative.

If IUZOPT = 1, prints time, ground-water head, and thickness of unsaturated zone, and cumulative volumes of infiltration, recharge, storage, change in storage and ground-water discharge to land surface.

If IUZOPT = 2, same as option 1 except rates of infiltration, recharge, change in storage, and ground-water discharge also are printed.

If IUZOPT = 3, prints time, ground-water head, thickness of unsaturated zone, followed by a series of depths and water contents in the unsaturated zone.

ITEMS 9-16 VARIABLES:

NUZF1—is an integer value for reusing or reading infiltration rates that can change each stress period. If NUZF1 > 0, then infiltration rates at land surface are specified. If NUZF1 < 0, then infiltration rates from the previous stress period are used. This variable only is used during steady-state stress periods for integrated simulations. A dummy value must be specified for stress periods that are not steady state.

FINF—is an array of positive real values used to define the infiltration rates (L/T) at land surface for each vertical column of cells. This variable only is used during steady-state stress periods.

NUZF2—is an integer value for reusing or reading ET demand rates that can change each stress period. If NUZF2 > 0, then ET demand rates are specified. If NUZF2 < 0, then ET demand rates from the previous stress period are used. {This variable is not used in GSFLOW simulations because the ET demand below the soil-zone base is calculated by PRMS. However, if ET is to be simulated below the soil-zone base (IETFLG > 0), NUZF2 should be set to a value less than zero.}

PET—is an array of positive real values used to define the ET demand rates (L/T) within the ET extinction depth interval for each vertical column of cells. {**PET** is not used for GSFLOW simulations.}

NUZF3—is an integer value for reusing or reading ET extinction depths that can change each stress period. If NUZF3 > 0, then ET extinction depths are specified. If NUZF3 < 0, then depths from the previous stress period are used. {If ET is to be simulated below the soil-zone base (IETFLG > 0) NUZF3 must be specified.}

EXTDP—is an array of positive real values used to define the ET extinction depths {relative to the altitude of the soil-zone base for GSFLOW simulations. If ET is to be simulated below the soil-zone base (IETFLG > 0) EXTDP must be specified for the first stress period at a minimum.}

NUZF4—is an integer value for reusing or reading the extinction water content that can change each stress period. If NUZF4 > 0, then extinction water contents are specified. If NUZF4 < 0, then the extinction water contents from the previous stress period are used. {If ET is to be simulated below the soil-zone base (IETFLG > 0) NUZF4 must be specified.}

EXTWC—is an array of positive real values used to define the extinction water content below which ET cannot be removed from the unsaturated zone {below the soil-zone base.} {If ET is to be simulated below the soil-zone base (IETFLG > 0) EXTWC must be specified for the first stress period at a minimum.}

Gage Package

The input file for specifying stream and lake gaging-station locations is read if file type (Ftype) GAGE is included in the MODFLOW Name File. A particular stream reach or lake can be designated for a gaging or monitoring station. At each designated stream reach or lake, the time, stream or lake stage, and streamflow out of the reach or lake volume will be written to a separate output file to facilitate model output evaluation and graphical post processing of the calculated data. Several options are available to also print additional information about a stream reach or lake. The output file will contain two header lines that provide relevant information and the text will be contained within quotes.

FOR EACH SIMULATION

1.   NUMGAGE

FOR EACH GAGING STATION

Items 2a and 2b are used to specify information for stream or lake gaging stations, respectively. The total number of lines specified for items 2a and 2b must equal NUMGAGE. Items 2a and 2b can be entered in any arbitrary order, and it is permissible to interleaf items for stream-gaging stations with those for lake gaging stations if NUMGAGE > 1.

FOR EACH STREAM GAGING STATION

2a.   GAGESEG   GAGERCH   UNIT   OUTTYPE

FOR EACH LAKE GAGING STATION

2b.   LAKE   UNIT   [OUTTYPE]

Explanation of Variables Read by the GAGE Package:

NUMGAGE—is the total number of stream and lake gaging stations.

GAGESEG—is an integer value that is the stream segment number where the gage is located.

GAGERCH—is an integer value that is the stream reach number where the gage is located.

LAKE—is a negative integer value that is the lake number where the gage is located.

UNIT—is an integer value that is the unit number of the output file for this gage. A unique unit number must be specified for each gage and the unit numbers must match data file types and file names in the MODFLOW Name File (see Harbaugh and others, 2000, p. 42-44).

OUTTYPE—for a stream gage, OUTTYPE is an integer value that is a flag for the type of expanded listing desired in output file. For a stream gage the options are:

OUTTYPE = 0, use standard default listing of time, stream stage, and outflow.

OUTTYPE = 1, default values plus depth, width, flow at midpoint, precipitation, evapotranspiration, and runoff. Computed runoff from the Unsaturated-Zone Flow (UZF) Package is added when the UZF Package is active.

OUTTYPE = 2, default values plus streambed conductance for the reach, head difference across streambed, and hydraulic gradient across streambed.

OUTTYPE = 3, {This option is not used in GSFLOW.}

OUTTYPE = 4, all of the above.

OUTTYPE = 5, use for diversions to provide a listing of time, stage, flow diverted, maximum assigned diversion rate, and flow at end of upstream segment prior to diversion.

OUTTYPE = 6, use for unsaturated flow routing to provide a listing of time, stream stage, ground-water head beneath stream, streambed seepage, change in unsaturated zone storage, and recharge to ground water.

OUTTYPE = 7, use for unsaturated flow routing to provide a listing of time and the unsaturated-water content profile beneath the stream. Two profiles are printed. The first is the volume averaged water content of all unsaturated zone cells (multiple unsaturated zone cells are allowed) beneath a stream reach. The second profile is the volume averaged water content beneath the low-flow channel only. The two profiles are identical when only one unsaturated zone cell is assigned beneath a stream reach.

For a lake gage, OUTTYPE is an integer value that is a flag for type of expanded listing desired in output file. It is read only when UNIT is a negative value. Options are:

OUTTYPE = 0, standard default listing of time, lake stage, and lake volume.

OUTTYPE = 1, default values plus all inflows to and outflows from lake (as volumes during time increment), and total lake conductance. Computed runoff from the UZF Package is added whenever the UZF Package is active.

OUTTYPE = 2, default values plus changes in lake stage and lake volume.

OUTTYPE = 3, all of the above.

Note: Total lake conductance (OUTTYPE = 1 and 3 for a lake gaging station) is the sum of the conductances of each seepage interface for each lake. Changes in lake stage and volume (OUTTYPE = 2 or 3) are listed as incremental changes from previous time increment and as cumulative change since the start of the simulation.

Strongly Implicit Procedure Package

Input to the Strongly Implicit Procedure (SIP) Package is read from the file that is type SIP in the Name File. All numeric variables are free format if the option FREE is specified in the Basic Package input file; otherwise, all the variables have 10-character fields. The SIP Package is documented in Harbaugh (2005) and users of GSFLOW are encouraged to review that document when using the SIP Package.

FOR EACH SIMULATION

0. [#Text]
    Item 0 is optional—# must be in column 1. Item 0 can be repeated multiple times.
1. MXITER   NPARM
2. ACCL   HCLOSE   IPCALC   WSEED   IPRSIP

Explanation of Variables Read by the SIP Package:

Text—is a character variable (199 characters) that starts in column 2. Any characters can be included in Text. The # character must be in column 1. Lines beginning with # are restricted to the first lines of the file. Text is written to the Listing File.

MXITER—is the maximum number of times through the iteration loop in one time step in an attempt to solve the system of finite-difference equations.

NPARM—is the number of iteration variables to be used. Five variables generally are sufficient.

ACCL—is the acceleration variable, which must be greater than zero and is generally equal to one. If a zero is entered, it is changed to one.

HCLOSE—is the head change criterion for convergence. When the maximum absolute value of head change from all nodes during an iteration is less than or equal to HCLOSE, iteration stops.

IPCALC—is a flag indicating where the seed for calculating iteration variables will come from.

    0—the seed entered by the user will be used.

    1—the seed will be calculated at the start of the simulation from problem variables.

WSEED—is the seed for calculating iteration variables. WSEED is always read but is used only if IPCALC is equal to zero.

IPRSIP—is the printout interval for SIP. IPRSIP, if equal to zero, is changed to 999. The maximum head change (positive or negative) is printed each iteration of a time step whenever the time step is an even multiple of IPRSIP. This printout also occurs at the end of each stress period regardless of the value of **IPRSIP**.

Preconditioned Conjugate-Gradient Package

Input to the Preconditioned Conjugate-Gradient (PCG) Package is read from the file that is type PCG in the Name File. All numeric variables are free format if the option FREE is specified in the Basic Package input file; otherwise, all variables have 10-character fields. The PCG Package is documented in Harbaugh (2005) and users of GSFLOW are encouraged to review that document when using the PCG Package.

FOR EACH SIMULATION

0. [#Text]
   Item 0 is optional—# must be in column 1. Item 0 can be repeated multiple times.
1. MXITER   ITER1   NPCOND
2. HCLOSE   RCLOSE   RELAX   NBPOL   IPRPCG   MUTPCG   DAMPPCG   [DAMPPCG2]

Explanation of Variables Read by the PCG Package:

Text—is a character variable (199 characters) that starts in column 2. Any characters can be included in Text. The # character must be in column 1. Lines beginning with # are restricted to the first lines of the file. Text is written to the Listing File.

MXITER—is the maximum number of outer iterations (Hill, 1990). For a linear problem MXITER should be 1, unless more than 50 inner iterations are required, when MXITER could be as large as 1000.

ITER1—is the number of inner iterations (Hill, 1990). For nonlinear problems, ITER1 usually ranges from 10 to 30; a value of 30 will be sufficient for most linear problems.

NPCOND—is the flag used to select the matrix conditioning method:

    1—is for Modified Incomplete Cholesky (for use on scalar computers)

    2—is for Polynomial (for use on vector computers or to conserve computer memory)

HCLOSE—is the head change criterion for convergence, in units of length. When the maximum absolute value of head change from all nodes during an iteration is less than or equal to HCLOSE, and the criterion for RCLOSE also is satisfied, iteration stops.

RCLOSE—is the residual criterion for convergence, in units of cubic length per time. The units for length and time are the same as established for all model data. (See LENUNI and ITMUNI input variables in the Discretization File.) When the maximum absolute value of the residual at all nodes during an iteration is less than or equal to RCLOSE, and the criterion for HCLOSE also is satisfied, iteration stops.

For nonlinear problems, convergence is achieved when the convergence criteria are satisfied for the first inner iteration.

RELAX—is the relaxation parameter used with NPCOND = 1. Usually, RELAX = 1.0, but for some problems a value of 0.99, 0.98, or 0.97 will reduce the number of iterations required for convergence. RELAX is not used if NPCOND is not 1.

NBPOL—is used when NPCOND = 2 to indicate whether the estimate of the upper bound on the maximum eigen value is 2.0, or whether the estimate will be calculated. NBPOL = 2 is used to specify the value is 2.0; for any other value of NBPOL, the estimate is calculated. Convergence is generally insensitive to this NBPOL. NBPOL is not used if NBPOL is not 2.

IPRPCG—is the printout interval for PCG. IPRPCG, if equal to zero, is changed to 999. The maximum head change (positive or negative) and residual change are printed for each iteration of a time step whenever the time step is an even multiple of IPRPCG. This printout also occurs at the end of each stress period regardless of the value of IPRPCG.

MUTPCG—is a flag that controls printing of convergence information from the solver:

> 0—is for printing tables of maximum head change and residual each iteration
>
> 1—is for printing only the total number of iterations
>
> 2—is for no printing
>
> 3—is for printing only if convergence fails

DAMPPCG—is the damping factor. A value of one indicates no damping. A value less than 1 and greater than 0 causes damping. {Because steady-state stress periods often require lower values of DAMPPCG for convergence, the option of using a separate damping factor for transient stress periods was added for GSFLOW. If DAMPPCG is specified as a negative value then an additional damping factor will be read for transient stress periods and DAMPPCG will be converted to a positive value and will only be used for steady-state stress periods.}

{DAMPPCG2—is the damping factor used for transient stress periods. DAMPPCG2 only is read if DAMPPCG<0. A value of one indicates no damping. A value less than 1 and greater than 0 causes damping.}

Direct Solver Package

Input to the Direct Solver (DE4) Package is read from the file that is type DE4 in the Name File. All numeric variables are free format. The DE4 Package is documented in Harbaugh (2005) and users of GSFLOW are encouraged to review that document when using the DE4 Package.

FOR EACH SIMULATION

0.   [#Text]
        Item 0 is optional—# must be in column 1. Item 0 can be repeated multiple times.
1.   ITMX   MXUP   MXLOW   MXBW
2.   IFREQ   MUTD4   ACCL   HCLOSE   IPRD4

Explanation of Variables Read by the DE4 Package:

Text—is a character variable (199 characters) that starts in column 2. Any characters can be included in Text. The # character must be in column 1. Lines beginning with # are restricted to the first lines of the file. Text is written to the Listing File.

ITMX—is the maximum number of iterations for each time step. Specify ITMX = 1 if iteration is not desired. Ideally, iteration would not be required for direct solution; however, iteration is necessary if the flow equation is nonlinear (see explanation for IFREQ = 3) or if computer precision limitations result in inaccurate calculations as indicated by a large water-budget error. For a nonlinear flow equation, each iteration is equally time consuming because [A] (eq. 2-27 in Harbaugh, 2005) changes each iteration and Gaussian elimination is required after each change. This is called external iteration. For a linear equation, iteration is substantially faster because [A] is changed at most once per time step; thus, Gaussian elimination is required at most once per time step. This is called internal iteration.

MXUP—is the maximum number of equations in the upper part of the equations to be solved. This value affects the amount of memory used by the DE4 Package. If specified as 0, the program will calculate MXUP as half the number of cells in the model, which is an upper limit. The actual number of equations in the upper part will be less than half the number of cells whenever no-flow and constant-head cells are included because flow equations are not formulated for these cells. The DE4 Package prints the actual number of equations in the upper part when it runs. The printed value can be used for MXUP in future runs to minimize memory usage.

MXLOW—is the maximum number of equations in the lower part of equations to be solved. This value affects the amount of memory used by the DE4 Package. If specified as 0, the program will calculate MXLOW as half the number of cells in the model, which is an upper limit. The actual number of equations in the lower part will be less than half the number of cells whenever no-flow and constant-head cells are included because flow equations are not formulated for these cells. The DE4 Package prints the actual number of equations in the lower part when it runs. The printed value can be used for MXLOW in future runs to minimize memory usage.

MXBW—is the maximum bandwidth plus 1 of the [AL] matrix. This value affects the amount of memory used by the DE4 Package. If specified as 0, the program will calculate MXBW as the product of the two smallest grid dimensions plus 1, which is an upper limit. The DE4 Package prints the actual bandwidth plus 1 when it runs. The printed value can be used for MXBW in future runs to minimize memory usage.

IFREQ—is a flag indicating the frequency at which coefficients in [A] change. This affects the efficiency of solution; much work can be avoided if the user knows that [A] remains constant all or part of the time.

IFREQ = 1 indicates that the flow equations are linear and that coefficients of simulated head for all stress terms are constant for all stress periods. To meet the linearity requirement, all model layers must be confined (which is specified in the Block-Centered Flow Package by setting LAYCON equal to 0 for all layers or in the Layer-Property Flow Package by setting LAYTYP equal to 0 for all layers), and formulations must not change based upon head (such as seepage from a river changing from head-dependent flow to a constant flow when head drops below the bottom of the riverbed). Examples of coefficients of simulated head for stress terms are riverbed conductance, drain conductance, maximum evapotranspiration rate, evapotranspiration extinction depth, and general-head boundary conductance.

IFREQ = 2 indicates that the flow equations are linear as described for IFREQ = 1, but coefficients of simulated head for some stress terms may change at the start of each stress period. Examples of coefficients of simulated head for stress terms are riverbed conductance, drain conductance, maximum evapotranspiration rate, evapotranspiration extinction depth, and general-head boundary conductance. For a simulation consisting of only one stress period, IFREQ = 2 has the same meaning as IFREQ = 1.

IFREQ = 3 indicates that a nonlinear flow equation is being solved, which means that some terms in [A] depend on simulated head. Examples of head-dependent terms in [A] are transmissivity for water-table layers, which is based on saturated thickness; flow terms for rivers, drains, and evapotranspiration if they convert between head-dependent flow and constant flow; and the change in storage coefficient when a cell converts between confined and unconfined. When a nonlinear flow equation is being solved, external iteration (ITMX > 1) is normally required to accurately approximate the nonlinearities. Note that when nonlinearities caused by water-table calculations are part of a simulation, obvious signs may not be present in the output from a simulation that does not use external iteration to indicate that iteration is needed. In particular, the budget error may be acceptably small without iteration even though large error in head exists because of nonlinearity. Additional information about this issue is contained in Chapter 7 of Harbaugh (2005).

MUTD4—is a flag that indicates the quantity of information that is printed when convergence information is printed for a time step.

MUTD4 = 0 indicates that the number of iterations in the time step and the maximum head change each iteration are printed.

MUTD4 = 1 indicates that only the number of iterations in the time step is printed.

MUTD4 = 2 indicates no information is printed.

ACCL—is a multiplier for the computed head change each iteration. Normally this value is 1. A value greater than 1 may be useful for improving the rate of convergence when using external iteration to solve nonlinear problems (IFREQ = 3). ACCL should always be 1 for linear problems. When ITMX = 1, ACCL is changed to 1 regardless of the input value; however, a value must always be specified.

HCLOSE—is the head change closure criterion. If iterating (ITMX > 1), iteration stops when the absolute value of head change at every node is less than or equal to HCLOSE. HCLOSE is not used if not iterating, but a value must always be specified.

IPRD4—is the time step interval for printing out convergence information when iterating (ITMX > 1). For example, if IPRD4 is 2, convergence information is printed every other time step. A value must always be specified even if not iterating.

Geometric Multigrid Solver Package

Input to the Geometric Multigrid Solver (GMG) Package is read from the file that is type GMG in the Name File. All variables in the input list are free format. The GMG Package is documented in Wilson and Naff (2004) and users of GSFLOW are encouraged to review that document when using the GMG Package. The GMG Package is based on the PCG Package algorithm.

FOR EACH SIMULATION

0. [#Text]
    Item 0 is optional—# must be in column 1. Item 0 can be repeated multiple times and between items 1 through 4.
1. RCLOSE   IITER   HCLOSE   MXITER
2. DAMP   IADAMP   IOUTGMG
3. ISM ISC
4. RELAX

Explanation of Variables Read by the GMG Package:

Text—is a character variable (199 characters) that starts in column 2. Any characters can be included in Text. The # character must be in column 1. Lines beginning with # are restricted to the first lines of the file. Text is written to the Listing File.

RCLOSE— is the residual convergence criterion for the inner iteration. The PCG algorithm computes the l2-norm of the residual and compares it against RCLOSE. Typically, RCLOSE is set to the same value as HCLOSE. If RCLOSE is set too high, then additional outer iterations may be required due to the linear equation not being solved with sufficient accuracy. Conversely, too restrictive setting for RCLOSE for nonlinear problems may force an unnecessarily accurate linear solution. This may be alleviated with the IITER parameter or with damping.

The convergence criteria for the GMG Package may be quite different from the other solver packages. Please refer to the "Convergence Criteria" section of Wilson and Naff (2004, p. 9).

IITER— is the maximum number of PCG iterations for each linear solution. A value of 100 is typically sufficient. It is frequently useful to specify a smaller number for nonlinear problems so as to prevent an excessive number of inner iterations.

HCLOSE— is the head-change convergence criterion for nonlinear problems. After each linear solve (inner iteration), the max-norm of the head change is compared to HCLOSE. HCLOSE can be set to a large number for linear problems; HCLOSE is ignored if MXITER=1.

MXITER— is the maximum number of outer iterations. For linear problems, MXITER can be set to 1. For nonlinear problems, MXITER needs to be larger, but rarely more than 100.

DAMP— is the value of the damping parameter. For linear problems, a value of 1.0 should be used. For nonlinear problems, a value less than 1.0 but greater than 0.0 may be necessary to achieve convergence. A typical value for nonlinear problems is 0.5. Damping also helps control the convergence criterion of the linear solve to alleviate excessive PCG iterations [see equation 20 in Wilson and Naff (2004, p. 9)].

IADAMP— is a flag that controls adaptive damping. The possible values of IADAMP and their meanings are as follows:

If IADAMP = 0, then the value assigned to DAMP is used as a constant damping parameter. time of the simulation. This results in use of the first value in arrays FLWRAT, SBHED, and AuxVar for each respective boundary cell.

If IADAMP ≠ 0, then the value of DAMP is used for the first nonlinear iteration.

[see "Nonlinear Solution" section in Wilson and Naff (2004, p. 8)]. The damping parameter is adaptively varied on the basis of the head change, using Cooley's method as described in Mehl and Hill (2001), for subsequent iterations.

IOUTGMG— is a flag that controls the output of the GMG solver. The possible values of IOUTGMG and their meanings are as follows:

0—is for printing only the solver inputs.

1—is for printing of each linear solve, the number of PCG iterations, the value of the damping parameter, the l2-norm of the residual, and the maxnorm of the headchange and its location (column, row, layer). At the end of a time/stress period, the total number of GMG calls, PCG iterations, and a running total of PCG iterations for all time/stress periods are printed.

2—is for printing the convergence history of the PCG iteration that includes the l2-norm of the residual and the convergence factor for each iteration.

3—is the same as 1, except output is sent to the terminal instead of the MODFLOW list file.

4—is the same as 2, except output is sent to the terminal instead of the MODFLOW list file.

ISM— is a flag that controls the type of smoother used in the multigrid preconditioner. The possible values for ISM and their meanings are as follows:

If ISM = 0, then ILU(0) smoothing is implemented in the multigrid preconditioner. This smoothing requires an additional vector on each multigrid level to store the pivots in the ILU factorization.

If ISM = 1, then Symmetric GaussSeidel (SGS) smoothing is implemented in the multigrid preconditioner. No additional storage is required for this smoother; users may want to use this option if available memory is exceeded or nearly exceeded when using ISM =0. Using SGS smoothing is not as robust as ILU smoothing; additional iterations are likely to be required in reducing the residuals. In extreme cases, the solver may fail to converge as the residuals cannot be reduced sufficiently.

ISC— is a flag that controls semicoarsening in the multigrid preconditioner. The possible values of ISC and their meanings are given as follows:

If ISC = 0, then the rows, columns, and layers are all coarsened.

If ISC = 1, then the rows and columns are coarsened, but the layers are not.

If ISC = 2, then the columns and layers are coarsened, but the rows are not.

If ISC = 3, then the rows and layers are coarsened, but the columns are not.

If ISC = 4, then there is no coarsening.

Typically, the value of ISC should be 0 or 1. In the case that there are large vertical variations in the hydraulic conductivities, then a value of 1 should be used [see Remark 9 in Wilson and Naff (2004, p. 18). If no coarsening is implemented (ISC = 4), then the GMG solver is comparable to the PCG2 ILU(0) solver described in Hill (1990) and uses the least amount of memory.

RELAX— is a relaxation parameter for the ILU preconditioned conjugate gradient method. The RELAX parameter can be used to improve the spectral condition number of the ILU preconditioned system. The value of RELAX should be approximately one. However, the relaxation parameter can cause the factorization to break down. If this happens, then the GMG solver will report an assembly error and a value smaller than one for RELAX should be tried. This item is read only if ISC = 4.

Array Reading Utility Subroutines

The array reading utility subroutines provide a common way for all packages to read variables that have multiple values. The term "array" is simply a programming term for a variable that contains multiple values. There are three subroutines: U2DREL, U2DINT, and U1DREL. U2DREL reads real two-dimensional variables, U2DINT reads integer two-dimensional variables, and U1DREL reads real one-dimensional variables. All of these subroutines work similarly. They read one array-control line and, optionally, a data array in a format specified on the array-control line. Several alternate structures for the control line are provided. The original fixed-format control lines work as documented in McDonald and Harbaugh (1988), and four free-format versions have been added. The free-format versions are described first because they are easier to use.

FREE-FORMAT CONTROL LINES FOR ARRAY READERS:

Values in **_bold italics_** are keywords that can be specified as uppercase or lowercase. Each control line is limited to a length of 199 characters.

1. **_CONSTANT_**   CNSTNT
   All values in the array are set equal to CNSTNT.

2. **_INTERNAL_**   CNSTNT   FMTIN   IPRN
   The individual array elements will be read from the same file that contains the control line.

3. **_EXTERNAL_**   Nunit   CNSTNT   FMTIN   IPRN
   The individual array elements will be read from the file unit number specified by Nunit. The name of the file associated with this file unit must be contained in the Name File.

4. **_OPEN/CLOSE_**   FNAME   CNSTNT   FMTIN   IPRN
   The array will be read from the file whose name is specified by FNAME. This file will be opened on unit 99 just prior to reading the array and closed immediately after the array is read. This file should not be included in the Name File. A file that is read using this control line can contain only a single array.

FIXED-FORMAT CONTROL LINE FOR ARRAY READERS:

A fixed-format control line contains the following variables:
    LOCAT   CNSTNT   FMTIN   IPRN
These variables are explained below. LOCAT, CNSTNT, and IPRN are 10-character numeric fields. For U2DREL and U1DREL, CNSTNT is a real number. For U2DINT, CNSTNT is an integer and must not include a decimal. FMTIN is a 20-character text field. All four variables are always read when the control line is fixed format; however, some of the variables are unused in some situations. For example when LOCAT = 0, FMTIN and IPRN are not used.

Explanation of Variables in the Control Lines:

LOCAT—indicates the location of the array values for a fixed-format array control line. If LOCAT = 0, all elements are set equal to CNSTNT. If LOCAT > 0, it is the unit number specified in the MODFLOW NAME file for reading formatted lines using FMTIN as the format. If LOCAT < 0, it is the unit number for binary (unformatted) lines, and FMTIN is ignored. Also, when LOCAT is not 0, the array values are multiplied by CNSTNT after they are read.

CNSTNT—is a real-number constant for U2DREL and U1DREL, and an integer constant for U2DINT. If the array is being defined as a constant, CNSTNT is the constant value. If individual elements of the array are being read, the values are multiplied by CNSTNT after they are read. CNSTNT, when used as a multiplier and specified as 0, is changed to 1.

FMTIN—is the format for reading array elements. The format must contain 20 characters or less. The format must either be a standard Fortran format that is enclosed in parentheses, (FREE) which indicates free format, or (BINARY) which indicates binary (unformatted) data. When using a free-format control line, the format must be enclosed in apostrophes if it contains one or more blanks or commas. A binary file that can be read by MODFLOW may be created in only two ways. The first way is to use MODFLOW to create the file by saving heads in a binary file. This is commonly done when the user desires to use computed heads from one simulation as initial heads for a subsequent simulation. The other way to create a binary file is to write a special program that generates a binary file, and compile this program using a Fortran compiler that is compatible with the compiler used to compile MODFLOW. (FREE) and (BINARY) only can be specified in free-format control lines. Also, (BINARY) can be specified only when using U2DREL or U2DINT, and only when the control line is EXTERNAL or OPEN/CLOSE. When the (FREE) option is used, be sure that all array elements have a non-blank value and that a comma or at least one blank separates adjacent values.

IPRN—is a flag that indicates whether the array being read should be written to the Listing File after the array has been read and a code for indicating the format that should be used when the array is written. The format codes are different for each of the three array-reading subroutines as shown below. IPRN is set to zero when the specified value exceeds those defined. If IPRN is less than zero, the array will not be printed.

IPRN	U2DREL	U2DINT	U1DREL
0	10G11.4	10I11	10G12.5
1	11G10.3	60I1	5G12.5
2	9G13.6	40I2	
3	15F7.1	30I3	
4	15F7.2	25I4	
5	15F7.3	20I5	
6	15F7.4	10I11	
7	20F5.0	25I2	
8	20F5.1	15I4	
9	20F5.2	10I6	
10	20F5.3		
11	20F5.4		
12	10G11.4		
13	10F6.0		
14	10F6.1		
15	10F6.2		
16	10F6.3		
17	10F6.4		
18	10F6.5		
19	5G12.5		
20	6G11.4		
21	7G9.2		

Nunit—is the unit for reading the array when the EXTERNAL free-format control line is used.

Examples of Free-Format Control Lines

The following examples use free-format control lines for reading an array. The example array is a real array consisting of 4 rows with 7 columns per row:

CONSTANT 5.7	This sets an entire array to the value 5.7.
INTERNAL 1.0 (7F4.0) 3	This reads the array values from the
1.2 3.7 9.3 4.2 2.2 9.9 1.0	file that contains the control line.
3.3 4.9 7.3 7.5 8.2 8.7 6.6	Thus, the values immediately follow the
4.5 5.7 2.2 1.1 1.7 6.7 6.9	control line.
7.4 3.5 7.8 8.5 7.4 6.8 8.8	
EXTERNAL 52 1.0 (7F4.0) 3	This reads the array from the formatted file opened on unit 52.
EXTERNAL 47 1.0 (BINARY) 3	This reads the array from the binary file opened on unit 47.
OPEN/CLOSE test.dat 1.0 (7F4.0) 3	This reads the array from the file named "test.dat".

List Utility Subroutine (ULSTRD)

Subroutine ULSTRD reads lists that are any number of repetitions of an input item that contains multiple variables. Examples of packages that make use of this subroutine are the General-Head Boundary, Well, and Streamflow-Routing Packages.

1. [***EXTERNAL*** IN] or [***OPEN/CLOSE*** FNAME]
   If Item 1 is not included, then the list is read from the package file. Item 1 must begin with the keyword EXTERNAL or the keyword OPEN/CLOSE (not both).
2. [***SFAC*** Scale]
3. List

Explanation of Variables Read by the List Utility Subroutine:

IN—is the unit number for a file from which the list will be read. The name of the file associated with this file unit must be contained in the Name File, and its file type must be DATA in the Name File.

FNAME—is the name of a file from which the list will be read. This file will be opened on unit 99 just before reading the list and closed immediately after the list is read. This file should not be included in the Name File.

Scale—is a scale factor that is multiplied times the value of one or more variables within every line of the list. The input instructions that define a list, which will be read by ULSTRD, should specify the variables to which ***SFAC*** applies. If item 2 is not included, then Scale is 1.0. If item 2 is included, it must begin with the keyword ***SFAC***. The values of the list variables that are printed to the listing file include the effect of Scale.

List—is a specified number of lines of data in which each line contains a specified number of variables. The first three variables are always layer, row, and column. The other fields vary according to which package is calling this subroutine.

Appendix 2. Definitions of Symbols.

Symbol	Equation	Definition
$A_{aq}^{m,n}$	92	Area of the lakebed covering a finite-difference cell, in length squared
A_{dwn}^{msf-1r}	82	Cross-sectional area at the downstream end of a stream reach for the previous, streamflow-routing time step $msfr$-1, in length squared
A_{fdc}	40, 42, 108	Top area of the finite-difference cell, in length squared
A_{fdca}	42, 60	Top area of finite-difference cell, in acres
A_{GVR}	42, 45, 56	Area of gravity reservoir, in acres
$A_{GVR,L}$	46	Area of gravity reservoir L, in acres
A_{HRU}	22, 31, 32, 33, 37, 39, 45, 46, 58, 67a	Area of the HRU, in acres
$A_{HRU,dwn}$	36	Area of the downslope HRU, in acres
$A_{HRU,up}$	36	Area of the upslope HRU, in acres
A_J	77, 90	Area of HRU J, in acres
$a_{i,j,k,N}$	112a, 112b	Flow from the Nth source into cell i, j, k, in cubic length per time
A_{lksurf}^{m-1}	93	Surface area of the lake for the last iteration of time step m-1, in length squared
$A_{lakeHRU}$	91	Area of the lake HRU, in acres
A_{perv}	37, 38a, 38b, 39	Pervious area of the HRU, in acres
A_{up}^{msfr}	82	Cross-sectional area at the upstream end of a stream reach for the current streamflow-routing time step $msfr$, in length squared
A_{up}^{msfr-1}	82	Cross-sectional area at the upstream end of a stream reach for the previous streamflow-routing time step $msfr$-1, in length squared
ap_{HRU}^m	15a	Degree-day based ratio of actual to potential shortwave radiation for the HRU during time step m, dimensionless
Asc_{HRU}^m	29, 32	Snow-covered area of the HRU determined by snow-cover areal depletion curve (Anderson, 1973), in acres
$awdth$	87	Empirical coefficients determined from regression methods, undefined dimension
B	16b	Constant used in the cloud-cover to solar-radiation relation, a value can be obtained from Thompson (1976, fig. 1), dimensionless
b_{month}	1	Monthly maximum (or minimum) daily temperature lapse rate representing the change in maximum (or minimum) air temperature per 1,000 feet or meters of altitude change for each month, January to December, in degrees Fahrenheit or Celsius per 1,000 length units
$b_{X,month}$ $b_{Y,month}$ and $b_{Z,month}$	3b, 3c	Maximum (or minimum) air-temperature regression coefficients for longitude, latitude, and altitude, respectively by month, starting with January, in degrees Fahrenheit or Celsius
$bwdth$	87	Empirical exponent determined from regression methods, dimensionless
C	105	Matrix of conductance values for the row, column, and layer directions (CR, CC, and CV) that are multiplied by the ground-water head in each variable-head cell and any coefficients multiplied by a variable-head cell to represent external sources and sinks, in length squared per time
C_j	107	Horizontal conductance for cell j in the row direction, in length squared per time
C_{j-1}	107	Horizontal conductance for cell j-1 in the row direction, in length squared per time
C_h	106	Horizontal conductance for a prism, in length squared per time
C_{ice}	25	Specific heat of ice, in calories per gram-degree Celsius

Symbol	Equation	Definition
C_{imper}^{m}	31	Water available for Hortonian runoff from the impervious part per unit area of the HRU during time step m, in inches
C_m	85, 86	Constant, which is 1.0 for cubic meters per second and 1.486 for cubic feet per second
$C_{mf2prms}$	40, 42	Conversion from MODFLOW-2005 length per time to PRMS inches per day
$C_{prms2mf}$	60	Conversion from PRMS inches per day to MODFLOW-2005 length per time
C_v	108	Vertical conductance, in length squared per time
$C1_{i,j,k,N}$	112a	Constant for head-dependent boundary conditions, in length squared per time
$C2_{i,j,k,N}$	112a	Constant for head-independent boundary conditions, in cubic length per time
$CC_{i-1/2}$	104	Conductance between nodes i-$1, j, k$ and i, j, k, in length squared per time
$CC_{i+1/2}$	104	Conductance between nodes i, j, k and i+$1, j, k$, in length squared per time
CC_{HRU}^{m}	16a, 16b	Decimal fraction of cloud cover on the HRU during time step m, dimensionless
$Cd_{lc}^{m,n}$	93	Lakebed conductance of cell face lc for time step m and iteration n, in length squared per time
Ci	52	Constant of integration, days
$cdpth$	87	Empirical coefficients determined from regression methods, undefined dimension
$Celthk$	41	Thickness of the finite-difference cell, in length
$Celtop$	41, 62	Top altitude of the finite-difference cell, in length
CF_{HRU}	7, 8	Monthly correction factor as a decimal fraction used to adjust rain (or snow) at the HRU
$CF_{HRU, rain}$	10a, 10d	Monthly snow correction factor as a decimal fraction of precipitation at the measurement station
CF_{xyz}	9a	Monthly correction factor as a decimal fraction used to adjust rain (or snow)
Ch_{bnd}	105	Vector of conductance values multiplied by ground-water head in each constant-head cell, in cubic length per time
CND_{sz}	41	Conductance across the soil-zone base equal to $\frac{K_v A_{fdc}}{0.5 celthk D_{usz}}\left(h_{fdc}^{m,n-1} - celtop + 0.5 D_{usz}\right)$, in length squared per time
$coefex_{HRU}$	59	Exponent in the equation used to compute gravity drainage from the gravity reservoir, dimensionless
$coeflin_{HRU}$	59	Linear coefficient in the equation used to compute gravity drainage from the gravity reservoir, in inches per day
$CR_{j-1/2}$	101, 102, 104, 107	Conductance between nodes i, j-$1, k$ and i, j, k, in length squared per time
$CR_{j+1/2}$	104	Conductance between nodes i, j, k and i, j+$1, k$, in length squared per time
crb_{month}	16a	Slope of the regression equation that relates cloud cover to daily minimum and maximum air temperature by month starting in January, dimensionless
cs_{HRU}	24	Snowpack settlement-time constant, per day
$CV_{k-1/2}$	104, 109, 110	Vertical conductance between cells i, j, k-1 and i, j, k, in length squared per time
$CV_{k+1/2}$	104, 109, 110	Vertical conductance between cells i, j, k and i, j, k+1, in length squared per time
D	114	Percent discrepancy, dimensionless
D_{CPR}^{m-1}	34b, 39, 63a	Volume per unit area of water in the capillary reservoir at the last iteration of time step m-1, in inches
$D_{CPR}^{m,n}$	39, 44, 46	Volume of water in the capillary reservoir at time step m and iteration n, in inches
$D_{CPR}^{m,n*}$	46	Revised volume per unit area of water in the capillary reservoir for time step m, iteration n, in inches
D_{GVR}^{m-1}	40, 54, 55, 56	Volume of water per unit area in the gravity reservoir at time step m-1, in inches
$D_{GVR}^{m,n}$	40, 48, 49, 50, 51, 52, 53, 54, 55, 58, 59	Volume of water per unit area in the gravity reservoir at time step m and iteration n, in inches

Symbol	Equation	Definition
$D_{GVR}^{m,n-1}$	45	Volume per unit area of water in the gravity reservoir for time step m and iteration n-1, in acre-inch
$D_{GVR}^{m,n*}$	43, 58	Volume per unit area in a gravity reservoir after slow interflow and before gravity drainage for time step m and iteration n, in inches
$D_{GVR}^{m,n**}$	43	Revised volume per unit area in a gravity reservoir after gravity drainage and before Dunnian runoff for time step m and iteration n, in inches
$D_{GVR,tot}^{m,n**}$	64	Volume of water in all gravity reservoirs after interflow and gravity drainage in an HRU, divided by the area of the HRU for time step m and iteration n, in inches
D_{imper}^{m}	33	Impervious storage as volume per unit area for the HRU, during time step m, in inches
D_{imper}^{m-1}	31, 32, 33	Impervious storage for the last iteration of time step m-1, as volume per unit area for the HRU, in inches
$D_{mdpt}^{m,n}$	86, 87, 89	Depth of water at the midpoint of a stream reach for time step m and iteration n, in length
$D_{mf2GVR}^{m,n}$	42, 43	Volume per unit area from the finite-difference cell to a connected gravity reservoir for time step m and iteration n, in inches
D_{PFR}^{m}	67	Volume of water in the preferential-flow reservoir per unit area for time step m, in inches
$D_{PFR}^{m,n}$	48	Volume per unit area in the preferential-flow reservoir for time step m, iteration n, in inches
$D_{PFR}^{m,n-1}$	64	Volume per unit area of water in the preferential-flow reservoir before fast interflow for time step m and iteration n-1, in inches
D_{PFR}^{m-1}	48, 67	Volume per unit area in the preferential-flow reservoir at the last iteration of time step m-1, in inches
$D_{rej}^{m,n}$	43	Gravity drainage volume per unit area that is rejected by the finite-difference cell, in inches
$D_{SZfc}^{m,n}$	64, 65	Volume of water per unit area of the soil zone above field capacity for time step m and iteration n, in inches
D_{usz}	41, 62	Depth of undulations at soil-zone base; in length
$Ddeficit_{CPR}^{m,n}$	44, 45	Volume of water per unit HRU area required to replenish the capillary reservoir to the field-capacity threshold for time step m and iteration n, in inches
$Dexcess_{CPR}^{m,n}$	40	Volume of water per unit area of excess water in the capillary reservoir for time step m and iteration n, in inches
$Dfif_{J}^{m,n}$	77, 90	Fast interflow from preferential flow reservoirs per unit area of HRU J for time step m and iteration n, in inches
$Dfct_{HRU}$	44, 46, 47, 63a, 65	Maximum volume of water per unit area in the capillary reservoir, in inches
$dfsub_{HRU}$	30	Decimal fraction of potential evapotranspiration that is sublimated from the snow surface, dimensionless
$Dfup_{HRU}^{m,n}$	39	Volume of fast interflow per unit area into the capillary reservoir from all contributing HRUs at time step m and iteration n, in inches
$Dhru_{PFR}^{m}$	38a	Volume of water per unit area added to all connected preferential-flow reservoirs for time step m, in inches
$Dhru_{PFR}^{m-1}$		Volume of water per unit area added to all connected preferential-flow reservoirs for time step m-1, in inches
$Dimx_{imper}$	31	Maximum retention storage for HRU impervious area, in inches
DM^{m}	12, 13a, 14b, 14c	Solar declination for time step m, angular degrees
Dmx_{HRU}	59	Maximum amount of gravity drainage from the gravity reservoir, in inches
$Dpft_{HRU}$	47, 48	Preferential-flow threshold as volume per unit area, in inches
$DR_{CPR}^{m,n}$	63	Soil-water content ratio in the capillary reservoir for time step m and iteration n, dimensionless

Symbol	Equation	Definition
$Drem_{CPR}^m$	38b, 39	Volume of water per unit area of the pervious part of the HRU that infiltrates into the capillary reservoir at time step m, in inches
$Drpl_{GVR}^{m,n}$	45	Volume of water per unit area that is removed from the gravity reservoir to replenish the capillary reservoir during time step m and iteration n, in inches
$Drpl_{GVR,L}^{m,n}$	46	Volume of water per unit area that is removed from gravity reservoir L to replenish the capillary reservoir during time step m and iteration n, in inches
Ds_{HRU}^m	24, 25	Total snowpack depth for the HRU during time step m, in inches
$Dsat_{HRU}$	47, 65	Maximum volume of water per unit area in the soil zone, in inches
$Dsif_{HRU}^{m,n}$	57	Volume per unit area of slow interflow from the gravity reservoirs of the HRU for time step m, iteration n, in inches
Dse_{HRU}^m	24, 25	Snowpack liquid water equivalent depth for the HRU at time step m, in inches
$Dslup_{HRU}^{m,n}$	39	Volume of slow interflow per unit area into the capillary reservoir from all contributing HRUs at time step m and iteration n, in inches
Dup_{HRU}^m	34a	Antecedent volume per unit area of water in the capillary reservoir that is available for evaporation during time step m, in inches
$Dupmx_{HRU}$	34a	Maximum volume per unit area of water in upper capillary reservoir, in inches
$D(\theta)$	70	Hydraulic diffusion coefficient, in length squared per time
e	19, 56	Exponential function constant (~2.7182818), dimensionless
E^m	11	Obliquity of the Sun's ecliptic for time step m, angular degrees
e_t	115b	Budget error at time t, in cubic length per time
$EAPR_{CPR}^{m,n}$	63	Actual to potential evapotranspiration ratio in the capillary reservoir for time step m and iteration n, dimensionless
EC	11	Eccentricity of the Earth's orbit (~0.01671), radians
eic_t	116d	Volumetric flow rate of evaporation of intercepted precipitation at time t, in cubic length per time
eim_t	116d	Volumetric flow rate of evaporation from impervious areas at time t, in cubic length per time
eP_t	116d	Volumetric flow rate of evapotranspiration from pervious areas at time t, in cubic length per time
esp_t	116d	Volumetric flow rate of snowpack sublimation at time t, in cubic length per time
ET_{GVR}^m	63e	Actual evapotranspiration removed from the capillary reservoir for time step m and iteration n, in inches per day
$Evap_{imper}^m$	32, 33	Evaporation from the impervious part of the HRU for time step m, in inches
ex_t	116c	Volumetric flow rate of ground-water discharge at time t, in cubic length per time
$F_{J,sr}$	77	Decimal fraction of the total area of HRU J that contributes runoff and interflow to a particular stream reach, dimensionless
$F_{J,lakeHRU}$	90	Decimal fraction of the total area of HRU J that contributes runoff and interflow to a lake HRU, dimensionless
f_1^{uzf}	112b	Recharge received from the unsaturated zone that was stored in the unsaturated interval through which the water table rose during time step m, in length cubed per time
f_2^{uzf}	112b	Recharge that flowed across the surface defined by the water table in the cell at time m-1, in cubic length per time
$face_{j+1/2}$	95, 96	Cell face between cells j and $j+1$, dimensionless
$face_{j-1/2}$	95, 96	Cell face between cells $j-1$ and j, dimensionless
$Fcontrib_{HRU,up}$	36	Decimal fraction of area in the upslope HRU that contributes Hortonian runoff to the downslope HRU, dimensionless
$Fden_{HRU}$	38a, 38b, 47	Decimal fraction of the soil zone available for preferential flow, dimensionless
$fdpth$	87	Empirical exponent determined from regression methods, dimensionless

Symbol	Equation	Definition
Fmn_{HRU}	34a	Minimum possible contributing area for pervious runoff, as a decimal fraction of HRU area, dimensionless
Fmx_{HRU}	34a, 34b	Maximum possible contributing area for pervious runoff, as a decimal fraction of HRU area, dimensionless
$Fperv_{HRU}^{m}$	34a, 34b, 35	Decimal fraction of the surface-runoff-contributing area of the pervious parts in the HRU for time step m, dimensionless
$Frain_{HRU}^{m}$	6	Decimal fraction of total precipitation occurring as rain on an HRU for time step m, dimensionless
$fstcoef_{lin}$	67a	Linear flow-routing coefficient for fast interflow, per day
$fstcoef_{sq}$	67a	Nonlinear flow-routing coefficient for fast interflow, in day per inch
gz_{t}	116a	Volume of water in gravity reservoirs of the soil zone at time t, in cubic length
gz_{t-1}	116b	Volume of water in gravity reservoirs of the soil zone at time t-1, in cubic length
h	94, 95, 96, 97	Potentiometric head, in length
$\overline{h}$	105	Vector of ground-water heads for all variable-head cells, in length
$h_{i,j-1,k}$	101	Head at node $i, j-1, k$, length
$h_{i,j,k}$	101, 112a	Head at node i, j, k, in length
$h_{i,j,k}^{m}$	103, 104, 111, 112b	Head at node i, j, k, at end of time step m
$h_{i-1,j,k}^{m}$	104	Head at node i-1, j, k, at end of time step m
$h_{i+1,j,k}^{m}$	104	Head at node i+1, j, k, at end of time step m
$h_{i,j-1,k}^{m}$	104	Head at node i, j-1, k, at end of time step m
$h_{i,j+1,k}^{m}$	104	Head at node i, j+1, k, at end of time step m
$h_{i,j,k-1}^{m}$	104	Head at node i, j, k-1, at end of time step m
$h_{i,j,k+1}^{m}$	104	Head at node i, j, k+1, at end of time step m
$h_{aqfdc}^{m,n}$	92	Ground-water head in the finite-difference cell adjacent to the lake at the cell node for time step m and iteration n, in length
$h_{fdc}^{m,n}$	41, 62, 78	Ground-water head in the finite-difference cell for time step m and iteration n, in length
$h_{fdc}^{m,n-1}$	41	Ground-water head in the finite-difference cell for time step m, iteration n-1; in length
$h_{lake}^{m,n}$	92, 93	Lake stage for time step m and iteration n, in length
h_{lake}^{m-1}	93	Lake stage for time step m-1, in length
$h_{afdc,lc}^{m,n-1}$	93	Ground-water head in finite-difference cell associated with cell face lc for time step m and iteration n-1, in length
$h_{str}^{m,n}$	78	Stream head at the midpoint of the stream reach for time step m and iteration n, in length
HC_{HRU}	18	Hamon monthly air-temperature coefficient, in inch-cubic meter per gram
Hc_{HRU}^{m}	28	Convective or sensible heat at air-snow interface during time step m, in calories
He_{HRU}^{m}	28	Latent heat (sublimation and condensation) at the air-snow interface during time step m, in calories
HF	29	Specific latent heat of fusion to melt 1-inch of water-equivalent ice at 0°Celsius, 203.2 calories per inch
Hg_{HRU}^{m}	28	Heat gained from the ground during time step m, in calories
Hm_{HRU}^{m}	28, 29	Energy available for snowmelt during time step m, in calories

Symbol	Equation	Definition
Hp_{HRU}^m	28	Heat gained from precipitation during time step m, in calories
Hq_{HRU}^m	28	Heat required for internal state change during time step m, in calories
Hs_{HRU}^m	28	Energy gained due to shortwave radiation during time step m, in calories
Ht_{HRU}^m	25	Daily heat transferred from the surface layer to the lower layer of the snowpack during time step m, in calories per square centimeter
i	68, 69b, 69c, 76a, 76b	Evapotranspiration rate beneath the soil-zone base per unit depth, in length per time per length
ic_t	116a	Volume of intercepted precipitation in plant canopy reservoirs at time t, in cubic length
ic_{t-1}	116b	Volumes of intercepted precipitation in plant canopy reservoirs at time t-1, in cubic length
im_t	116a	Volume of water in impervious reservoir at time t, in cubic length
im_{t-1}	116b	Volume of water in impervious reservoir at time t-1, in cubic length
in	114	Total inflow to the model, cubic length per time
in_t	115b, 116c	Total inflow at time t, in cubic length per time
J	77, 90	Counter for the HRU number, dimensionless
jd	11	Julian day number (3 is subtracted as the solar year begins on December 29), days
JH_{HRU}	20a	Jensen-Haise air-temperature coefficient for the HRU, in degrees
JH_{month}	20a	Monthly Jensen-Haise air-temperature coefficient, per degrees Fahrenheit
JJ	77, 90	Total number of HRUs that contribute surface runoff and interflow to a particular stream reach (77) or lake (90)
k	57	Total number of gravity reservoirs in the HRU, dimensionless
K_{aq}	92	Horizontal or vertical hydraulic conductivity of aquifer finite-difference cell adjacent to lake cell, in length per time
K_{lkbd}	92	Hydraulic conductivity of the lakebed, in length per time
K_s	61a, 61b, 74	Vertical saturated hydraulic conductivity of the unsaturated zone, in length per time
K_{strbed}	78, 89	Hydraulic conductivity of the streambed, in length per time
K_v	41, 108	Vertical hydraulic conductivity of the finite-difference cell, in length per time
ke_{HRU}^m	25, 26	Effective thermal conductivity of the snowpack, in calories per second-gram-degree Celsius
K_{xx}	94, 95, 96	Hydraulic conductivity tensor aligned with the x coordinate axes, in length per time
K_{yy}	94	Hydraulic conductivity tensor aligned with the y coordinate axes, in length per time
K_{zz}	94	Hydraulic conductivity tensor aligned with the z coordinate axes, in length per time
$K(\theta)$	68, 69a, 70, 71, 74, 75	Unsaturated vertical hydraulic conductivity as a function of water content and is equal to the vertical flux, in length per time
$K(\theta_{z_1})$	71, 73, 76a	Unsaturated vertical hydraulic conductivity at depth z_1, in length per time
$K(\theta_{z_2})$	71, 73, 76a	Unsaturated vertical hydraulic conductivity at depth z_2, in length per time
$KR_{j-1/2}$	102	Hydraulic conductivity between nodes i, $j-1$, k and i, j, k, in length per time
L	46, 60, 66	Counter for the gravity reservoir number, dimensionless
lat	13a	Latitude of the horizontal surface (basin centroid, HRU or equivalent-slope surface), positive values are in the northern hemisphere, negative values are in the southern hemisphere, in radians
lat'_{HRU}	14b, 14c	Latitude of the equivalent-slope surface of the HRU, in radians
lc	93	Counter used to refer to individual finite-difference cell faces in contact with the submerged lakebed
LC	93	Number of finite-difference cell faces in contact with the submerged lakebed
$Ldist_i$	5b	Inverse distance between HRU centroid and measurement station i

Symbol	Equation	Definition
$Length_{fdc}$	106	Length of the prism, in length
$length_{str}$	78, 89	Length of the stream reach, in length
LL	46, 60, 66	Total number of gravity reservoirs, dimensionless
$long'_{HRU}$	14c	Longitude offset between the equivalent-slope surface and the HRU, in radians
$npstas$	10a	Number of precipitation stations, dimensionless
$ntstas$	4, 5a	Number of air temperature measurement stations, dimensionless
out	114	Total outflow from the model, cubic length per time
out_t	115b, 116d	Total outflow at time t, in cubic length per time
P^m_{HRU}	7, 9a, 10a, 22, 23	Precipitation at the HRU during time step m, in inches
$P^m_{lakeHRU}$	91	Precipitation on the lake HRU for time step m, in inches
P^m_{sta}	7, 10a	Measured precipitation at the station during time step m, in inches
p_t	116c	Volumetric flow rate of precipitation on modeled region at time t, in cubic length per time
$\overline{P}_{base}$	8	Mean monthly precipitation at the base station, in inches per day
$\overline{P}_{lapse}$	8	Mean monthly precipitation at the lapse station, in inches per day
$\overline{P}_{sta,month}$	10c, 10d	Mean monthly precipitation at each measurement station, length
$\overline{P}^m_{stas}$	9c	Mean precipitation of all stations during time step m, in inches
pw_t	116a	Volume of water in snowpack storage at time t, in cubic length
pw_{t-1}	116b	Volume of water in snowpack storage at time t-1, in cubic length
paf_{base}	8	Mean monthly factor used to adjust rain or snow lapse rate (usually 1.0), dimensionless
$Pancoef_{month}$	21	Monthly pan-evaporation coefficient, dimensionless
$Panevap^m_{sta}$	21	Pan evaporation at each measurement station for time step m, length
PET^m_{HRU}	18, 20a, 21, 30, 32, 63e	Potential evapotranspiration for the HRU during time step m, in inches
$pmixaf_{month}$	6	Monthly rain adjustment factor for a mixed precipitation event (usually 1.0), dimensionless
$pptadj_{HRU}$	15b	Precipitation-day adjustment factor, dimensionless
$Pnet^m_{HRU}$	23, 31, 32, 33, 34b, 35, 37	Precipitation that reaches the ground during time step m, in inches
Ps^m_{HRU}	24	Net snowfall rate, as a liquid water equivalent, on the HRU during time step m, in inches per day
Ptf^m_{HRU}	22, 23	Precipitation throughfall on the HRU during time step m, in inches
$q^{m,n}_{gd,pot}$	59	Potential gravity drainage per unit area for time step m and iteration n, in inches per day
$q^{m,n}_{GVR,in}$	49, 51, 52, 56	Volumetric inflow rate per unit area to the gravity reservoir, in inches per day
$q^{m,n}_{GVR,sif}$	49, 50, 56, 58	Slow interflow rate per unit area from the gravity reservoir, in inches per day
$q^{m,n}_{i,sif}$	57	Slow interflow from gravity reservoir i for time step m, iteration n, in inches per day.
$q^{m,n}_{gdc,net}$	61a, 61b, 62	Net gravity drainage to the connected finite-difference cell for time step m, and iteration n, in length per time
$q^{m,n*}_{gdc,net}$	62	Net gravity drainage to the saturated zone for time step m and iteration n, in length per time[*]
$q^{m,n}_{gdc,pot}$	60	Potential gravity drainage for time step m and iteration n from all gravity reservoirs connected to a finite-difference cell, in length per time

Symbol	Equation	Definition
$q_{gd,L}^{m,n}$	60	Potential gravity drainage per unit area of gravity reservoir L for time step m and iteration n, in inches per day
$q_{gd,L}^{m,n-1}$	60	Potential gravity drainage per unit area of gravity reservoir L for time step m and iteration n-1, in inches per day
Q_{in}	113	Total inflow to the modeled region during a time step, in cubic length per time
$q_{i+1/2}$	98, 99	Volumetric flow rate between cells i and i+1, in cubic length per time
$q_{i-1/2}$	98, 99	Volumetric flow rate between cells i-1 and i, in cubic length per time
$q_{j+1/2}$	96, 98, 99	Volumetric flow rate between cells j and j+1, in cubic length per time
$q_{j-1/2}$	96, 98, 99, 101	Volumetric flow rate between cells j-1 and j, in cubic length per time
$q_{k+1/2}$	98, 99	Volumetric flow rate between cells k and k+1, in cubic length per time
$q_{k-1/2}$	98, 99	Volumetric flow rate between cells k-1 and k, in cubic length per time
Q_L	99, 100	Volumetric flow rate into a cell (negative value is a volumetric flow out of cell), in cubic length per time
$Q_{lakeHRUtolakemf}^{m,n}$	91	Volumetric flow rate from a lake HRU to a lake represented in MODFLOW-2005 for time step m and iteration n, in cubic length per time
$Q_{lakeHRUtomf}^{m,n}$	93	Sum of precipitation, evaporation, surface runoff, and interflow calculated in PRMS, in cubic length per time
$Q_{lakeleak}^{m,n}$	92	Volumetric-flow rate across the lakebed to center of aquifer finite-difference cell for time step m and iteration n and is outward or downward leakage when positive and ground-water discharge when negative, in cubic length per time
$Q_{lateral}^{m,n}$	77, 79, 84, 88	Volumetric flow rate into a stream reach from all connected HRUs for time step m and iteration n, in cubic length per time
$Q_{mdpt}^{m,n}$	84, 85, 86, 87	Volumetric-flow rate at the midpoint of a stream reach for time step m and iteration n, in cubic length per time
Q_{out}	113	Total outflow from the modeled region during a time step, in cubic length per time
$Q_{mdpt}^{m,n}$	84	Volumetric-flow rate at the midpoint of a stream reach for time step m and iteration n, in cubic length per time m
$Q_{srdvr}^{m,n}$	79, 84	Volumetric flow rate diverted from the end of the stream reach for time step m and iteration n, in cubic length per time
$Q_{srevp}^{m,n}$	79, 84, 88	Specified evaporation rate from the stream surface multiplied by the wetted plan-view area for time step m and iteration n, in cubic length per time
Q_{srin}^{m}	79, 84, 88	Specified volumetric flow rate at the beginning of a stream reach for time step m, in cubic length per time
$Q_{srleak}^{m,n}$	78, 84, 89	Volumetric-flow rate across the streambed for time step m and iteration n and is downward leakage when positive and ground-water discharge to the stream when negative, in cubic length per time
$Q_{srout}^{m,n}$	79	Volumetric flow rate from the end of a stream reach for time step m and iteration n, in cubic length per time
$Q_{srpp}^{m,n}$	79, 84, 88	Specified precipitation rate on the stream surface multiplied by the wetted plan-view area for time step m and iteration n, in cubic length per time
$Q_{srup}^{m,n}$	79, 84, 88	Sum of the volumetric flow rate that enters the stream reach from outflow of upstream reaches for time step m and iteration n, in cubic length per time
$Q_{srleak,mx}^{m,n}$	88	Total amount of water available for stream leakage to ground water for time step m and iteration n, in cubic length per time
$Q_{totstrin}^{m,n}$	93	Sum of all tributary stream inflows for time step m and iteration n, in cubic length per time
$Q_{totstrout}^{m,n}$	93	Sum of lake outflows to streams for time step m and iteration n, in cubic length per time

Symbol	Equation	Definition
Q_{wd}^{m}	93	Specified withdrawal from a lake (negative value adds water) for time step m, in cubic length per time
Q_{dwn}^{msfr}	83	Volumetric flow at the downstream end of the stream reach for the current streamflow-routing time step $msfr$, in cubic length per time
Q_{dwn}^{msfr-1}	83	Volumetric flow at the downstream end of the stream reach for the previous streamflow-routing time step $msfr$-1, in cubic length per time
Q_{up}^{msfr}	83	Volumetric flow at the upstream end of the stream reach for the current streamflow-routing time step $msfr$, in cubic length per time
Q_{up}^{msfr-1}	83	Volumetric flow at the upstream end of the stream reach for the previous streamflow-routing time step $msfr$-1, in cubic length per time
q_{str}	80	Sum of all lateral inflows and outflows per unit length of stream listed on the right side of equation 79, in length squared per time
Q_W	105	Volumetric flow rate added to variable-head cells that represents external sources and sinks to the ground-water system, in cubic length per time
gif_t	116d	Volumetric flow rate of interflow leaving modeled region at time t, in cubic length per time
qce_{HRU}^{m}	30	Evaporation loss from interception storage for the HRU during time step m, in inches
qsi_{perv}^{m}	37, 38a, 38b	Soil infiltration over the pervious part of the HRU for time step m, in acre-inch
$qsnmx_{HRU}$	37	Daily maximum snowmelt infiltration into the soil zone for the HRU, in inches
qso_t	116d	Volumetric flow rate of surface runoff leaving the modeled region at time t, in cubic length per time
$R_{hydraulic}$	85	Hydraulic radius of the stream, which is equal to the stream area divided by the wetted perimeter, in length
rad	11	Revolution speed of the Earth (~0.0172), radians per day
Rah_{HRU}^{m}	15a, 15b, 16b, 17	Measured or computed as the horizontal plane shortwave radiation on the HRU during time step m, in calories per square centimeter per day
$\overline{Rain}_{HRU,month}$	10c	Mean monthly rain on each HRU that can be obtained from National Weather Service's spatial distribution of mean annual precipitation for the 1971–2000 climate normal period, length
$Rasw_{HRU}^{m}$	15b, 17, 20a	Computed shortwave radiation on the HRU during time step m, in calories per square centimeter per day
$ROd_{HRU}^{m,n}$	65, 66	Volume of Dunnian runoff per unit area from an HRU for time step m and iteration n, in inches
$ROd_{J}^{m,n}$	77, 90	Dunnian runoff per unit area from HRU J for time step m and iteration n, in inches
$ROd_{L}^{m,n}$	66	Dunnian runoff from gravity reservoir L for time step m, iteration n, in inches
$ROdup_{HRU}^{m,n}$	39	Volume of Dunnian runoff per unit area into the capillary reservoir from all contributing HRUs at time step m and iteration n, in inches
$ROh_{HRU,dwn}^{m}$	36	Volume per unit area of Hortonian runoff to a downslope HRU or stream segment during time step m, in inches
ROh_{imper}^{m}	31, 32, 33	Hortonian runoff from the impervious part of the HRU per unit area during time step m, in inches
$ROh_{impr,up}^{m}$	36	Runoff per unit area from the impervious part of the upslope HRU for time step m, in inches
$ROh_{J}^{m,n}$	77, 90	Hortonian runoff per unit area from HRU J for time step m and iteration n, in inches
ROh_{perv}^{m}	35, 37	Runoff per unit area from the pervious part of the HRU for time step m, in inches
$ROh_{perv,up}^{m}$	36	Runoff per unit area from the pervious part of the upslope HRU for time step m, in inches
$ROhup_{HRU}^{m}$	31, 32, 33, 35, 37	Sum of Hortonian runoff from all upslope contributing HRUs as a volume per unit area of the HRU for time step m, in inches

Symbol	Equation	Definition
$Roughness$	85, 86	Manning's roughness coefficient, dimensionless
Rsp_{HRU}^{m}	14a, 15a, 16b	Potential solar radiation on the HRU during time step m, in calories per square centimeter per day
S_s	94, 97	Specific storage of the porous rock, per length
S_t	115a, 116a	Storage at time t, in cubic length
S_{t-1}	115a, 116a	Storage at the previous time step, in cubic length
S_y	61a	Specific yield that is used to approximate θ_s-θ_r, dimensionless
sc^m	14a	60-minute period solar constant for time step m, in calories per square centimeter per hour
sh_{HRU}^{m}	13, 14b, 18	Daylight length on the HRU for time step m, hours
$Slope_0$	81, 85, 86	Slope of the channel in the longitudinal profile, dimensionless
$Slope_f$	81	Friction slope of the channel in the longitudinal profile, dimensionless
$slope_{HRU}$	15b, 17	Slope of the HRU, dimensionless
$slwcoef_{lin}$	50, 51, 52, 53, 54, 56	Linear flow-routing coefficient for slow interflow, per day
$slwcoef_{sq}$	50, 51, 52, 53, 54, 56	Nonlinear flow-routing coefficient for slow interflow, in day per inch
sm_t	116a	Volume of water in capillary reservoirs of the soil zone at time t, in cubic length
sm_{t-1}	116b	Volume of water in capillary reservoirs of the soil zone at time t-1, in cubic length
Smc_{HRU}	34b	Coefficient used to calculate decimal fraction of pervious surfaces, dimensionless
$Smex_{HRU}$	34b	Coefficient used to calculate the decimal fraction of pervious surfaces, per inch
$Smidx_{HRU}^{m}$	34b	Soil moisture index of the capillary reservoir for time step m, in inches
$\overline{Snow}_{HRU,month}$	10d	Mean monthly snow on each HRU that can be obtained from National Weather Service's spatial distribution of mean annual precipitation for the 1971–2000 climate normal period, length
Spc_{HRU}^{m}	22	Storage in the plant canopy (summer or winter) on the HRU during time step m, in acre-inch
$Spca_{HRU}^{m}$	22	Available storage in the plant canopy of the HRU during time step m, in acre-inch
$Spcmx_{HRU}$	22	Maximum storage in the plant canopy (snow, winter rain, or summer rain) of the HRU, in inches
sr^m	13b	Hour angle of sunrise, measured from the local meridian of a horizontal surface (HRU or equivalent-slope surface) for time step m, radians
sr_{HRU}^{m}	13c, 14c	Hour angle of sunrise, measured from the local meridian of a horizontal surface (HRU or equivalent-slope surface) for time step m, radians
SS	97, 98, 100,	Volume averaged specific storage, per length
$SS_{i,j,k}$	104, 111	Volume averaged specific storage for cell i, j, k, per length
ss^m	13a, 13b	Hour angle of sunset, measured from the local meridian of a horizontal surface (HRU or equivalent-slope surface) for time step m, in radians
ss_{HRU}^{m}	13c, 14c	Hour angle of sunset on the sloped surface of the HRU for time step m, in radians
sub_{HRU}^{m}	30, 32	Sublimation from the HRU during time step m, in inches
t	68, 69a, 69b, 70, 71, 73, 80, 82, 94, 97	Time
t_*^{m}	103, 111	Time at the end of time step m
t_*^{m-1}	103, 111	Time at the end of time step m-1
T_{base}^{m}	2	Measured maximum (or minimum) daily temperature at the base station assigned to an HRU for time step m, degrees Fahrenheit or Celsius

Symbol	Equation	Definition
T_{HRU}^m	1, 2, 3a, 5a	Maximum (or minimum) daily temperature at each HRU for time step m, in degrees Fahrenheit or Celsius
T_i^m, T_{i+1}^m	4, 5a	Measured maximum (or minimum) temperature at measurement station i and $i+1$ for time step m, degrees Fahrenheit or Celsius
T_{lapse}^m	4, 5a	Basin-average maximum (or minimum) temperature lapse rate for time step m, degrees Fahrenheit or Celsius per length
T_{lower}^m	25	Temperature of the snowpack lower layer during time step m, in degrees Celsius
T_{sta}^m	1	Measured maximum (or minimum) daily temperature at the station for time step m, in degrees Fahrenheit or Celsius
T_{surf}^m	25	Temperature of the snowpack surface layer during time step m, in degrees Celsius
$\bar{T}_{HRU}$	19, 20a, 20b	Mean daily temperature on the HRU, in degrees Celsius
$\bar{T}_{stas}^m$	3c	Mean measured maximum or (minimum) daily temperature of all stations for time step m, in degrees Fahrenheit or Celsius
taf_{HRU}	1, 2, 3a, 5a	Maximum (or minimum) daily HRU temperature adjustment factor, which is estimated on the basis of slope and aspect, in degrees Fahrenheit or Celsius
$thck_{CB}$	110	Thickness of the semi-confining unit, in length
$thck_{i,j,k}$	109, 110	Thickness of cell i, j, k, in length
$thck_{i,j,k+1}$	109, 110	Thickness of cell $i, j, k+1$, in length
$thick_{strbed}$	78, 89	Streambed thickness in the stream reach, in length
$thick_{aq}$	92	Horizontal or vertical distance from lakebed to the center of the adjacent finite-difference cell, in length
$thick_{lkbd}$	92	Lakebed thickness, in length
Tmn_{HRU}^m	6, 16a	Minimum air temperature assigned to the HRU for time step m, degrees Fahrenheit or Celsius
Tmx_{HRU}^m	6, 16a	Maximum air temperature assigned to the HRU for time step m, degrees Fahrenheit or Celsius
$Tmxsnow_{HRU}$	6	Monthly maximum air temperature at which precipitation is all snow for the HRU, degrees Fahrenheit or Celsius
Tr	106	Transmissivity of the cell, in length squared per time
u_s	73	Velocity of a sharp wetting front, in length per time
V	100, 104	Volume of the cell, in cubic length
$V_{lakeHRU}^{m,n}$	90	Volume of surface runoff and interflow routed to a lake HRU from contributing HRUs for time step m and iteration n, in acre-inch
$v(\theta)$	69a, 69c, 75	Characteristic velocity restricted to the downward (positive z) direction, in length per time
$VK_{i,j,k}$	109, 110	Vertical hydraulic conductivity of cell i, j, k, in length per time
$VK_{i,j,k+1}$	110	Vertical hydraulic conductivity of cell $i, j, k+1$, in length per time
$VKCB_{i,j,k}$	110	Vertical hydraulic conductivity of the semi-confining unit beneath cell i, j, k, in length squared per time
Vsm_{HRU}^m	29, 31, 32, 33, 37	Volume of snowpack melted during time step m, in acre-inch
w	60	Weighting factor used to average gravity drainage between iterations, set to 0.5
W	94, 98, 104	Volumetric flow rate per unit volume representing sources and/or sinks of water, $W < 0.0$ for flow out of the ground-water system, and $W > 0.0$ for flow into the system, per time
$w_{mdpt}^{m,n}$	87	Stream width at reach midpoint, in length
w_{str}	86	Steam width, in length

Symbol	Equation	Definition
$wetper_{str}$	78, 89	Wetted perimeter of the streambed, in length
$Width_{fdc}$	106	Width of the prism perpendicular to the direction for which the conductance is defined, in length
x	94, 95, 96, 97	Horizontal coordinate axis normal to the y and z coordinate axis, in length
$x_{channel}$	80, 83	Distance along the stream channel, in length
X_{HRU}	5b, 10b	Longitude of the HRU centroid, length
$\overline{X}_{HRU}$	3b, 9b	Normalized longitude of the HRU, dimensionless
$\overline{X}_{sta}$	3c, 9c	Mean normalized longitude of all stations, dimensionless
$Xppt_i$	10b	Longitude of each precipitation measurement station, length
$Xtemp_i$	5b	Longitude of each air temperature measurement station, length
y	94, 95, 97	Horizontal coordinate axis normal to the x and z coordinate axis, in length
Y_{HRU}	3b, 9b	Normalized longitude of the HRU, dimensionless
$\overline{Y}_{HRU}$	3b, 9b	Normalized latitude of the HRU, dimensionless
$\overline{Y}_{sta}$	3c, 9c	Mean normalized latitude of all stations, dimensionless
$Yppt_i$	10b	Latitude of each precipitation measurement station, in length
$Ytemp_i$	5b	Latitude of each air temperature measurement station, in length
z	68, 69a, 70, 71, 72, 94, 95, 97	Altitude in the vertical direction, in length
z_1	70, 71, 72, 76a	Point above the wetting front, in length
z_2	70, 71, 72, 76a	Point below the wetting front, in length
Z_{base}	2, 8	Altitude of the base station, in length
z_f	72, 73	Depth of the sharp front, in length
Z_{HRU}	1, 2, 5a, 8	Mean land-surface altitude of the HRU, in length
$Z_{tsta,i}$, $Z_{tsta,i+1}$	4, 5a	Altitude of measurement stations i and $i+1$, length
Z_{lapse}	2, 8	Altitude of the lapse station, in length
$\overline{Z}_{tsta}$	3c, 9c	Mean normalized altitude of all stations, dimensionless
Δc_i	95, 96, 97, 98, 102, 109, 110, 111	Length of finite difference cell along the column direction, in length
$\dfrac{\Delta h}{\Delta t_*}$	97, 98, 100	Change in head over a specified time interval
$\dfrac{\Delta h_{i,j,k}^m}{\Delta t_*}$	103	Change in head at cell i, j, k at the end of time step m over a specified time interval
Δr_j	95, 97, 98, 109, 110, 111	Length of finite difference cell along the row direction, in length
$\Delta r_{j-1/2}$	102	Distance between nodes $i, j-1, k$ and i, j, k, in length
Δs_t	115a, 115b	Total storage change of the time step ending at time t, in cubic length
ΔS_{total}	113	Total storage change in all reservoirs from beginning to end of time step, in cubic length
$\Delta storage$	114	Total storage change, in cubic length
Δt_{snow}	25	Snow computation time step, in 43,200 seconds (half-day interval)
Δt	40, 42, 55, 56, 58, 67, 76b, 77, 91, 113, 114	GSFLOW time step, in one day
Δt_{mf}	93	MODFLOW-2005 time step, in time

Symbol	Equation	Definition
Δt_{sfr}	82	Time-step for routing streamflow, in time
Δv_k	95, 96, 97, 98, 102, 111	Length of finite difference cell along the layer (vertical) direction, in length
$\dfrac{\Delta V}{\Delta t}$	111	Rate of accumulation of water in the cell, in cubic length per time
Δx	83	Spatial increment for routing streamflow, equal to the reach length
Δz	108	Cell thickness, or, for unconfined conditions, the thickness of the saturated region of the cell, in length
$\Delta \theta$	75	Change in water content between two adjacent locations along a trailing wave, volume of water per volume of rock
ε	61a, 74	Brooks-Corey exponent, dimensionless
θ	68, 69a, 69b, 70, 71, 72, 74	Volumetric water content, in volume of water per volume of rock
θ^{mn}	61a, 61b	Volumetric water content at the top of the unsaturated zone for time step m, and iteration n, dimensionless
θ_{z_1}	72, 73	Volumetric water content at depth z_1 in volume of water per volume of rock
θ_{z_2}	72, 73	Volumetric water content at depth z_2, in volume of water per volume of rock
θ_τ	76b	Water content of trailing waves above the evapotranspiration extinction depth after time τ, in volume of water per volume of rock
$\theta_{\tau+\Delta t}$	76b	Water content of trailing waves above the evapotranspiration extinction depth after time $\tau + \Delta$, in volume of water per volume of rock
θ_r	61a, 74	Residual water content of the unsaturated zone, in volume of water per volume of rock
θ_s	61b, 74	Saturated water content of the unsaturated zone, in volume of water per volume of rock
λ_{HRU}	20a, 20b	Latent heat of vaporization on the HRU for time step m, in calories per gram
π	13c, 14c, 25	Constant pi (~3.1415926535898), dimensionless
ρ_{HRU}	18, 19	Saturated water-vapor density (absolute humidity), in grams per cubic meter
ρ'_{HRU}	22, 23	Plant canopy density (winter or summer) as a decimal fraction of the HRU area, dimensionless
ρs	24	Density of new-fallen snow, as a decimal fraction, dimensionless
$\rho\,snow^m_{HRU}$	25	Volumetric density fraction of the snowpack for the HRU during time step m, dimensionless
ρ''^m_{HRU}	26, 27, 30	Volumetric density fraction of the snowpack for the HRU during time step m, dimensionless
ρsm	24	Average maximum snowpack density, as a decimal fraction of the liquid water equivelent, dimensionless
ω	83, 93	Time-weighting factor that ranges between 0.5 and 1, dimensionless

www.ingramcontent.com/pod-product-compliance
Lightning Source LLC
Chambersburg PA
CBHW081438170526
45166CB00008B/2240